Archimedes

Volume 12

Archimedes

NEW STUDIES IN THE HISTORY AND PHILOSOPHY OF SCIENCE AND TECHNOLOGY

VOLUME 12

Archimedes has three fundamental goals; to further the integration of the histories of science and technology with one another: to investigate the technical, social and practical histories of specific developments in science and technology; and finally, where possible and desirable, to bring the histories of science and technology into closer contact with the philosophy of science. To these ends, each volume will have its own theme and title and will be planned by one or more members of the Advisory Board in consultation with the editor. Although the volumes have specific themes, the series itself will not be limited to one or even to a few particular areas. Its subjects include any of the sciences, ranging from biology through physics, all aspects of technology, broadly construed, as well as historically-engaged philosophy of science or technology. Taken as a whole, *Archimedes* will be of interest to historians, philosophers, and scientists, as well as to those in business and industry who seek to understand how science and industry have come to be so strongly linked.

Universities and Science in the Early Modern Period

Edited by

MORDECHAI FEINGOLD
California Institute of Technology, U.S.A.

and

VICTOR NAVARRO-BROTONS
University of Valencia – CSIC, Spain

Springer

A C.I.P. Catalogue record for this book is available from the Library of Congress.

ISBN-10 1-4020-3974-3 (HB)
ISBN-13 978-1-4020-3974-4 (HB)
ISBN-10 1-4020-3975-1 (e-book)
ISBN-13 978-1-4020-3975-1 (e-book)

Published by Springer,
P.O. Box 17, 3300 AA Dordrecht, The Netherlands.

www.springer.com

Printed on acid-free paper

Printed in the Netherlands.

TABLE OF CONTENTS

LIST OF CONTRIBUTORS

Diana Soto Arango is a professor of the history of education at the Universidad Pedagógica y Tecnológica de Colombia.

Luis Carlos Arboleda is an emeritus professor of history of mathematics at the Institute of Education and Pedagogy, Universidad del Valle, Cali, Colombia.

Antonio García Belmar teaches the history of science at the University of Alicante, Spain.

José Ramón Bertomeu Sánchez teaches history of science at the University of Valencia, Spain.

Ugo Baldini is a professor of history at the University of Padua.

Luis García Ballester was a professor of research of the CSIC (Spanish Council for the Research) in the Institution "Milà i Fontanals" (Barcelona) and, from 1996 until his death in 2000, professor of history of science in the University of Cantabria.

Maria Teresa Borgato is an associate professor of didactics and foundations of mathematics at the University of Ferrara.

Luís Miguel Carolino is a researcher at the Centro de Estudos de História da Ciência, University of Évora, Portugal.

José Chabás teaches at the Universitat Pompeu Fabra, Barcelona, Spain.

Piero del Negro teacher of modern history at the University of Padua.

Mordechai Feingold is a professor of history at the California Institute of Technology.

Maija Kallinen is a professor of history of science at the University of Oulu.

Henrique Leitao is a research associate at the Center for the History of Science at the University of Lisbon.

José María López Piñero is an emeritus professor of history of medicine at the University of Valencia.

Víctor Navarro-Brotóns is a professor of history of science at the University of Valencia, Spain.

Agustí Nieto-Galan is a lecturer in the history of science at the *Universitat Autònoma de Barcelona*, Spain.

Luigi Pepe is a professor of the history of mathematics at the University of Ferrara.

José Luis Peset is a research professor in the Institute of History in the Spanish Council for Scientific Research.

Antoni Roca-Rosell is lecturer of history of science and technology at the Universitat Politècnica de Catalunya, Barcelona, Spain..

Grazyna Rosińska is a member of the Institute for the History of Science in Polish Academy of Sciences, Warsaw and professor of history at the M. Curie-Sklodowska University, Lublin.

Cristina Sendra Mocholí teaches history of science at University of Valencia.

Paul Wood is a director of the Humanities Centre and a Professor in the Department of History at the University of Victoria.

INTRODUCTION

This book includes most of the contributions presented at a conference on "Universities and Science in the Early Modern Period" held in 1999 in Valencia, Spain. The conference was part of the "Five Centuries of the Life of the University of Valencia" (*Cinc Segles*) celebrations, and from the outset we had the generous support of the "Patronato" (Foundation) overseeing the events.

In recent decades, as a result of a renewed attention to the institutional, political, social, and cultural context of scientific activity, we have witnessed a reappraisal of the role of the universities in the construction and development of early modern science. In essence, the following conclusions have been reached: (1) the attitudes regarding scientific progress or novelty differed from country to country and follow different trajectories in the course of the early modern period; (2) institutions of higher learning were the main centers of education for most scientists; (3) although the universities were sometimes slow to assimilate new scientific knowledge, when they did so it helped not only to remove the suspicion that the new science was intellectually subversive but also to make science a respectable and even prestigious activity; (4) the universities gave the scientific movement considerable material support in the form of research facilities such as anatomical theaters, botanical gardens, and expensive instruments; (5) the universities provided professional employment and a means of support to many scientists; and (6) although the relations among the universities and the academies or scientific societies were sometimes antagonistic, the two types of institutions often worked together in harmony, performing complementary rather than competing functions; moreover, individuals moved from one institution to another, as did knowledge, methods, and scientific practices.

The objectives of the conference were: (1) to examine and assess the current state of the problem; (2) to provide new data, methods, and perspectives on it; and (3) to compare historical approaches carried out in various countries—especially in the Iberian world, an area that is rarely considered in studies of the subject. The articles presented here cover the entire early modern period, from the 15th to the 18th centuries. The majority of the papers concentrate on one of the three great periods that historians of science generally recognize—the Renaissance, the Scientific Revolution, and the Age of Enlightenment—although two studies, both devoted to Italy, cover all three periods. At the same time, the study of the university as an institution permits the examination of elements contributing to the question of continuity and change in early modern science.

M. Feingold and V. Navarro-Brotóns (eds.), Universities and Science in the Early Modern Period, 1–7.

The contributions are distinguished by their geo-cultural and geo-political diversity; however, those relating to the Iberian world (Spain, Portugal, and the Viceroyalty of New Granada in colonial America) clearly predominate. Other countries represented in the volume include Italy, Switzerland, Poland, England, and France. Papers dedicated to Holland and Mexico were presented at the conference but are not included in the volume. Concerning the themes studied in the essays, one group takes up activities connected with the scientific disciplines taken as a whole, including teaching and research, while another group examines particular disciplines such as mathematics and its applications, natural philosophy, botany, medicine, and astronomy.

The importance of the Renaissance universities, particularly in areas such as Italy and Germany, is a well-established fact. The articles dealing with this period included here also confirm the importance of the universities within an area less well known, Iberia, and bring new precision and information relating to the Italian and Polish universities. The study by José Chabás emphasizes the importance of the University of Salamanca as a principal center of astronomical activity in the Iberian peninsula in the 15th century. Chabás also underscores the collaboration between Jews and Christians and the contacts between the academic and non-academic worlds, a subject exemplified in the life and work of the outstanding Salamancan astronomer Abraham Zacuto.

The late Luis García Ballester, utilizing recently discovered manuscript documentation as well as printed sources, was able to reconstruct the teaching of medicine at Salamanca during this period. His contribution emphasizes that at Salamanca there existed a movement that approached Galenism in a manner similar to prominent European faculties of medicine such as Paris, Montpellier, and various Italian centers.

More famous than Salamanca was the University of Cracow, which in the second half of the 15th century arose as "an international center for the study of astronomy", in the words of Birkenmaier. Examining the teaching activities and the diffusion of the works of Martinus Rex and Giuseppe Bianchini, Grazina Rosinska contributes new information about the renovation of the teaching of mathematics applied to astronomy at Krakow and elsewhere. Since Bianchini (who was the first western mathematician to use purely decimal tables of trigonometric functions) was not a university professor, by studying the diffusion of his writings, Rosinska is able to provide a notable example of the reciprocal influence among the university and extrauniversity sites.

Returning to the Iberian world, thirty new university centers were created in the 16th century, some of which, such as Valencia and Alcalá, rivaled Salamanca. Valencia's faculty of medicine became the most important in 16th-century Spain. José Maria López Piñero shows that Valencia's faculty incorporated the principal novelties in the teaching of medicine that were introduced into Europe: the study of anatomy according to the methods introduced by Vesalius, the teaching of botany applied to medicine with an associated botanical garden, and the teaching of surgery. Moreover, the clinical observations and environmentalist-oriented public health studies were put into the foreground, while a chair of 'secret remedies' with a Paracelsian orientation was created, although it lasted for only one academic term.

In the same period, the universities of Valencia and Alcalá introduced chairs of mathematics, as Salamanca had already done in the 15th century, thereby considerably expanding the repertory of subjects. At the same time, the content of teaching was reformed and renewed, on the one hand by the introduction of lessons on the practical applications of mathematics to geography, cartography, topography, nautical science, and artillery, and on the other hand by the introduction of new mathematical techniques and discussions of cosmological questions (including those relating to the heliocentric theory, to the nature of the nova of 1572 and to comets). All of these are evident in the activities of the outstanding mathematician and humanist Jerónimo Muñoz and his disciples, who are the subjects of Víctor Navarro's article.

Cosmography and navigation had great importance in the Iberian world. The teaching of mathematics oriented to these subjects is the main characteristic of the Portuguese case, here studied by Henrique Leitao and Luis Miguel Carolino. These authors emphasize the role of Jesuit teaching in the "Aula de Esfera" (Chamber of the Sphere) of the College of San Antonio in Lisbon, which was oriented to the training of cosmographers and navigators, whose professor proceeded from the Collegio Romano and other European centers. The practical orientation of mathematics created a divorce of mathematics from natural philosophy. Nevertheless, the teaching of philosophy was far from closed to new scientific ideas. Its professors attempted to integrate non-Aristotelian ideas, such as those of Tycho concerning the celestial localization of comets, into the Aristotelian framework.

The attempt to assimilate astronomical and cosmological novelties into traditional systems is well represented in the work of Jacques de Chevreul, a University of Paris professor between the 1620s and the 1640s. Roger Ariew analyzes de Chevreul's treatise on the sphere and shows how he used the new astronomical observations of Galileo and other authors. Nevertheless, in the most controversial astronomical–cosmological issues, such as the incorruptibility of the heavens and the question of the lunar surface, de Chevruel resorted to a probabilistic language in order to avoid publicly favoring any particular interpretation.

In countries such as Switzerland, the university was the main center of scientific activity in the seventeenth and the eighteenth centuries. Founded in 1477 and encouraged by successive monarchs, by 1620 the University of Uppsala became a major seat of learning. The teaching of the mathematical disciplines was advanced by virtue of its practical applications and by the necessities of civic life. Maija Kallinen describes the activities of the most relevant figure of the period, the physician Olof Rudbeck, a true polymath, who counted among his various activities the teaching of mixed mathematics. Kallinen emphasizes that Rudbeck's semi-private lessons are striking examples of how semi-official activities could, within limitations, significantly complement more official forms of instruction. In Italy as well, where some of the first European universities were created, scientific activity in the early modern period was strongly linked to this institution. As Luigi Pepe and Maria Teresa Borgato demonstrate, this would explain both the successes of science in Italy and its limitations. The works of these two authors cover the entire period for the Italian case. Pepe discusses the weaknesses of the Italian academies, their complementary character in relation to the universities and the interactions and conflicts among them during this period. He dedicates one part of his paper to the Napoleonic reforms, which inaugurated a new epoch for the Italian universities as a consequence of the intense geopolitical changes that took place on the Italian peninsula. Borgato, focusing on mathematics, studies the links between mathematical research and the university, pointing out how in the 18th century the revival of mathematical research in Italy took place essentially within the university environment.

The impact of the Scientific Revolution and the profound social transformations of the period had repercussions in the European universities. Indeed, the last decades of the 17th century and the first decades of the 18th century might be classified as a period of crisis and reform in the university: Crisis as the traditional university confronted new philosophical and scientific currents, and reform as it adapted to the new currents. This process occurred in a paradigmatic manner in countries such as Spain, which had scarcely participated in the philosophical and scientific changes of the 17th century. As José Luis Peset points out, in the second half of the 18th century, a series of reforms took place in Spain with the support of the monarchy. New texts were introduced (in large part editions and translations of foreign works) and, in accordance with the ideas of the epoch, the teaching of science was oriented toward practical applications. A large number laboratories, cabinets for experimental demonstration, botanical gardens, natural history collections, and astronomical observatories were either created or expanded in the peninsular universities.

In the case of the University of Valencia, the various curricula culminated in the so-called Blasco Plan, which created chairs of chemistry and botany. The study of various reports and plans relative to the teaching of botany has enabled Cristina Sendra to analyze the position of particular professors in relation to these educational reforms. The professors who are the subjects of Sendra's article debated whether or not botany might be taught in the university courses on medicine as well as what should be considered to be the science of botany. The ideas oulined by the authors of these reports included the recognition of botany as a discipline with its own methods and patterns of work, separate from the medical profession.

While it is clear that the Bourbons patronized the university reforms in Spain, Agustín Nieto and Antonio Roca demonstrate that Cataluña was a special case in point. As a consequence of the War of Spanish Succession, the new dynasty that came to power in Spain suppressed all of the Catalonian universities and created a new one at Cervera. In this university, certain Jesuit professors attempted to introduce elements of the new physics in the philosophy chairs. At the same time, the absence of a university was replaced by the emergence of alternative institutions, principally in Barcelona, such as the Academia Militar de Matemáticas, the Real Colegio de Cirugía, the Academia Médico-Práctica, the Real Academia de Ciencias Naturales y Artes, and the Colegio de Cordelles. In addition, the technical schools of the Junta de Comercio were established to educate craftsmen and early industrial workers.

The Barcelona Junta created a workable system of higher education that lasted for decades. The Barcelona network of schools had explicit utilitarian objectives and most of their syllabuses were based on contemporary European scientific and technological developments.

The university reforms sponsored by the Spanish Bourbons were also extended to the universities of the Viceroyalty of New Granada and to those of the Audiencia of Caracas. Luis Carlos Arboleda and Diana Soto investigate these reforms, focusing in particular on the teaching of physics and Newtonian cosmology. The Copernican and Newtonian theories began to circulate in these countries in relation to activities such as the geodesic expeditions, the establishment of borders, and the construction of fortifications. In teaching these subjects, the Jesuits typically adopted an eclectic style, considering cosmological theories as useful hypotheses. In the second half of the century, various professors introduced the systematic teaching of the new physics, although not without polemics and difficulties. Arboleda and Soto consider these teachings to be manifestations of the institutionalization of enlightened ideas in the university classrooms, which were not completed in the colonial era. The Creole elite attempted to assimilate the new scientific currents, integrating them into their own socio-political and cultural project.

One of the most dynamic centers of Enlightenment thinking, as is well known, was Scotland. The Scottish universities participated in and contributed to Enlightenment thought in significant ways. In his contribution to this volume, Paul Wood explores the reasons why the study of natural knowledge flourished in the Scottish universities during this period. Wood's paper focuses on issues relating to the role that the universities played in improving the ethos that fostered science, the impact of the ideal of "politeness" on the transformation of the Scottish university curricula in the 18th century, the ways in which the related notion of the "gentleman" shaped the teaching of mathematics and natural sciences, the interplay between religion and natural knowledge, the specific institutional features of the Scottish universities that affected the cultivation and transmission of natural knowledge and the significant part played by the universities in the creation of the public sphere. One of Wood's main conclusions is that in Scotland public culture was to a considerable extent an offshoot of academe.

Piero del Negro and Ugo Baldini deal, respectively, with the universities of Padua and Rome. In his study, Del Negro revises the historiography concerning the politics guided by the Reformatiori dello Studio di Padova. Many indicators suggest that the 18th century was an age of progress for the University of Padua. If formal enrollment decreased, the number of "real" students grew. New institutions for scientific research were created, including physics laboratories, a natural history museum, an astronomical observatory, a school of obstetrics, medical, and surgical clinics in the hospital and a chair of architecture. In addition, many new courses were added. Also at Padua, as in other European universities, what characterized the Enlightenment reform was its utilitarian inspiration, the belief that scientific knowledge could be exploited for the benefit of society as a whole.

The reforms of the Enlightenment also reached the University of Rome "La Sapienza", although the changes were not as extensive or effective as in other Italian universities. At least until the 1780s the distance between the Sapienza and the other universities increased. Ugo Baldini offers one of the first comprehensive studies of this university's activities in relation to science. In addition to describing the reforms, he analyzes the limited degree of their application in the teaching of different disciplines: philosophy; mathematics, medicine, physiology, anatomy, chemistry, botany and natural history. Around 1780, the practice of science in Rome approached the level that for decades had already been reached in the Universities of Padua, Pavia, Turín, Bologna, and Pisa. This development did not occur in opposition to or independently of the intentions of the Papal government. Rather, it took place thanks to initiatives emanating at least partly from the Papacy. Thus it is not an exaggeration to characterize this as a period of "Papal reform" in science. The case of the Roman Sapienza is similar to the other Italian universities, whose role in the cultural history of cities like Naples, Messina, Catania, or Palermo is underestimated, evaluated too generally, or subjected to biased judgments.

The study of the subject matter taught in the universities, as well as the methods and teaching instruments used, requires that special attention be given to textbooks and to the processes by which they were created. These are the themes that concern Antonio García Belmar and Jose Bertomeu Sánchez, who underscore the need to revise the idea that teaching was merely the transmission of knowledge. They insist that communication is one of the practices by which scientific knowledge is constructed. To illustrate this thesis, García Belmar and Bertomeu Sánchez examine the characteristics of chemistry textbooks elaborated in Spain and France in the late eighteenth and early nineteenth centuries, as well as the conditions of their production, evolution, and change. Their study emphasizes the creative labor of the authors of teaching manuals in spite the multiple institutional, academic, political, and economic conditions that they had to confront.

In summary, a long-road lies ahead to reach an adequate and well-founded understanding of the universities in the scientific culture of the early modern era. Nevertheless, we think that this collection of studies, besides contributing new data, raises new questions and suggests new lines of investigation, confirming the demand for a reevaluation of the role of the universities to which we referred at the outset.

We wish to express our gratitude to the following institutions for making the conference possible: the Patronat "Cinc Segles" of the University of Valencia, the Valencian Government (Generalitat Valenciana), the Spanish Agency for International Cooperation (Agencia Española de Cooperación Internacional), the Diputació de València, the Catalan Society for the History of Science and Technology and the Instituto de Historia de la Ciencia y Documentación "López Piñero". We also wish to thank the members of the Organizing Committee, whose work was essential in preparing the conference: Vicente L. Salavert, Jesús I. Catalá, Tayra Lanuza and Mavi Corell. We are also grateful to the members of the Scientific Committee: Ugo

Baldini, Luis García Ballester, Luigi Pepe, José Luis Peset, and José María López Piñero. We also wish to thank Jehane Kuhn for her exemplary editorial work.

Finally, we would like to thank those participants whose works could not be included in the volume, as well as to those who papers have been included but who were unable to attend the conference, the late Luis García Ballester and Luis Carlos Arboleda.

GRAZYNA ROSIŃSKA

"MATHEMATICS FOR ASTRONOMY" AT UNIVERSITIES IN COPERNICUS' TIME: MODERN ATTITUDES TOWARD ANCIENT PROBLEMS

INTRODUCTION

From the Middle Ages until the end of the 15th century and even later, students attending universities were introduced into philosophy and natural sciences by way of the four "liberal arts" of the *quadrivium* which included arithmetic and geometry. Those who subsequently chose to specialize in astronomy, particularly at universities such as Cracow, which had a chair in the subject from the beginning of the 15th century, were gratified with supplementary portion of mathematics included into the lectures on astronomy.[1] It was the case of the lectures on the *Canones tabularum* or on the composition of astronomical instruments or else on various *Theoricae planetarum*. In addition, there existed special treatises on "mathematics for astronomy" or special commentaries on the commonly used "ordinary" textbooks on mathematics. All these types of mathematical texts devoted to the training of astronomers will be referred to in what follows.

The period we are interested in is limited to some 60 years, between the early 40s of the 15th century and the beginning of the 16th century. It was as early as the first half of the 15th century that the algebra was introduced in an exposition of mathematical astronomy (the case of Bianchini's *Flores Almagesti*) and arithmetic of decimal fractions was invented for use in astronomical computations. From that period come the tables of decimal trigonometric functions developed subsequently in Regiomontanus', Copernicus', and Rheticus' works. Our considerations are focused, therefore, on the achievement of mathematicians whose activity began in the first half of the 15th century: Giovanni Bianchini of Ferrara (ca. 1400–1470) and Martinus Rex of Żurawica (or of Premislia, ca. 1420–1453). While Bianchini is the author of the modern form of trigonometric tables, Martinus Rex, professor of mathematics and astronomy, first at Cracow university and then at universities in Prague and Bologna, created scientific milieus prepared to accept Bianchini's accomplishments. (Although Peurbach and Regiomontanus were their partial contemporaries, it should be noted that in 1440 Peurbach was 17 and Regiomontanus only 4).

Bianchini, a brilliant autodidact, spent almost all his adult life in Ferrara as an administrator of the estate of the d'Este family.[2] His treatises, however, written in Latin and filled with references to Euclid and Ptolemy, were intended as university textbooks in mathematics and astronomy. Bianchini's scientific writings come from around

M. Feingold and V. Navarro-Brotóns (eds.), Universities and Science in the Early Modern Period, 9–28.

1440–1460, and according to research on the 15th century astronomical manuscripts preserved in various collections, were known during his author life all over Europe.[3] Martinus Rex, not as creative as Bianchini, studied at Cracow probably from 1438. His most important work, the *Summa super tabulas*, written in 1444, certainly influenced the teaching of mathematics at Cracow University in the second half of the 15th century.[4]

The aim of our study is the presentation of some arithmetical aspects of 15th century trigonometry that were relevant, as it seems, to the evolution of the concept of number that originated in the early 40s of the 15th century and continued into the 16th and 17th centuries, hand in hand with the development of early modern physics. Therefore, after presenting Martinus Rex' and Giovanni Bianchini's expositions of trigonometry, and the transmission of their achievement to the second half of the 15th century, this study will continue with a view of the mathematics taught at universities at the end of the century (Copernicus enrolled at Cracow university in 1491) and it will conclude with Copernicus' trigonometric tables, preserved in the so-called *Uppsala codex* and in *De revolutionibus.*

Among the "ancient problems" that were still present in 15th century expositions of "mathematics for astronomy", we shall consider the inadequacy of the Pythagorean concept of number for the (quantitative) description of reality. In Antiquity as well as in the Middle Ages, the term *number* itself referred only to the whole, positive numbers and thus it did not correspond to the requirements of mathematical astronomy, among other applications, working as a rule with incommensurable quantities. I am speaking here of the concept of number as adopted in "scientific arithmetic", based on Boethius' *De institutione arithmetica*, and present in the university textbooks. In "vulgar arithmetic", *logistika*, it was admitted the use of what could now be called the "naïve concept of real number", sufficiently large to include irrationals and numerical expressions of negative quantities, and thus adequate to perform all arithmetical operations.

As it results from Bianchini's *Flores Almagesti*, the 15th century "mathematics for astronomers" used overtly "real numbers" as well as methods of calculation borrowed directly from the *scuole d'abbaco*. Two treatises open the *Flores Almagesti*, the *Arithmetica*, and the *Algebra*. The *Arithmetica* teaches astronomers, among others, the method of "composition", known in the mercantil arithmetic as *infilzamento* (*infilzare*) and then the four operations with negative numbers, together with the rule of signes; furthermore, it gives the principle of the extension of the decimal positional system of numbers to decimal positional fractions.[5] The *Algebra* assumes the doctrines exposed in the *Arithmetica*. Both treatises, written in medieval Latin and accompanied with "geometrical proofs" constructed with reference to Euclid, certainly contributed to the mathematical training of 15th century astronomers, Peurbach and Regiomontanus included.[6] In Cracow, Bianchini's mathematical (trigonometric) and astronomical tables were systematically copied, rearranged, and adapted to Cracow meridian, to begin with Martinus Rex' students and to finish with Copernicus, and with his younger colleagues in Cracow.[7]

In the first half of the 15th century, the formal difficulties with the "mathematics for astronomers" were sometimes consciously ignored or avoided—apparently—by the lecturers on astronomy. Such is the case of Martinus who believes that the difficulties can be circumvented thanks to a special interpretation of the relations between *matter* and *number*, and considers the trigonometric functions of tangent and cotangent (*umbra recta and umbra extensa*) as ratios of numbers rather than as lines—or as lines that express the *ratios* of numbers. As for Bianchini, he changes the concept of a trigonometric function using "numbers", decimal positional fractions, for its expression.

RENAISSANCE "MATHEMATICS FOR ASTRONOMY"

Renaissance "mathematics for astronomy" has not yet been systematically studied, though valuable studies have been published aiming, for instance, at establishing Paolo Toscanelli's and above all Regiomontanus' contribution to mathematics.[8] Generally speaking, "mathematics for astronomers" is characterized by:

a. The use of *fractiones physicae*: sexagesimal positional fractions, which had met the needs of astronomy since Babylonian times.
b. Arithmetical operations with large numbers. Special arithmetical techniques were invented to deal with large numbers.
c. The developed theory of proportions applied to plane and spherical trigonometry, and approximations of incommensurables to numbers (fractions).
d. Particular interest in the development of tables of sine, versed sine, and "umbrae" (tangent and cotangent).

All these characteristics had been present in mathematical astronomy since Ptolemy's times. Renaissance mathematics (for astronomy) was enriched in the course of time by developments in the Muslim East that had begun in the ninth century, and in the Latin West, particularly during the 14th century, in Paris. In the 15th century, a specific enrichment of "mathematics for astronomy" came from the *scuole d'abbaco* flourishing in Italy.

The dimensions of the achievement of the "abbacists" have become evident thanks to research conducted by historians of mathematics in the last decades of the 20th century. Practical mathematics has been studied especially by Italian scholars, including Gino Arrighi,[9] Rafaella Franci, and Laura Toti-Rigatelli.[10] Subsequently, Jacques Sessiano[11] and Warren Van Egmond[12] have revealed new dimensions of the late medieval and the early renaissance practical mathematics. Finally, efforts have been made to illustrate the impact of mathematics from the *scuole d'abbaco* upon the achievement of 16th century mathematicians, Bombelli included.[13]

In my own work on the 15th century "mathematics taught at universities to future astronomers", I seek to present a new concept of number emerging from the treatises on mathematics composed by Giovanni Bianchini. Bianchini extended the concept of number to negative numbers and to irrationals. Then, he introduced decimal positional fractions applying them to astronomical computations, and particularly to approximation of irrationals to numbers. Finally, he used the unit segment for the geometrical expression of arithmetical operations with irrationals.[14]

The concept of number, used by Bianchini's predecessors, resulted of accumulative principle applied to its construction. Nicomachus' of Geraza, second century, definition of number runs as follows, according to d'Ooge translation: Number is a limited multitude or a combination of units... and Boethius in sixth century writes following Nicomachus' definition: *Numerus est cumulatio unitatum....*[15] The same idea persists in definitions of number present in the university textbooks used in 15th century: *Numerus est multitudo ex unitatibus profusa*... and *unitas non est numerus sed principium numeri.*[16] In contrast with the earlier tradition, Bianchini considers the unity as a number, and not as a "principle of numbers, itself not being a number". Moreover, the way of explaining certain arithmetical operations confirms that he bases the concept of number on a multiplicative principle rather that on the cumulative one, and considers "number" as a *ratio* of a quantity to its unit. This idea permits him to use the inverse of number in the operation of division and perform division by means of multiplication by the inverse of the divisor: *Quando integri dividuntur per fractiones, multiplicentur integri secundum proportionem unitatis ad illasmet fractiones divisoris. Quando fractiones dividuntur per integra, multiplicentur fractiones secundum proportionem unitatis ad numerum divisorem.*[17]

All these issues, the irrationals and negative values considered as numbers, the decimal positional fractions, and the multiplicative principle underlying the concept of number, are present in Bianchini's mathematical and astronomical writings, beginning with his *Arithmetica*, which I am inclined to date to about 1440, and his *Compositio instrumenti*, dated to 1442.[18]

In what follows, Martinus' lecture on trigonometry, delivered at Cracow university in 1444 and included into his *Summa super tabulas*, will be presented, and then Bianchini's *Canones super tabulas sinuum* and his *Compositio instrumenti* will be considered, together with his *Arithmetica* and his decimal trigonometric tables. In 1444 Martinus was probably unaware of the fact that at about the same time, in Italy, Bianchini had composed decimal trigonometric tables, first for $R = 60 \times 10^n$, and some years later for $R = 10^n$. There is no evidence that Martinus and Bianchini met, though they could have done so during Martinus' professorship in Bologna in the years 1447–1449.

As for the philosophical attitudes of Martinus and Bianchini toward the relation between magnitude and number, and between number and matter, underlying trigonometry as applied to astronomy, it was Martinus who posed in his lecture the problem of the correspondence (or rather of the lack of correspondence) between number and Nature; more exactly between mathematics and nature of matter. The approach to the solution of this problem would be made by Giovanni Bianchini. It should be emphasized, however, that Bianchini considers the problem of the correspondence between number and magnitude (matter) on a methodological rather than philosophical level. He is concerned with the formal correctness (or rather the lack of correctness that he finds in the works by his contemporaries) of the numerical expression of the *ratios* between two heterogeneous entities: arcs (measured in degrees) and lines in a circle.

MARTINUS REX' EXPOSITION OF TRIGONOMETRY

Martinus, who was trained at Cracow University in the years 1438 (or 1440)–1445, thought of himself as a heir to a tradition of trigonometric tables going back to Ptolemy, and further developed in the 14th century by the moderns (*moderni*), among whom he mentions by name only John of Lineris.

Among textbooks on mathematics used in Cracow at the time of Martinus' studies were Sacrobosco's *Algorismi* (*Algorismus de integris* and *Algorismus de minutiis*), the *Tractatus de proportionibus* of Jordanus Nemorarius, and the *Geometria speculativa* of Thomas Bradwardine. It is through the *Geometria speculativa* that students come in touch with Archimedes' *De mensura circuli* and through the *De curvis superficiebus* (attributed by Martinus to Geber) with Archimedes' *De sphera et cylindro*.[19] It is very possible that Richard of Wallingford's *Quadripartitum* was also known. The basic arithmetic of sexagesimal fractions was taught in Cracow using treatises like the common *algorismi minutiarum* or a short treatise that begins *Circulus obliquus qui signifer nuncupatur,* with explication of the eight [!] arithmetical operations: *Sunt autem octo species huius negocii scilicet: reduccio, addicio, subtraccio, duplacio, multiplicacio, divisio et radicum extractio, tam in numeris cubicis quam quadratis.*[20] In addition, treatises by contemporary Cracow lecturers on mathematics were used, for instance the commentary to *Algorismus minutiarum*, written in 1430 by Sandivogius of Czechlo.[21] Martinus himself was the author, apart from the *Summa super tabulas*, of mathematical treatises such as an *Algorismus minutiarum* and a *Geometria practica.*[22]

Martinus on the mathematical origins of astronomy and on the relation: numerus—materia (natura materiae)

Martinus' *Summa*, addressed first to Cracow students, but then, taught most probably also in Prague, touches on virtually all questions dealing with astronomy, including geometry of a sphere, the composition of trigonometric and astronomical tables, and the construction of some of the astronomical instruments, together with rules the use of them in some simple observations, then the introduction to astronomical tables (the Alphonsine and the de Lineris tables) and above all, the basics of "mathematics for astronomy". In fact, at the beginning of the *Summa* Martinus declares: "*Summa tamen hec nonnisi ex mathematica consideracione ortum habet.*" [This Summa takes its origins exclusively from mathematical considerations] (250r).[23]

This radical statement in favor of the mathematical origins of the *Summa* (and thus of astronomy?) is motivated by the fundamental relation that exists, according to Martinus' philosophical views, between "materia" (or "natura" or else "natura materiae") and mathematics. The link between these aspects of the reality is so deep that it permits us to construct astronomy on a mathematical basis, even when number (mathematical speculations involving number) cannot describe matter (the nature of matter) exactly. The fragment that points to the limitation of mathematics with respect to matter runs as follows:

> [Thabulae astronomicae] calculate sunt per speculaciones mathematicas, que speculaciones mathematice quandoque alio modo capiunt materiam quam habet se in sui natura. [The astronomical tables] are calculated by means of mathematical speculations, and these speculations sometimes grasp matter differently from [the way] the matter is in its own nature]. (275v)

Thus, even if mathematics is able to grasp matter, the mathematical (geometrical or numerical) expression may not always correspond exactly to the nature of matter. It seems that the text can be interpreted in two ways: either that there is an occasional lack of correspondence of the mathematical models to celestial reality, caused by the distortion of this reality by mathematics, or that in Nature there is "something" that still remains inaccessible through mathematical *formulae*. In the second case, it is a question of the incompleteness of mathematical models with respect to the nature of matter, rather than of the falsehood of these models.

If the second reading (incompleteness) is correct, it may be concluded that Martinus is offering here the philosophical foundations of astronomy considered as a *scientia mixta*. The context of Martinus' statement shows that the "mixture" concerns at least two elements that make up astronomy, namely the observations and the numbers (mathematical speculations) that express the results of observations, presented in the geometrical (kinemtaical) models and in the astronomical tables. It has to be noted, however, that the idea of the inherence of number in the Nature, that permits the "mathematical speculations" about the Nature, is completed by the idea of the transcendence of the Nature with regard to number (and to these "speculations").

Incidentally, the same relation between Nature and number (mathematical speculations), but considered from another point of view, seems to be sometimes indicated by the term *mathematica practica*—practical mathematics—used in 15th century taxonomies of knowledge. "Practical mathematics" would be understood here not in the sense of "mathematics that serves for practical purposes" but rather "mathematics that has to be verified through practice" or "mathematics that has to be justified by the experience of Nature". And, consequently, the "mathematical models" of the reality that underlie astronomical previsions, from which not only coherence but also correspondence to the Nature is required.

"Commensurability" or "Incommensurability"? The expression of irrational relations by "numbers": minutiae physicae

Problems with the mathematical description of matter (the celestial spheres) begin with the exposition of the measurement of the circle. The geometry of the circle obliges Martinus to deal with incommensurate magnitudes. Normally, the discussion of the *ratio* of the circumference of a circle to its diameter should appear here together with a discussion of the expression by numbers of the *ratios* between the incommensurable magnitudes. Martinus, however, tries to avoid the problem of incommensurability of magnitudes rather than address it, conscious, *as it seems, of its difficulty*. In fact, he uses the term *commensuratio* "commensuration" instead of "incommensuration", when speaking of the *ratio* of the circumference of the circle to its diameter:

> Commensuratio circumferencie et diametri. Habita... commensuracione dyametri ad circumferenciam..." [Commensuration of circumference and diameter... Having commensuration of the diameter to circumference...] (275v).

Also, he gives the numerical value of π as 22/7, citing Archimedes but in fact at variance with Archimedes' approximation which is $3\frac{10}{71} < \pi < 3\frac{1}{7}$. Martinus considers only the second value and treats it as a fraction, at least he does so at the beginning of the lecture:

> Si enim nota est dyameter et circumferencia nota erit, et econtra. Demonstrat enim Geber [!] "De curvis superficiebus" sicut se habet 22 ad 7 sic se habet circumferencia ad dyametrum circuli, ut patet ad sensum, per circinum probando. [If a diameter is known, the circumference will also be known and vice versa. Geber [!] in De curvis superficiebus proves that the ratio of a circumference to its diameter is like that of 22 to 7. This is apparent to sense, when [the question] is checked using compass] (276r).

Is this a simple lack of precision or a deliberate procedure? Is it possible that Martinus, having in mind the approximations of the *ratio* of the circumference of a circle to its diameter, considers the problem of incommensurable quantities as solved by means of further approximations of irrationals to fractions? Or it is rather that he continues to consider the rough value of $\pi = 22/7$ as correct?

> Rursum, si quis haberet circumferenciam et vellet habere dyametrum multiplicet circumferenciam per 7 et productum dividat per 22. Et quociens est dyameter. Verbi gracia sit circumferencia 44. Multiplicetur per 7 et veniunt 308, que dividantur per 22 et proveniunt 14 dyameter. Et per istum modum poterint commensurari omnes circumferencie suis dyametris et econtra. [Thus, if one knows a circumference [of a circle] and would like to know its diameter, multiply the circumference by 7 and divide the result by 22. Let be circumference equal to 44. Multiplied by 7 results in 308, which divided by 22 results in a diameter of 14. And in this way all circumferences may be commensurated to their diameters and vice versa] (276r).

When Martinus transposes plain trigonometry into the reality of the celestial sphere and tries to express the length of the diameter in degrees, the question of a number or of a ratio of incommensurable numbers—"inherent" in matter returns:

> ...Sed cum totum celum dividitur in 360 gradus, si igitur vis habere dyametrum in gradibus celi, multiplica circumferenciam per 7 et divide per 22. Multiplicando vero per 7 veniunt 2520 et dividendo per 22 veniunt 114 et remanent 12 vicesimesecunde que significant minuta unius gradus. Quod patet: semper sit 60 primus numerus et operando secundum regulas 12 vicesimesecunde fere 32 minuta faciunt. [But, since the heavens are divided into 360 degrees, thus if

you want to know the diameter in the degrees of heavens, multiply the circumference by 7 and divide by 22. Through the multiplication by 7 you will receive 2520, which divided by 22 will give 114 and 12 twentytwo seconds which mean minutes of a degree. So it is clear: let 60 always be the first number and, following the rules, 12 twentytwo seconds will be [equal to] about 32 minutes]. (284v)

Two points are to be noted in Martinus' text: first, the return of the problem of incommensurability of the circumference of a circle and its diameter and, expressed—this time—by a rough approximation to minutes (...*fere 32 minuta faciunt*) and second, the expression of a straight line (diameter) in degrees. Both of them were commonly accepted by mathematicians of that period, with the exception of Bianchini who will try to abandon the old tradition, and to introduce in his concept of trigonometric functions the unitary radius of a circle. In that way, he always expresses the π by number in his trigonometric tables, and gives the values of trigonometric functions in decimal positional fractions.

Trigonometric functions and the knowledge of Nature

Let us come back to Cracow and to consider the section of Martinus' *Summa* devoted to the construction of the table of sines. It opens with the statement:

...Cognicio sinuum est causa cognicionis motuum in orbibus celstibus non solum tabulas de Lineris cernentibus, verum et Alfoncji, et instrumentorum astronomicalium, videlicet albeonis et similium. Ex speculacione enim sinuum et veri motus sunt computati. Nam noscenti sinus et cordas faciliter adveniet corrigere equaciones tabularum, supposita solum ecentricitate alicuius planete in ordine ad totum semidiametrum. [The knowledge of sines is the cause of the knowledge of movements of the celestial spheres. And this concerns not only the tables [of the movements of the heavenly bodies] of Lineris, but also of Alphonsus, and it [concerns equally] astronomical instruments, the Albion and the like. For the true movements [of planets] are also calculated through the computations of sines. To those who know sines and chords, it is easier to correct the equations [given in] the tables, knowing only the eccentricity of a planet related to the whole semidiameter]. (286v).

Et tota veritas thabularum in cognicione sinuum et cordarum pendet. Et non solum in his que supercelestia sunt, sed et in geometricis. [And the whole truths of tables depends on the knowledge of sines and chords. And not only in supra-celestial things, but also in geometry] (284v).[24]

Subsequently, Martinus refers to the 14th century table of sines, based on the Ptholemean table of chords:

Semidiameter autem qualibet in 60 partes dividitur, quare dyameter in 120, ex quorum cognicione in cognicionem sinuum devenitur; ex sinibus autem in cognicionem thabularum verorum motuum planetarum et revolucionis firmamenti. [Each semidiameter is divided into 60 parts, thus the diameter in 120. From this knowledge one proceeds to the knowledge of sines; and from sines into the knowledge of the tables of the true movement of planets and of the revolutions of the firmament.] (286v)

The status of sines and of the umbrae *(tangent and cotangent): are they lines or ratios?*

In his geometrical introduction to trigonometry, Martinus refers to the Pythagorean theorem and then considers sines and chords as lines:

> Et hec linea quandoque sortitur nomen sinus, quandoque nomen corde. Ex quibus sinibus et cordis omnes thabule [sunt] composite . . . [And to that line sometimes is attributed the name sine and sometimes chord. And of these sines and chords are composed all tables . . .] (284v)

As for the character of these lines, Martinus expresses an opinion that sines and chords, as considered on the celestial sphere, do not exist as a reality but are the result of the imagination of mathematicians (*linee imaginarie*).

Apart from the concept of sines as lines, real or imaginary, in Martinus' lecture there is also a definition of the astronomical functions conceived as ratios and as proportions:

> [. . .] Quibus sic stantibus ad quantitatem corde et sinus i.e. proporcionem cum quolibet arcu, secundum quam proporcionem omnes thabule de veris motibus sunt composite, taliter accede. [. . . All considered, proceed in this way to obtain the quantity of chords and sines i.e. to [their] proportion to each arc. According to this proportion all tables of the true movement (of the celestial bodies) have been composed.] (285r)

The concept of proportionality, expressed through the numerical values, is even more stressed when the composition of the tables of *umbrae* is considered. According to the Muslim tradition, going back to the ninth century, the *umbra recta* and *umbra versa* result from the *ratio* of sines to cosines and vice versa. Since, obviously, Martinus does not use the term "cosine", he recommends using the table of sines for the composition of the tables of *umbrae* (*recta* and *versa*) taking for the composition of the table of tangents: tg α = sine α/sine $(90 - \alpha)$, for an $\alpha = 1^\circ, 2^\circ, 3^\circ$ etc. His table is composed for the length of gnomon equal 12 parts.

> Componitur autem thabula in hunc modum: [. . .] capias sinum rectum correspondentem uni [sic] gradui ex presenti thabula et reduc usque in secunda. Deinde capias sinum rectum eciam in thabula presenti, positum contra 89 gradus . . . [Proceed in this way in the composition of the table: [. . .] you will take the sine that corresponds to one degree from the present table and reduce it to seconds. Then take the sine, from the same table, disposed against 89 degrees . . .] (286v–287r)

Thus, when Martinus uses the concept of "proportion" for tabulating tangents (*umbra recta*) he is no longer thinking of proportionality between lines or between lines and arcs but of proportionality of numerical expressions (in degrees, minutes, etc. reduced to seconds) of sines to cosines.

GIOVANNI BIANCHINI'S EXPOSITION OF TRIGONOMETRIC FUNCTIONS. THE SINE CONSIDERED AS A LINE AND AS A *RATIO* EXPRESSED BY NUMBER: A DECIMAL FRACTION

The first step: trigonometric tables computed for $R = 60 \times 10^n$

Bianchini, in contrast to Martinus Rex who acknowledges the contribution of the 14th century mathematicians to trigonometry, refers himself directly to Ptolemy. He does it simply to demonstrate that any progress, in the matter of trigonometric tables since Ptolemy's times, is due to his own inventions. These inventions, it seems, result from Bianchini's research into the proper numerical expression of the *ratios* of magnitudes. Furthermore, Bianchini was particularly concerned by the fundamental inconsistency, inherent, according to him, in earlier trigonometrical tables and resulting from the fact that the *ratios* of "heterogeneous magnitudes", such as "arcs" and "lines", were tabulated in them. Moreover, he perceived it as incoherent that sines, being lines, should be expressed by degrees, since this was the unit of measure of angles and arcs. This inconsistency, according to Bianchini, is resolved by the use of numbers of a special status, namely by decimal fractions, suited to express all the ratios of all magnitudes, even of the heterogeneous ones.

Bianchini's *Declaratio tabularum compositarum per Johannem Blanchinum de sinu et arcu* (cap. 1) begins as follows:

> Primo notandum est quod [...] Ptholomeus et alii posteriores operati sunt per tabulas de sinibus seu de cordis medietatis [arcus] compositis per gradus et minuta cum suis fractionibus sexagenariis, que in operationibus multiplicandi et dividendi, seu in radicibus extrahendis, oportet omnia reducere ad minores fractiones; que [operationes] certe tediose et laboriose sunt et de facili per operantes comittitur error—nec videtur consonare quando arcus, cuius quantitas per gradus et minuta denominatur, quod corde ipsius, ex lineis rectis composite, per gradus et minuta determinantur. Ex quibus proposui in tabulis meis sinus per numeros componere, presupponendo dyametrum circuli 120, prout Ptholomeus proposuit, sed sibi addendo tres fractiones ad invicem decenarias. Erit ergo totus dyameter 120 000 et per consequens semidyameter 60 000. [First, it is to be noted [...] that Ptolemy and other later [authors] have operated using tables of sines or of chords of half the arc, composed with degrees and minutes and their sexagesimal fractions. In multiplications, divisions and extractions of roots it was necessary to reduce everything to the smallest fractions. Certainly, these operations are tedious and laborious and an error is easily committed. [Moreover] it does not seem consistent when chords, being straight lines, are denoted

> with degrees and minutes [which are proper to express] the quantity of an arc.] Therefore, I have proposed in my tables to compose sines by numbers. I posit the diameter to be 120, according to Ptolemy, but I add to it three fractions that are decimals of one another. Thus, the whole diameter will be 120 000, and consequently the semidiameter 60 000.] (ms BJ 556, f.11va–11vb).[25]

Since almost all values of trigonometrical functions express *ratios* between incommensurable magnitudes, it is not surprising that Bianchini's ("naive") concept of a real number appears in the trigonometrical part of his work.[26] The "modern tool" is used for the purpose, namely, of decimal positional fractions. The advantages of Bianchini's tables manifest themselves in the application of them for astronomical computations. In chapter 3 of the *Canones* mentioned above the rule of multiplication of sines is given:

> Quando contingeret multiplicare sinus per sinum, generaliter productum fractionum semper in fractionibus augetur. Resecande sunt quatuor figure ultime ad dextram, quia sunt fractiones fractionum. [. . .] In reliquis autem remanentibus, ut supra dictum est, tres ex ipsis ad dexteram sunt fractiones millenarie **graduum**, et si alie supersunt ad sinistram sunt **gradus** integri seu **partes ex 120 partibus** totius dyametri. [When it occurs to multiply sine by sine, [since] generally the product of fractions always augments in fractions, the four last numerals [digits] on the right are to be cut off [from the product] because they are fractions of fractions. But of those that remain, as was said above, three of those at the right are millenary fractions of **degrees**, and if others remain to the left, they are whole **degrees**, or **parts of the 120 parts** of the total diameter].

This procedure, involving degrees combined together with decimal fractions, is also presented in Book III of the *Flores Almagesti* where, as Bianchini states:

> [. . .] per 12 capitula potissime per figuras geometricas et regulas arismetrice [sic] declaravi omnes [11ra] cordas seu sinus subtensos [. . .] super quibus composui tabulas ad propositum, per numeros **discretos** et **continuos**. [. . . I have vigorously declared in 12 chapters all that concerns chords and sines, by means of diagrams and rules of arithmetic, [. . .] and I have constructed tables for that purpose, with **discrete** and **continuous** numbers.]

The "discrete numbers" are obviously degrees, the relics of Ptolemean tables still persisting in Bianchini's. However, "continuous numbers" are Bianchini's own invention; they are, according to his own terminology "fractions as reciprocally decimal". Manuscript sources show that very soon after this, Bianchini makes a further step toward the arithmetization of trigonometric tables: he abandons the remnants of sexagesimal calculus and compiles purely decimal tables of tangents (calculated for $R = 10^3$) and of cosecants ($R = 10^4$). Both are included in the set of the *Tabulae magistrales*:

1. *Tangents:* Tabula magistralis quarta. Numerus divisionis sinus primi cuiuslibet gradus quarti circuli per sinum eius secundum. [Magistral table number four. The quotient of the division of the prime sine of each degree of the quarter of a circle by its second sine] (BJ 556, 52r–52v).[27]
2. *Cosecants:* This table (untitled, BJ 556, 53r–53v) gives values calculated for $R = 10^4$. Bianchini refers to this table in the *Additiones Canonum primi mobilis*.[28] All Bianchini's trigonometric tables are calculated for intervals of 10 min.

The process of the extension of the decimal number system to decimal positional fractions

Bianchini was the first mathematician in the West to use purely decimal tables of trigonometric functions. It most probably occurred not later than in early 40s, since a complete exposition of the arithmetic of decimal positional fractions was ready as early as 1442 (the year of the dedication to Leonello d'Este of the treatise "Compositio instrumenti"). (A little earlier similar tables drawn up by Al-Kashi (died 1429), had appeared in the East.) In 1467, more than 20 years after Bianchini's tables, Regiomontanus produced his decimal table of tangents calculated for $R = 10^5$ and intervals of 1°, and then his decimal tables of sine/cosine for $R = 10^7$ and 1 s intervals.

Bianchini's idea of decimal positional fractions evolved. In the *Arithmetica*, where the principle of the potential extension of the decimal positional number system from integrals to fractions is given, Bianchini considers decimal fractions in Simon Stevin's way. For him, as for Stevin, "decimal fractions" are integrals.[29] He gives a precise definition of such a "decimal fraction" when he declares:

> [...] omnis figura firmata in ordine numerorum denotat fractionem decenariam loci figurae immediate sequentis ad sinistram, ut puta 342. Dico quod 2 denotat duo decimi unius decenae, et 4 sunt 4 decimi unius centenarii etc. [... each figure, fixed in the order of numbers, denotes a decimal fraction of the figure that follows it immediately on its left. For example, [in the number] 342, I say that 2 denotes two tenths of a ten, and 4 are 4 tenths of a hundred etc.][30]

But, subsequently, Bianchini goes further than Stevin will go a 100 years later. The next step is found in the *Compositio instrumenti.* This treatise, dedicated to Leonello d'Este, of which the manuscript copy is preserved in Modena, was recently published by Paolo Garuti.[31] It is devoted to the construction and use of a surveying instrument called "biffa". In his description of the scaling of this instrument, Bianchini uses fractions of the tenth progress. Then, he explains the four arithmetical operations with decimal positional fractions, as applicable to the numerical solutions of the typical surveying problems (based on the similarity of triangles and the use of the Pythagoras theorem).

Obviously, the complete doctrine of decimal positional fractions, exposed in the treatise on surveying, is meant by Bianchini as generally applicable to the solution of computational problems. We learn from the *Compositio instrumenti* of two ways of writing down decimal fractions. The first occurs when readings are taken from the decimal scale of the instrument. Here, the number is presented in the metrological structure, accompanied by the units of measure. Bianchini writes: "*pedes* 0 *untie* 7 *minuta* 4 *et secunda* 6."

The same quantity, however, expressed in seconds, functions as a number, a decimal positional fraction, in the following notation: .746. *secunde* (understood as 0.746 *pedes*).[32]

The number—decimal fraction—liberated from the units of measure, will be used subsequently in arithmetical operations. The points that are seen at the beginning and at the end of number mark its fractional nature. The first point signals the "zero" of integrals. A clear distinction of the fractional part of the number from its integral part, using the decimal point, confirms that Bianchini is really dealing with decimal fractions in the *Compositio instrumenti*. Moreover, it is to be noted that in the multiplication and division of fractions by fractions, Bianchini places the decimal point in the product or in the quotient using the formulae: $10^m \times 10^n = 10^{m+n}$ and $10^m : 10^n = 10^{m-n}$.[33]

MATHEMATICS AT CRACOW UNIVERSITY AT THE END OF THE 15TH CENTURY: JOHANNES GLOGOVIENSIS, LECTURER ON MATHEMATICS IN CRACOW AND IN VIENNA AND THE LEGACY OF BIANCHINI AND REGIOMONTANUS. THE REAPPEARANCE OF THE MEDIEVAL ARITHMETIC

The time of Copernicus' university studies in Cracow, the last decade of the 15th century, is particularly interesting because of the reception of earlier contributions, made by Bianchini and Regiomontanus, to mathematical astronomy. From the 60s on, the new generation of astronomical tables was used in Cracow: first, Bianchini's monumental *Tabulae primi mobilis* accompanied by the *Tabulae magistrales*, and then Regiomontanus' *Tabulae directionum* with the table of tangents—*Tabula fecunda*. But in the same period, the interest in mathematical and astronomical tables was not balanced by an interest in arithmetic. The teaching of "mathematics for astronomy", both in Cracow and in Vienna, was at that time still at the level of 13th century elementary arithmetic. This is true in the case of two eminent professors, Albertus de Brudzewo, professor in Cracow in the 80s and Johannes de Glogovia (Glogoviensis), professor at the end of the 15th and 16th centuries in Cracow and in Vienna.[34] The mathematical (arithmetical) introduction into Brudzewski's *Commentary* on Peurbach's *Theorice nove planetarum* teaches the basics of computation with sexagesimal fractions. As for Glogoviensis, he comments, in Cracow and in Vienna, on Sacrobosco's *Algorismi* as well as the *Algorismus novus de integris* composed by Georg Peurbach in Sacrobosco's tradition. In all these textbooks, number is defined according to Boetian tradition as composed of units (*numerus est cumulatio unitatum* ...) and several pages are devoted to the explanation of the geometrical nature of numbers:

> Ponenda autem est hec divisio numerorum: alius linealis, alius superficialis, alius quadratus, alius cubicus sive solidus sive corporalis. [The following division of

numbers is to be posited: line-number, which is different from a superficial-number, and a square number, and a cube or solid or corporeal number].[35]

The codex BJ 1840, written by Johannes Glogoviensis and owned first by him then by Michael de Wratislavia, is a good example. It contains the following treatises and lectures intended to train astronomers (in Cracow and in Vienna): *Algorismus novus de integris*, *Algorismus de minutiis vulgaribus*, *Algorismus de minutiis physicalibus*, *De regula proportionum sive aliter Regula Mercatorum dicta*, *Opus Algorismi iocundissimi*, *Algorismus novus de integris* mgri Georgii Peurbachii, and *Canon multiplicacionis*. They are followed by Glogoviensis' commentary on Sacrobosco's *Omnia que a primeva ...*, the Vienna lecture by Glogoviensis also based on Sacrobosco, finally the Enigmata: *Si vis scire quot dudum habeat aliquis in bursa ...*, and the *Tabula algoristica de numeris quadratis, cubicis et eorum radicibus.*[36]

In the official lectures on mathematics there is not a trace of Bianchini's *Arithmetica* with the exposition of the principle of decimal fractions, nor of his *Algebra*, both works meant to provide students with mathematical tools for astronomical computations.

GIOVANNI BIANCHINI'S PATTERNS OF COPERNICUS' TRIGONOMETRIC TABLES

Of the four tables of trigonometric functions linked with the name of Copernicus—because they were calculated by him, or only copied by him, or else attributed to him by historians for a time[37]—two are pertinent to our studies: the table of sines for $R = 10^5$ and 10' intervals and the table of secants for $R = 10^4$ and 1° intervals. Both exist in autograph.

The first one is incorporated in the *De revolutionibus*, Book I, chapter 12. Several questions about it remain unsolved (such as the origins of Copernicus' numerical errors, and the differences between the errors in the holograph and the first two editions, etc.). As for the question of the model followed by Copernicus, it still remained unanswered in 1974, when the English version of the *De revolutionibus* was published with Edward Rosen's Commentary.[38] By the end of the 70s, I pointed out the possibility of the influence on Copernicus of Bianchini's tables of decimal trigonometric functions, then recently discovered (first published in 1981). In fact, the arrangement of Copernicus' table corresponds to that of almost all Bianchini's mathematical and astronomical tables, the intervals of 10 s included.

The second of Copernicus' decimal trigonometric tables, the table of secants composed for $R = 10^4$ and 1° intervals, was inscribed by Copernicus between the columns of Regiomontanus' printed *Tabula fecunda*, the table of tangents, calculated on the same base, $R = 10^4$. In fact, Copernicus used the skeleton of Regiomontanus' table, inscribing numerical values, computed by himself, in front of degrees running from 1 to 90 in Regiomontanus' table, as can be seen in numerous editions of the table and its facsimile reproductions.[39] Here, the question of the possible inspiration of Copernicus concerns not so much the decimal base of the table and the 1° intervals, (Copernicus, when inscribing the data, follows simply the pattern of Regiomontanus' table), but rather the introduction of a trigonometrical function other than sine, cosine, and tangent. In the absence of textual evidence, I would not like to suggest that Copernicus' idea of calculating a table of secants was inspired by Bianchini's decimal table of cosecants. But, as I pointed out above, the Cracow astronomers did have at their disposal the set of Bianchini's *Tabulae magistrales*, the table of cosecants included, as early as the 60s of the 15th century.

15TH CENTURY MATHEMATICS AND THE HISTORY OF EARLY MODERN SCIENCE

The first (successful) attempts at the extension of the concept of number, to include negative values and incommensurable ones, are commonly attributed to 16th century mathematicians: Stifel accepted negative and irrational numbers, Bombelli was seen by historians (E. Bortolotti) as the forerunner of Descartes in the use of a "unit segment" for the geometrical expression of number, and Stevin as the inventor of the decimal positional fractions. Some of these ideas and techniques of computation, however, were expounded about a 100 years earlier by Bianchini in treatises intended by their author to help in the university teaching of "mathematics for astronomy".

Not all these inventions were adopted by astronomers in the course of the 15th century. The idea of the table of sines calculated on the pattern 60×10^n was almost immediately followed by Peurbach. By the time of Regiomontanus' stay in Venice, in the early 1460s, Bianchini's *Flores Almagesti*, as well as his *Tabulae primi mobilis* (together with the *Tabulae magistrales*, including the decimal tables of tangent and cosecant), was circulating in Italy. As for the shape of the trigonometric tables of Copernicus, it seems to me that the legacy of Bianchini, known in Cracow since the second half of the 15th century, permits us to see in Copernicus' achievement the reminiscence of Bianchini's tables of decimal trigonometric functions.

The evolution of the concept of number that occurred in the 15th century scientific milieus, from Boethius' version of the Pythagorean idea of number toward the "justified" use of a (naive) concept of real number, in fact was motivated by the requirements of the university teaching of astronomy. It is not surprising, therefore, that the extension of the concept of number to negative values—a consequence of the use of algebra in astronomical computations—and to irrationals approximated to decimal fractions, was first accomplished within the framework of "mathematics for astronomy". Within the same framework initiated the evolution of the concept of trigonometric functions, from the lines—sides of a right triangle—toward the ratios of numbers in a circle with a unitary radius. Though the tables of decimal trigonometric functions contributed to the simplification of computations, they have been not invented, according to Bianchini, for only practical purposes: Bianchini was concerned with the problem of the proper expression of the *ratio* between "heterogeneous" magnitudes, arcs, and lines in a circle. According to him, numbers, the decimal positional fractions, were the right solution to the problem. Among the consequences of the introduction of the decimal base to the trigonometric tables will be the simplification of the formula for the circumference of a circle from $2\pi r$ to 2π, with the numerical value of π approximated to a decimal fraction. The next step will consist in the expression of trigonometric functions by means of π. We are here at the origins of the concept of circular trigonometric functions, developed further in the 17th century mathematics.

ACKNOWLEDGMENTS

Studies on the 15th century astronomical sources preserved in Italy were possible thanks to the University of Harvard Center for Italian Renaissance Studies, Villa I Tatti, Florence, and the Centro Studie Incontri Europei P. G. Frassati, Rome. The Institute for the History of Science (Polish Academy of Sciences, Warsaw) facilitated me an examination of the astronomical manuscripts in Cracow, Biblioteca Jagiellonska, and the Swedish Royal Academy of History and Antiquities (Stockholm) the examination of Copernicus' manuscripts preserved in Uppsala, the University Library. It is my pleasure to thank Mrs Jehane Kuhn for the improvement of my English.

NOTES

[1] A chair of mathematics and astronomy was founded at Cracow University in the first decade of the 15th century by a certain Stobnerus, citizen of Cracow. The document of the foundation (a copy from 1472) is still preserved in the Archives of the Jagellonian University, ms. 44, f. 45r. Professor "Stobnerianus" was responsible for mathematical and astronomical formation of students of the *quadrivium*, particularly for their knowledge of astronomical tables and their skills in predicting eclipses of the Sun and Moon: "... collegiatus domini Stobneri duos actus faciat. Pro uno legat in mathematica hoc ordine, videlicet Euclidem, perspectivam, arismetricam et musicam et *Theoricam planetarum*, demum *Tabulas Alphoncii*, premisso *Algorismo minuciarum*. [...] Pro secundo actu practicet et publicet notoria [...] eclipses, almanach et minuciones [sanguinis] pro honore Universitatis". Cf. ms. in the Archiwum Uniwersytetu Jagiellońskiego, nr. 68, p. 18. The collection of astronomical manuscripts, still preserved in the Jagellonian Library, permits us to reconstruct the activity of Cracow mathematicians and astronomers, from as early as the end of the 14th century. It was the subject of studies exemplified by the editions (or at least partial editions) of mathematical and astronomical texts by L. A. Birkenmajer, "Krakowskie tablice syzygiów dla r. 1379 i 1380. Przyczynek do dziejów astronomii w Polsce w XIV wieku", in *Rozprawy Akademii Umiejętności, Wydział matematyczno-przyrodniczy,* seria II, t.1. (Kraków, 1891), pp. 261–285. J. Dobrzycki, "New sources for the history of calendar reform", in Proceedings No. 2. XIV International Congress of the History of Science, Tokyo & Kyoto, 1974, pp. 35–36. J. Dobrzycki, "The Tabulae resolutae", in M. Comes, R. Puig, and J. Samsó (eds.), *De astronomia Alphonsi Regis* (Barcelona, 1987), pp. 71–77. R. L. Kremer and J. Dobrzycki, "Alphonsine Meridians: Tradition Versus Experience in Astronomical Practice c. 1500", *Journal for the History of Astronomy* 29(2):187–199. G. Rosińska, "Une table astronomique de Laurent de Raciborz. Le commentaire qui l'accompagne", *Mediaevalia Philosophica Polonorum* XIX:141–147 (1974) (there are some doubts concerning the attribution to the table "Radices ad meridianum Cracoviensem A.D. 1420 completo et valent ad tabulas sequentes pro instrumentis Campani" to Laurent of Raciborz, but there is no doubt that this table was compiled in Cracow about 1420). G. Rosińska, "Sandivogius de Czechel et l'école astronomique de Cracovie vers 1430", *Organon* 9:217–229 (1973). G. Rosińska, "Instrumenty astronomiczne na Uniwersytecie krakowskim w XV wieku" ("Astronomical instruments at Cracow University in the 15th century"). *Studia Copernicana*, Vol. XI. Wrocaw, Ossolineum 1974 (in Polish and Latin, English summary).

[2] G. F. Vescovini, "Bianchini Giovanni", in *Dizionario Biografico Degli Italiani*, Vol. X (1968), pp. 194–196, where the references to manuscript sources and early prints are given.

[3] L. Thorndike, "Giovanni Bianchini in Paris Manuscripts", *Scripta Mathematica*, Vol. 16, 1950, pp. 5–12 and 169–180. L. Thorndike, "Giovanni Bianchini in Italian manuscripts", *Scripta Mathematica*, Vol. 19, 1953, pp. 5–13. E. Poulle, *La bibliothèque scientifique d'un imprimeur humaniste au XV*[e] *siècle. Catalogue des manuscrits d'Arnaud de Bruxelles à la Bibliothèque Nationale de Paris* (Librairie Droz: Genève, 1963), pp. 38–44, pp. 59–60, pp. 73–74. G. Rosińska, Scientific Writings and Astronomical Tables in Cracow. A Census of Manuscript Sources. *Studia Copernicana*, vol. XXII, Wrocław 1984. Nrs. 218, 429, 485, 875, 1702 and II. 1–67. W. Kokott signals the adaptation of Bianchini's planetary tables to the latitude of Leipzig (51°): "Syzygia as Pivots: An Unusual Mid-Fifteenth-Century Working Ephemeris", *Journal for the History of Astronomy* 29: 129.

[4] Martinus' Summa is preserved in Cracow, Jagellonian Library (Biblioteka Jagiellońska), ms. BJ 1927, ff. 250r–318r and in Oxford, Bodleian Library, ms. Can. misc. 499, ff. 222ra–258vb. Marcin Rex (Król) de Premislia (alias de Żurawica) see L. A. Birkenmajer, "Marcin Bylica z Olkusza oraz narzędzia astronomiczne, które zapisał Uniwersytetowi Jagiellońskiemu w roku 1493". Kraków 1892, pp. 21–27 (Cap. III. Marcin z Żórawic, jeden z profesorów Marcina Bylicy). Z. Kuksewicz, "Marcin Król z Żurawicy. Stan Badań", in *Materiały i Studia Zakładu Historii Filozofii Starżytnej i Średniowiecznej PAN. Seria A: Materiały do historii filozofii średniowiecznej w Polsce*, Vol. I, pp. 118–140. Zathey, "Biblioteka Jagiellońska w latach 1364–1492", in *Historia Biblioteki Jagiellońskiej* (Kraków, 1966), pp. 108–109 where a list of Martinus' treatises preserved in the Jagellonian Library is given. J. Dianni, "Pierwszy znany traktat rękopiśmienny w literaturze matematycznej w Polsce. Algorismus minutiarum Martini Regis, de Premislia". ("*Algorismus minutiarum* by Marcin Król". in Polish, English summary), *Kwartalnik Historii Nauki i Techniki* (*Quarterly Journal of the History of Science and Technology*) 12:269–288 (1967, nr. 2. G. Rosińska, "Nieznany traktat astronomiczny Marcina Króla z Żurawicy" ("An unknown astronomical treatise by Martinus Król of Żurawica" (in Polish, Latin, English summary), *Kwartalnik Historii Nauki i Techniki* 17:227–233, nr. 2, and G. Rosińska, "Scientific Writings and Astronomical Tables in Cracow..." *op. cit.* nrs. 751, 767, 688, 1203, 1211, 1213, 1228, 2228, 2243. According to L. A. Birkenmajer, Martinus was in Prague in 1445 and in Bologna in 1447–1449, where he taught astronomy in 1449: L. A. Birkenmajer, "Marcin Bylica z Olkusza oraz narzędzia astronomiczne które zapisał Uniwersytetow Jagiellońskiemu w roku 1493." Kraków, 1892, pp. 22–27 and pp. 113–119. U. Dallari, "I Rotuli dei Lettori Legisti e Artisti dello Studio Bolognese dal 1384 al 1799". Vol. 1, Bologna 1888, p. 26, col. 2: "Ad lecturam Astronomie: D. M. Johannes de Fondis arcium et medicine doctor, D. M. Martinus de Polonia arcium doctor". Martinus died in Cracow in 1452, see M. Kowalczyk, "Przyczynki do biografii Henryka Czecha i Marcina Króla z Żurawicy", in *Biuletyn Biblioteki Jagiellońskiej*. Vol. 22 (1971), pp. 87–91.

[5] G. Rosińska, "A Chapter in the History of the Renaissance Mathematics: Negative Numbers and the Formulation of the Law of Signs (Ferrara, Italy ca. 1450)". *Kwartalnik Historii Nauki i Techniki* 41:53–70 (1996), nr.3–4. G. Rosińska, "Decimal Positional Fractions. Their Use for the Surveying Purposes (Ferrara, 1442)". *Kwartalnik Historii Nauki i Techniki* 40:17–31 (1995), nr. 4.

[6] A. Gerl, "Trgonometrisch-astronomisches Rechnen kurz vor Copernicus. Der Briefwechsel Regiomontanus-Bianchini". Boethius Bd. XXI, Stuttgart 1989, where the author seeks to present Bianchini as a (possible) partner of Regiomontanus in discussions of problems of the spherical astronomy. See p. 26, pp. 265–269, pp. 331–335.

[7] L. A. Birkenmajer, Mikołaj Kopernik (Kraków 1900), pp. 21–11, p. 25, p. 28, pp. 60–64, p. 162, pp. 228–229.

[8] J. L. Jervis, "Cometary Theory in Fifteenth-Century Europe". (*Studia Copernicana*, vol. XXVI). Wrocław 1985. Appendix B: "Toscanelli's Mathematical Computations", pp. 162–169.

M. Folkerts, "Regiomontanus als Mathematiker". *Centaurus* 21(3–4):214–245 (1977). E. Glowatzki and H. Göttsche, "Die Tafeln des Regiomontans: Ein Jahrhundertwerk". *Algorismus*, Heft 2, Münich 1990. W. Kaunzner, "Über Regiomontanus als Mathematiker", in Wien (ed.), *Regiomontanus-Studien*, (1980), pp. 125–145. W. Kaunzner, "Zum Stand der Westeuropaeischen Mathematik zur Zeit der Entdeckung Americas", in S. S. Demidov et al. (eds.), *Amphora. Festschrift fur Hans Wussing zu seinem 65. Geburtstag.* (Basel: Boston, Berlin, 1992), pp. 362–374. N. M. Swerdlow, "Regiomontanus's Concentric-Sphere Models for the Sun and Moon", *Journal for the History of Astronomy* 30(Part 1): 1–23.

[9] G. Arrighi (ed.), "Piero della Francesca, Trattato d'abaco. Dal Codice Ashburnhamiano 280 (359*–291*) della Biblioteca Medicea Laurenziana di Firenze", a cura e con introduzione di Gino Arrighi. Pisa, Domus Galilaeana 1970, where a bibliography of principal works by Arrighi is given on pp. 13–13, notes 9–10.

[10] R. Franci and L. Toti Rigatelli, "Introduzione all'aritmetica mercantile del Medioevo e del Rinascimento", Urbino 1982; L. Toti Rigatelli, "Matematici fiorentini del tre-quattrocento", in *Symposia mathematica*, Vol. 27, 1968, pp. 3–67.

[11] J. Sessiano, "On an Algorithm for the Approximation of Surds From a Provencal Treatise", in C. Hay (ed.), *Mathematics from Manuscript to Print* (Oxford: Clarendon Press, 1988), pp. 30–55.

[12] W. V. Egmond, "Practical Mathematics in the Italian Renaissance. A Catalog of Italian Abacus Manuscripts and Printed Books to 1600", *Supplemento agli Annali dell'Istituto di Storia delle Scienze di Firenze*; Firenze 1980, fasc. 1.

[13] E. Giusti, "L'algebra del Trattato d'abaco di Piero della Francesa: Osservazioni e Congetture", *Bollettino di Storia delle Scienze Matematiche* Vol. XI, 1991, pp. 55–83. S. A. Jaywardene, "The Influence of Practical Arithmetic on the Algebra of Rafael Bombelli", *Isis* 64:510–523 (1973).

[14] G. Rosińska, "The 'Fifteenth-Century Roots' of Modern Mathematics. The Unit segment. Its Function in Bianchini's De Arithmetica, Bombelli's L'Algebra and Descartes' La Géometrie", *Kwartalnik Historii Nauki i Techniki*. Vol. 41:59–62 (1996), nr. 3–4.

[15] Nicomachus of Gerasa "Introduction to Arithmetic". Translated into English M. L. D'Ooge. With Studies in Greek Arithmetic by F. E. Robbins and L. C. Karpinski. (New York, 1926), p. 190.

[16] Ms. BJ 1840, ff. 19r, 37r, 43r–44r.

[17] "When integers are divided by fractions they are multiplied by unity in proportion to the divisor. When fractions are divided by integers they are multiplied by unity in proportion to the number divisor..." G. Rosińska, "The 'fifteenth century roots' of modern mathematics..." *op. cit.* pp. 58–60.

[18] The dating of Bianchini's *Arithmetica* to ca. 1440 results from the fundamental character of this work with the respect to other Bianchini's mathematical and astronomical writings (in some of which the Arithmetica is quoted). Furthermore, Bianchini accquired his training in practical mathematics before 1427 (the year of Bianchini's coming to Ferrara). As for the dating of Bianchini's *Compositio instrumenti*, essential for the dating of the introduction of the arithmetic of decimal fractions into European mathematics, I refer to the arguments in favor of the year 1442 given in G. Rosińska, "Decimal positional fractions..." *op. cit.* pp. 28–29, note 5.

[19] G. Rosińska, "Kwadratura kola i 'liczba π' w nauczaniu matematyki na Uniwersytecie krakowskim w pierwszej polowie XV wieku. Recepcja Archimedesa *De mensura circuli* poprzez Tomasza Bradwardina *Geometria speculativa*". ("The quadrature of circle and 'number π' in the teaching of mathematics at Cracow university in the first half of the 15th century. Reception of Archimedes' *De mensura circuli* by means of Thomas Bradwardinus' *Geometria speculativa*"). In Polish and Latin. English summary. *Kwartalnik Historii Nauki i Techniki*. Vol. 42, nr. 2, 2000, pp. 49–62.

[20] Treatise *Circulus obliquus qui signifer nuncupatur dividitur in sex signa*... is preserved in the Jagellonian mss.: BJ 459, ff. 3ra–4vb; BJ 563, ff. 240ra–242rb; BJ 602, ff. 131r–133r. All mss. come from the 20s of the 15th century.
[21] BJ 1929, ff. 151r–183v.
[22] *Algorismus minutiarum*: mss. BJ 1844, ff. 275r–295v; BJ 1859, ff. 41r–56r; BJ 1927, ff. 189r–211v. J. Dianni, "Pierwszy znany traktat rękopiśmienny..." *op. cit.* pp. 269–289. *Geometria practica*: mss. BJ 1865, ff. 39r–51v; BJ 1968, ff. 1r–10v. L. A. Birkenmajer [ed.], "Marcina Króla z Przemyśla Geometria praktyczna" (Warszawa, 1895).
[23] All quotations of Martinus' *Summa* come from the ms. BJ 1927.
[24] From the point of view of paleography both readings, *veritas* and *varietas*, are admissible; *veritas*, however, seems to be justified by the context. I consulted also the ms. Cm 499 f. 241vb.
[25] A similar text from the Flores Almagesti, ms. BJ 558, f. 17r was published by L. A. Birkenmajer, "Flores Almagesti. Ein angeblich verlorengegangener Traktat Giovanni Bianchini's, Mathematikers und Astronomen von Ferrara aus dem XV. Jahrhundert" (Extrait du Bulletin de l'Académie des Sciences de Cracovie). Cracovie 1911. Birkenmajer was the first to consider Bianchini's authorship of the sine table ($R = 60 \cdot 10$), on the base of the 15th century note in the margin of the table in the ms. BJ 600 p. 268.
[26] An historical perspective on the foundations of the "pragmatist" concept of real number, present in astronomy since Babylonian times, is given in N. Bourbaki, "Éléments d'histoire des mathématiques". Nouvelle édition revue, corrigée et augmentée. (Paris: Hermann 1974), pp. 184–195, pp. 184–191.
[27] Fragments of the Tables of tangent and cosecant was published by G. Rosińska, "Tables trigonométriques de Giovanni Bianchini", *Historia Mathematica*. Vol. 8, 1981, pp. 49–50.
[28] Ms. BJ 556, f. 5ra–7va: *Addiciones Canonum primi mobilis ordinate per dominum Iohannem Blanchinum.*
[29] In fact, Stevin states in the title of his work (the Franch version, Leyde 1585) as follows: "La Disme. Enseignant facilement expedier par nombres entiers sans rompuz, tous comptes se rencontrans aux affaires des Hommes". See E. J. Dijksterhuis' comment on this subject in E. J. Dijksterhuis, "Simon Stevin. Science in Netherlands around 1600", in R. Hooykas and M. G. J. Minnaert (eds.), The Hague (Martin Nijhoff, 1970), p. 19.
[30] In translations of Bianchini's *Arithmetica* I refere to my critical edition of the Latin original, forthcomming in the Studia Copernicana series. I have established the final text on the base of the six still extant 15th century manuscripts preserved in Cracow, Paris, Bologna, Perugia, and Vatican.
[31] G. Bianchini, "*Compositio instrumenti.* (Cod. Lat. α. T. 6. 19) della Biblioteca Estense di Modena". A cura di Paolo Garuti con introduzione di Gino Arrighi, in *Rendiconti di Classe Lettere e Scienze Morali e Storiche* Vol. 125 (1), 1991, pp. 95–127. (*Istituto Lombardo Accademia di Scienze e Lettere,* Milano 1992).
[32] Ms. Modena B.E. Lat. 145, f.6 . Ed. P. Garuti, *op. cit.* p. 116. See G. Rosińska, "Decimal..." *op. cit.* p. 24.
[33] Ms. Modena B.E. Lat. 145, f.6 . Ed. P. Garuti, *op. cit.* p. 116. See G. Rosińska, Decimal..." *op. cit.* p. 24.
[34] L. A. Birkenmajer, "Stromata Copernicana". Kraków 1924, pp. 103–126.
[35] Ms. BJ 1840, f. 58v.
[36] Ms. BJ 1840, f. 1r: *Algorismus novus de integris*; f. 8r: *Algorismus de minutiis vulgaribus*; f. 10r: *Algorismus de minutiis phisicalibus*; 10v: *De regula proportionum sive aliter Regula Mercatorum dicta*; f.11v: *Tercia pars est de proporcionibus. Unde quinque sunt genera proportionum*; f.14r: *Opus Algorismi iucundissimi*; f. 19r: *Algorismus novus de integis mgri Georgii Pawrbachii* [!] *Wienensis*; f.37r: *Algorismus novus de integris compendiose studioseque more Italorum compilatus*; f.40v: *Canon multiplicationis*; Glogoviensis' Comentary on Sacrobosco's *Omnia que a primeva*...; f. 58r: *Inquit Ptholomeus in sapienciis Almagesti* [...] *Astronomie studium super Arismetrice et Geometrie demonstracionibus certissimis est fundatum*... with

Glogoviensis' Commentary on ff. 65r–69v; f. 80r: *Ex quo artes mathematice propter difficultatem Arismetrice . . .*; f. 92v: *Enigmata*; f. 94r: *Tabula algoristica de numeris quadratis, cubicis et de eorum radicibus.*

37 I have argued elswhere that the "fourth trigonometric table", linked with the name of Copernicus because ascribed to him or to Rheticus—namely the large table of sines/cosines, appended to Copernicus' "De lateribus et angulis triangulorum" [. . .] Vittembergae 1542—is essentially Regiomontanus' Table, calculated by him on the $R = 10^7$ and one minute intervals. It was first published in 1541 (Norimbergae, apud Iohannem Petreium). A 15th century manuscript copy of this Table is still preserved in ms. BJ 606, f. 171r–180r, where its title runs as follows: "Tabula sinuum nova Bude confecta [?] per magistrum Johannem de Regio monte 1468". See G. Rosińska, "Nie przypisujmy Rhetykowi dziela Regiomontana . . .". *Kwartalnik Historii Nauki i Techniki*. Vol. 28:615–619, 1983.

38 E. Rosen: "Copernican scholars have not yet discovered what model, if any, Copernicus followed in transforming Ptolemy's sexagesimal Table of Chords into an early form of the Modern Table of Natural Sines", in Nicholas Copernicus *Complete Works*. Vol. I. "On the Revolutions . . ." Ed. J. Dobrzycki, Translation and Commentary by E. Rosen. Warsaw-Cracow 1978, p. 363.

39 N. Copernicus. *Complete Works*. Vol. IV. "The Manuscripts of Nicholas Copernicus' Minor Works. Facsimiles". Warsaw-Cracow 1992, plate XXX. 62. See also P. Czartoryski, "The Library of Copernicus", in *Science and History. Studia in Honor of Edward Rosen*. (Studia Copernicana Vol. XVI) (Wrocław, 1978), p. 366.

JOSÉ CHABÁS

THE UNIVERSITY OF SALAMANCA AND THE RENAISSANCE OF ASTRONOMY DURING THE SECOND HALF OF THE 15TH CENTURY[1]

It is a well known fact that learning in the medieval universities was generally organized in four faculties: Arts, Law, Medicine, and Theology. All students wishing to attend any of the last three, the "higher faculties", had to go first through the Arts Faculty, which was thus considered as a preparation period for higher education.

Broadly speaking, the Arts Faculty was in charge of teaching the classical disciplines of the *trivium* and *quadrivium*. The three branches of the *trivium* were the verbal disciplines (*artes sermonicales*) of grammar, rhetoric, and logic, whereas the *quadrivium* was composed of the four mathematical disciplines: arithmetic, geometry, astronomy, and music. However, this classification of the liberal arts varied with time and from place to place. The fact remains that teaching of liberal arts, and particularly that of astronomy, was strongly conditioned by its role as a way to get to the higher faculties. Therefore, it is not surprising that astronomy was seen by most students as a tool for other disciplines, medicine in the first place, but not as a discipline in itself. Note also that the level of astronomy required for the purposes of medical astrology did not meet a high standard, and any basic astronomy would suffice. This is what astronomical textbooks indicate as well.

The most widely used textbook in this discipline during the late Middle Ages was a very simple treatise written in the 13th century by an Englishman called John of Holywood, better known as Sacrobosco. This book, entitled *De sphera*, survives in hundreds of manuscripts, which are found in libraries throughout Europe, thus giving an indication of its popularity among university students.[2] Sacrobosco's treatise, as well as similar less-diffused works, did not pay much attention to the motions of the planets. But this was a subject that by Sacrobosco's time had reached a very sophisticated level, mainly due to the compilation of astronomical tables in Muslim Spain. Then, in order to explain the foundations of planetary motions and as a complement to Sacrobosco's *Sphere*, another type of textbook appeared with the generic name of *Theoricae planetarum*.[3] Different versions of the *Theoricae planetarum* are also found in many miscellaneous codices, but the number of copies of all the versions together is much smaller than the number of copies of Sacrobosco's *Sphere*.

Needless to say, these textbooks were based on Ptolemy's astronomy, as presented mainly in his *Almagest*. This book lay behind the curriculum although it was rarely used in the university.[4] This is confirmed by the few copies of the *Almagest* preserved

M. Feingold and V. Navarro-Brotóns (eds.), Universities and Science in the Early Modern Period, 29–36.

in Latin manuscripts, despite the fact that Ptolemy's treatise was known in Latin since it was first translated from Arabic into Latin in Toledo by Gerard of Cremona in 1157.

So, for a long time astronomical activity continued to develop mainly outside the academic world, as was obviously the case before the founding of universities. Now, it is difficult to establish how academic astronomy shifted from an auxiliary discipline to a discipline in itself, and when and why separate chairs in astronomy were created.

We know that the Jagiellonian University of Cracow inaugurated a chair in the early 15th century. We also know that John of Gmunden, beginning in about 1420, was the incumbent of the chair of astronomy at the University of Vienna.[5] As for Spain, we have no information of that sort until about 1460. This seems to be the date the first chair of astronomy was created at the University of Salamanca. It is worthwhile noting that it was called "cathedra astrologia", for "astrologia" was the term most commonly used to designate mathematical astronomy, and it included those parts of astronomy not related to predictions or prognostications.

The first person to hold a chair of astronomy at the University of Salamanca, and most likely in any Spanish university, was Nicholaus Polonius.[6] As his name suggests, he probably came from Poland. Not much is known about him, but he is mentioned for the first time in the *Libro de Claustros* of the University of Salamanca for 1464–1474: on 22 March 1464 the vice-rector, Diego de Castro, comments on the decision to fill the chair of astrology after a long absence of Nicholaus Polonius: *per recessum et longam abscenciam vene[rabi]lis viri Nicholai polonij eiusdem cathedre ultimi cathedratici*.

As will be argued later, there is evidence that Polonius was in Salamanca no later than 1460. On the other hand, we do not know the reasons why he left his chair. As a matter of fact, there are many other important questions for which we do not have adequate answers: Why was it decided to create a chair of astrology/astronomy at the University of Salamanca? What were the social needs for it? Why a foreigner? What brought Polonius to Salamanca in the first place? Was not there any local astronomer available?

In his *Diccionario Hispano-Latino* (Salamanca *ca.* 1495), the humanist and grammarian Elio Antonio de Nebrija (1444–1522) wrote a dedication "Al mui magnifico e assi ilustre señor Don Juan de estuniga maestre dela cauelleria de alcantara dela orden del cister". In this text he mentions some of his professors at the University of Salamanca:

> "I dexando agora los años de mi niñez passados en mi tierra debaxo de bachilleres e maestros de gramatica e logica: dexando aquellos cinco años que en Salamanca oi en las mathematicas a Apolonio: en la filosofia natural a Pascual de aranda en la moral a Pedro de osma maestros cada uno en su arte mui señalado..." (f. a.ij.va)

The "Apolonio" mentioned by Nebrija has to be Nicholaus Polonius; there was no other professor of "mathematics" by that name. On the other hand, the 5 years spent by Nebrija at the University of Salamanca cover the period 1455–1460, just before his 10-year sojourn in Bologna. Thus, Nicholaus Polonius had to be present at the University no later than 1460.

What we definitely know about Polonius is that while in Salamanca he composed a set of astronomical tables based on the Alfonsine Tables. Polonius's first table has the heading "Tabule ad meridianum Salamantinum", and the tables that follow are indeed calculated for Salamanca with 1460 as epoch. They use zodiacal signs of 30°, rather than "physical" signs of 60° (as is the case for the Parisian version of the Alfonsine Tables), and the years begin in January 1 (more precisely, at noon of December 31). In fact, these tables seem to be one of the earliest examples for the use of the Alfonsine Tables in the Iberian Peninsula after their compilation in Christian Spain almost two centuries before.[7] Judah ben Moses ha-Cohen and Isaac ben Sid, the two main scientific collaborators working under the patronage of King Alfonso X of Castile, were responsible for the compilation of the Alfonsine Tables and the writing of the accompanying text explaining their use. It is also well known that the Alfonsine Tables compiled in Toledo spread throughout Europe and became the most important tool, if not the only one, for mathematical astronomy for a few centuries.[8] Incidentally, it is at least surprising that during these two centuries from, say, 1280–1460, a period that has been called the "post-Alfonsine era", there is so little evidence of Alfonsine astronomy in the Iberian Peninsula, despite the number of allusions to it found in astronomical works produced in Spain.

Besides his tables for Salamanca, Nicholaus Polonius also wrote some canons to explain their use. Both Polonius's text and tables are preserved in Lisbon, Torre do Tombo, MS 2115, as well as in a complex Latin manuscript of fundamental significance for the history of astronomy in Spain that is now at the Bodleian Library in Oxford (MS Can. Misc. 27). Polonius's canons were based on another text written about a decade before, in 1448, by a Polish astronomer, Andreas Grzymala of Poznan, for his students at the Jagiellonian University of Cracow, thus supporting the suggestion that Nicholaus Polonius came from Poland.[9] Moreover, Polonius's tables for Salamanca follow very closely a particular form of presenting the Alfonsine Tables called *Tabulae Resolutae*, extant in many 15th century copies from Central Europe, especially Poland.[10] Therefore, the University of Salamanca was in fact the gate through which some astronomical knowledge came in from Poland, although originally generated in Spain.

The arrival of Nicholaus Polonius to the University of Salamanca around 1460 and the use of the Alfonsine Tables apparently induced intensive astronomical activity to take place in Salamanca. This is attested by various Latin and Castilian manuscripts containing astronomical material specifically for that city. Among this material are two sets of tables of unknown authorship. They differ from those compiled by Polonius and can be designated as the *Tabule Verificate* and "Tables in Castilian".[11] Nevertheless, both sets share very specific features: they are both based in the Alfonsine Tables, and they are both computed for the beginning of 1461 as epoch, and for the meridian of Salamanca. What we just called "Tables in Castilian" is an extensive set of tables, uniquely preserved in a manuscript in Madrid, Biblioteca Nacional (MS 3385), with two interesting novelties: the headings of the tables are in Castilian and the entries are given in roman numerals. This is a clear sign that there was a sense at the time that a new period was beginning, for these two features greatly differ from those traditionally in use in astronomical tables compiled in the Iberian Peninsula.

These two sets of tables also include some tables composed in Central Europe at the beginning of the 15th century, as well as material derived from the medieval Jewish tradition in astronomy. This mixture, typically from a Spanish milieu at that time, remains a puzzle. It is not possible to assign authorship to either the *Tabule Verificate* or the "Tables in Castilian", even though there are several very good candidates. All of them held the chair of astronomy/astrology at the University of Salamanca in succession from 1464 to 1480: Juan de Salaya, Diego Ortiz de Calçadilla, and Fernando de Fontiveros. They all shared the same astronomical background, and had access to the tables of Nicholaus Polonius, the first incumbent of the chair.

Juan de Salaya succeeded Polonius and held the chair of astronomy at the University of Salamanca from 1464 to 1469; he was then appointed for the chair of logic, and had thus access to a better salary. In any event, he had the necessary skills to compile astronomical tables based on Polonius's tables. Salaya was certainly interested in astrology for he is mentioned in the testament of Don Gonzalo de Vivero, bishop of Salamanca, died in 1480. He gets three books: one on geomancy, one by Abû Ma^{c}shar (the famous oriental astrologer at the end of the ninth century), and yet another one not specified. But Juan de Salaya is mostly known for his translation into Castilian of an astronomical treatise in Hebrew, on which we will comment later.

Diego Ortiz de Calçadilla succeeded Salaya in the chair of astrology and held it until 1475. We have no information on his scientific activity and we only know that he was a follower of Juana "la Beltraneja" in her dispute over the succession to the crown of Castile with her aunt Isabel, who later became Queen of Castile after the death of King Enrique IV. Diego Ortiz then moved to Portugal in 1476 and entered the service of King Afonso V, who supported Doña Juana. The chair of astrology at Salamanca remained vacant for a while. It was later occupied by Fernando de Fontiveros, from May 1476 to March 1480. As was the case with Diego Ortiz, we have no information on the scientific work, if any, carried out by Fontiveros.

So far we have only dealt with a part of the astronomical activity in Salamanca in the second half of the 15th century, the Christian part, taking place at the University. There is a Jewish part as well, developed in a nonacademic milieu, represented by Abraham Zacut, who was to become the most prominent astronomer of his time in the Iberian Peninsula.

Abraham Zacut[12] was born in Salamanca in 1452 and was an outstanding intellectual figure in the Spanish Jewish community on the eve of the expulsion in 1492. His scientific work began in the 1470s, and continued in exile, in Portugal, North Africa, and ultimately in Jerusalem. In 1478, Zacut composed in Hebrew for a Jewish audience a voluminous set of astronomical tables called *ha-Ḥibbur ha-Gadol* (*The Great Composition*). The tables are preceded by a long introduction in 19 chapters where the author not only explains how to use the tables but gives in-depth information on theoretical astronomy and praises, or argues against, the works of his predecessors. The *Ḥibbur* has been preserved in various manuscripts in Hebrew and also in Latin.

But Zacut's fame rests mainly on a book published in Leiria, Portugal, in 1496, and known as the *Almanach Perpetuum*. Although printed in Portugal, this book was published in two versions, in Latin and Castilian. The *Almanach Perpetuum* has hitherto been considered as a summary of the *Ḥibbur*, translated from Hebrew, but a careful review of the evidence shows that Abraham Zacut had nothing to do with the printing of the *Almanach Perpetuum,* even though the vast majority of the astronomical tables in it where taken from his *Ḥibbur*.

According to various scholars, the bishop of Salamanca at the time, Don Gonzalo de Vivero, was a patron of Zacut. This assertion is largely based on a dedication preceding the *Almanach Perpetuum* and addressed to a bishop. It has been shown that this dedication was copied almost word-for-word from an entirely different book printed for the first time in 1490, written by Johannes Müller (better known as Regiomontanus), and dedicated to a different person, in an entirely different context. Regiomontaus's book is the *Tabulae directionum*, first printed in Augsburg in 1490, and the person to whom it is dedicated is a Hungarian archbishop called János Vitéz.[13]

However, Zacut did have a Christian patron, Juan de Zúñiga y Pimentel (d. 1504), maestro de la Orden de Alcántara, for whom Zacut wrote a *Treatise on the influence of the heavens* in 1486 in the city of Gata (Cáceres). As explained in the dedication, the purpose of that book was medical astrology:

> "E por esto el muy magnifico y de grãd linaje yllustre mi señor el maestre de alcantara don Ju*an* de çuñiga amador de todas las sciencias y sabidor en ellas que a su fama todos los sabios y letrados dexan sus t*ie*rras y su nascimi*ento* por buscar sosiego verdadero y p*er*fection conplida que a su causa se esfuerçan las sciencias y sus letrados y an rrefrigerio y remuneracion . . . Ouo por bien mãdar a mi Rabi abrahan zacut de salamãca astrologo su *cri*ado que conpusiese un tratado breve en las ynfluencias del cielo para que con este mas se ayudasen los medicos de su señoria".

Nevertheless, the main relationship between Zacut and his Christian environment is to be found in the astronomical contents of his works. The tables in the *Ḥibbur* were compiled by Zacut for the beginning of year 1473, and depend largely on the Alfonsine Tables, already available in Salamanca no later than 1460. Zacut was most probably aware of the existence of such tables in his hometown, and was ready to use the latest material at hand, quite abundant in Salamanca at the time.

The available material was not only produced and used by Christian astronomers, but it also derived from a long and rich astronomical tradition in Hebrew. Among these works is a Hebrew version of the Alfonsine Tables that is almost certainly related to Salamanca. It is extant in a manuscript now in St. Petersburg (Academy of Sciences, MS Heb. C-076). Two lists of radices for the planets are given; one is for the beginning of 1473. As this is the same date as that used by Zacut in his *Ḥibbur*, it is difficult to decide whether this manuscript is prior to Zacut or it contains material compiled by Zacut himself. In the *Ḥibbur* Zacut acknowledges the works of such distinguished astronomers as Levi ben Gerson, Immanuel ben Jacob Bonfils, Jacob ben David Bonjorn, and Judah ben Asher II (14th century); and Judah ben Verga (15th century), among others.

On the other hand, Zacut does not mention any of his Christian peers, who in turn do not mention Zacut anywhere in their writings. In fact, the only Christian author acknowledged by Zacut in the *Ḥibbur* is King Alfonso, called the Wise. Despite this silence, Zacut was actually in contact with Christian astronomers of Salamanca, and the evidence is given in a manuscript still in Salamanca (Biblioteca Universitaria, MS 2-163). It contains the translation of the *Ḥibbur* from Hebrew into Castilian by Juan de Salaya, a former professor of astronomy at the University between 1464 and 1469, as mentioned previously. The manuscript indicates that the translation was done with the help of Zacut himself, in 1481. This is a perfect example of the ongoing interaction between Christian and Jewish astronomers at the time, but also a nice example of the collaboration between academic and nonacademic scholars, for Zacut, a Jew, did not attend the university, either as a student or as a professor.

There is additional evidence for the interest on Zacut's *Ḥibbur* among Christian astronomers. This is attested by the translation into Latin of the tables themselves. Besides the three MSS in Hebrew containing the tables of the *Ḥibbur* (now in Munich, Lyon, and Warsaw), we have found three MSS in Latin, all of them extant in Spanish libraries (Madrid, Academia de la Historia; Madrid, Biblioteca Nacional, and Segovia, Cathedral). Judging from the quality of the translation, it was probably done by a Christian well aware of astronomy in Hebrew, but it cannot be excluded that Zacut himself played a role in the translation of the tables.

A contemporary of Abraham Zacut, Diego de Torres, was active at the University of Salamanca beginning in 1469. During the following 10 years he held various positions until he was nominated for the chair of astronomy in the 1480s. The date cannot be given more precisely, but there is evidence that he was professor of astronomy in 1485 and 1487.

Diego de Torres was the author of two works. The first is a short treatise on medical astrology, entitled: *Eclipse del Sol. Medicinas preseruativas y curatiuas y remedios contra la pestilencia que significa el eclipse del sol del año de mill cccc.lxxxv a xvi. de março.* The treatise was published in Salamanca in 1485. The data for this solar eclipse seem to derive from the table of eclipses in Zacut's *Ḥibbur*, in which case it would be the earliest attested use of a table in the *Ḥibbur* by a Christian scholar, and well before the publication of the *Almanach Perpetuum* in 1496.

The second work by Diego de Torres, *Obra astrológica*, was also written in Castilian, although its introduction and explicit are in Latin. This work has not yet been edited and the only known copy is in Madrid, Biblioteca Nacional. The explicit indicates that the author is "didacum de torres", master of arts and medicine, and professor of astronomy at the University of Salamanca. It also gives the date when the treatise was finished: May 25, 1487. The work is divided into four parts, and the first one is entirely devoted to mathematical astronomy. Some planetary positions, for an unspecified date, are given as examples. Recomputation of these positions show that they depend on the Alfonsine Tables, and it is therefore plausible that Diego de Torres took all these data from Abraham Zacut. Moreover, in chapter 1, Diego de Torres refers to a table for the solar declination, and mentions the entries for 30° and 60°: the values given are identical to those found in the table for the solar declination in Zacut's *Ḥibbur*. It should be noted, however, that Diego de Torres mentions Nicholaus Polonius, who had left the chair of astronomy more than 20 years previously, but he does not mention his contemporary and neighbor, Abraham Zacut, anywhere in his *Obra astrológica*. It is not surprising: that was the eve of the expulsion of the Jews by the rulers of Spain in 1492, a shameful episode in Spanish history which nowadays would be referred to as "ethnic cleaning". Zacut had to leave Salamanca, and moved to Portugal. When the practice of Judaism was declared illegal in Portugal, he settled in North Africa, and ultimately in Jerusalem, where he composed a few other astronomical tables based on his *Ḥibbur*. Abraham Zacut died in 1515, probably in Jerusalem or Damascus.

We have found other manuscripts in Latin containing astronomical material related to Salamanca, but unfortunately no names are mentioned in it. There is also new material that has appeared very recently concerning astronomical activity in Lisbon in the middle decades of the 15th century.[14] All of them attest that in the second half of the 15th century Salamanca became the leading place for astronomy in the Iberian Peninsula, and a center of production, not just consumption, of astronomy. Two events explaining this situation happened at that time: the creation of the first chair of astronomy/astrology at the University and the presence in that city of the foremost astronomer of that time, Abraham Zacut. Finally, it should be emphasized that astronomy became a meeting ground not only for scholars belonging to the Jewish and Christian communities, otherwise sharply separated at that time, but also for academic and nonacademic scholars, as shown by the examples given above.

NOTES

[1] Thanks are due to B. R. Goldstein (University of Pittsburgh) for his useful comments to this paper.

[2] For an edition of the Latin text and a translation into English, see L. Thorndike, *The* Sphere *of Sacrobosco and Its Commentators* (Chicago: The University of Chicago Press, 1949).

[3] The most widely diffused treatise of this kind was the *Theoricae novae planetarum* by Georg Peurbach (1423–1461), printed for the first time in Nuremberg in 1472, but basically written in 1454. For an edition, see, E. J. Aiton, "Peurbach's *Theoricae novae planetarum*: A Translation with Commentary", *Osiris* 3:5–44 (1987).

[4] J. D. North, *A History of the University in Europe*, p. 348 (1992).

[5] For a recent survey on John of Gmunden, see B. Porres, *Les tables astronomiques de Jean de Gmunden: edition et étude comparative*, Thèse de doctorat à l'École pratique des Hautes études, Paris (2003).

[6] On Nicholaus Polonius and the astronomical background in Salamanca, see J. Chabás, "Astronomy in Salamanca in The Mid-Fifteenth Century: The *Tabulae Resolutae*", *Journal for the History of Astronomy*, 29:167–175 (1998).

[7] As has been recently shown, the use of the Parisian version of the Alfonsine Tables is already attested *ca.* 1400 in Morella (Valencia): see J. Chabás, "Astronomía alfonsí en Morella a finales del siglo XIV", *Cronos. Cuadernos Valencianos de Historia de la Medicina y de la Ciencia*, 3:381–391 (2000).

[8] For the Castilian Alfonsine Tables and their diffusion in Europe, see J. Chabás and B. R. Goldstein, *The Alfonsine Tables of Toledo*, *Archimedes: New Studies in the History and Philosophy of Science and Technology* (Dordrecht and Boston: Kluwer Academic Publishers, 2003).
[9] B. de Porres and J. Chabás, "Los cánones de las *Tabulae Resolutae* para Salamanca: origen y transmisión", *Cronos. Cuadernos Valencianos de Historia de la Medicina y de la Ciencia*, 1:51–83 (1998).
[10] On the *Tabulae Resolutae*, see J. Dobryzci, "The *Tabulae Resolutae*", in M. Comes, R. Puig, and J. Samsó (eds.), *De Astronomia Alphonsi Regis* (Barcelona, 1987) pp. 71–77 ; J. Chabás, "The Diffusion of the Alfonsine Tables: The Case of the *Tabulae resolutae*", *Perspectives on Science*, 10:168–178 (2002).
[11] For an analysis of these tables, see J. Chabás and B. R. Goldstein, *Astronomy in the Iberian Peninsula: Abraham Zacut and the Transition from Manuscript to Print*, (Philadelphia : The American Philosophical Society, 2000).
[12] The most complete biography of Abraham Zacut is still that written by Francisco Cantera Burgos, *Abraham Zacut,* Madrid (1935); see also F. Cantera, "El judío salmantino Abraham Zacut", *Revista de la Academia de Ciencias Exactas, Físicas y Naturales de Madrid*, 27:63–398 (1931). On Zacut's astronomical activity and his scientific background, see Chabás and Goldstein (note 11).
[13] Chabás and Goldstein (see note 11).
[14] B. R. Goldstein, "The Astronomical Tables of Judah ben Verga", *Suhayl* 2:227–289 (2001).

LUIS GARCÍA BALLESTER

MEDICAL SCIENCE AND MEDICAL TEACHING AT THE UNIVERSITY OF SALAMANCA IN THE 15TH CENTURY*

INTRODUCTION

Universities were undoubtedly the most significant and lasting scientific institutions of the later Middle Ages. From the 13th century onward, they became the principal places throughout Christian Europe, although not the only ones, where the process of assimilating, developing, and transmitting scientific knowledge was carried out.[1] These institutions were decisive in order to explain the development of later medieval medical science, from both a quantitative and a qualitative point of view, and were fundamental for an understanding of particular socio-medical phenomena, such as the definition of the medical profession and its control by society.[2] To be more precise, medical faculties were gradually established in the course of the 13th century around an academic syllabus that was especially sensitive to the new material that came to their knowledge in the form of newly revealed works by Greek and Arab physicians. Such contributions conditioned university curricula, but at the same time also led to substantial changes in the scientific output of their members of staff.[3] Generally speaking, it can be stated that the closing years of the 13th century and the opening decades of the 14th century marked the peak of the scholastic form of medicine; this approach was full of fresh ideas and vitality, and also demonstrated itself to be efficient at providing solutions to the medical problems posed by the ruling élites of the time, as well as able to bring about an authentic process of intellectual seduction among learned non-Christian minority circles, such as those of the Jewish community.[4] In contrast, one of the characteristics of 13th-century Castilian medicine was the absence of the scholastic approach because of the weakness of the kingdom's university institutions. It was no coincidence that neither of the 13th-century physicians from the Iberian Peninsula who are known to have written scientific works, namely Petrus Hispanus (*c.*1210/1220–1277) and Arnau de Vilanova (*c.*1240–1311), was from Castile or León, nor that both of them were typical university products of the first wave of medical scholasticism: in the case of the former of Paris-Siena, and, in that of the latter, of Montpellier. It is nonetheless surprising that the Galenism produced in Montpellier or northern Italy at the turn of the 13th and 14th centuries should have taken so long to reach Salamanca with a minimal degree of continuity and documentary evidence. It was not until the 1411–1415 reform that documentation can be found in which it is recommended that the works of Arnau de Vilanova (which ones?) "... *et aliorum novorum super medicina*" should be acquired.[5] Was this recommendation fulfilled?

M. Feingold and V. Navarro-Brotóns (eds.), Universities and Science in the Early Modern Period, 37–64.

If it was, the works that were acquired, at least those by Arnau de Vilanova, have been lost, and there are hardly any traces of them in the limited number of extant medical works by 15th-century Salamanca professors or in works related to such university circles. Only in the closing years of the 15th century and the opening ones of the following century do they appear. What is more, no medical works are known to have been written at a Castilian university in the 14th century, and the extant examples by 15th-century professors lack both breadth of vision and complexity. It is quite evident that in no way can they be compared with the works produced by their counterparts at other southern European university centers.

In his classic work, Rashdall pointed to three general characteristics common to 13th-century Castilian *Studia* or universities: their close connection with the Crown, the strong influence that cathedral chapters wielded over them, the system of financial maintenance based on ecclesiastical tithes, and that each kingdom possessed "its" own university.[6] It should not be forgotten that under Alfonso X the former kingdoms of León and Castile coexisted with the extensive new territories of the Guadalquivir valley and the Kingdom of Murcia, in the lands of Castile; the last two were conquered or fell under Castilian sovereignty between 1243 (the capitulation of the Kingdom of Murcia) and 1248 (the conquest of Seville). In all this realm, in fact, there were only two universities, Salamanca and Valladolid, since that of Valencia can only be considered to have existed during a short period during the first half of the 13th century.[7] All the wide tract of territory lying to the south of the Plasencia–Madrid–Cuenca line, which included such cities as Toledo and such densely inhabited regions as Murcia and the Guadalquivir valley, lacked universities during the later Middle Ages.

There is no evidence available which might enable us to state that the Salamanca Faculty of Medicine was in existence in the 13th century.[8] When, then, was the transition made from the situation implied by the expression "I order . . . that there should be . . . two masters of physic", as laid down in Alfonso X's royal privilege (1254),[9] to a *facultas* with an administratively defined corporation, with premises, a sufficiently endowed library and a syllabus befitting the knowledge of the period? This does not seem to have taken place in the 13th century. On the other hand, in spite of the lack of documentary sources, a series of data leads one to accept that medical studies were being undertaken in the 14th century, at least during the second half. This is what is hinted at by a Bull of Clement V of 1313, in which the Archbishop of Santiago was informed that it would be advisable to carry out economic reforms that would guarantee the salaries of the professors of Salamanca, including those of medicine.[10] Nevertheless, the first indication that we have of a professor of medicine in Salamanca dates from shortly before 1363.[11] At approximately the same date the awarding of degrees in medicine, more precisely that of *magister in medicina*, can be confirmed.[12] To the end of the century there also belongs the royal warrant of Enrique III (September 4, 1391), issued in Valladolid, in which the king refers to certain complaints by the "masters of grammar and logic and philosophy and physic".[13] Everything seems to point to studies of medicine having been consolidated at the University of Salamanca in the second half of the 14th century.

Even though the volume of written source material does not increase in the 15th century—it should be borne in mind that the surviving cloister books only cover discontinuous periods in the second half of the century (1464–1481)—Amasuno's painstaking reconstruction[14] enables us to know the names of those who were teaching medicine from the two chairs in existence, that of Prime and that of Vespers, from 1405 to 1406 onward. It is a list of some 30 names of professors and assistants (young bachelors and graduates), who were responsible for the functioning of the complex teaching machinery of a scholastic faculty of medicine, ranging from ordinary and extraordinary normal classes, or revision classes, to the *quaestiones* and *disputationes* characteristic of the awarding of degrees.[15] None of them are known to have written an original academic medical work of any great importance: only brief treatises, some of which are no more than a very short list of prescriptions, on the plague in four cases (Gómez García of Salamanca, Licenciado Forés, Diego de Torres, and Fernán Álvarez Abarca) and a *Compendium* of medicine (by Gómez García of Salamanca, professor between 1433 and 1464, the year in which he died), which in fact is a list of recipes including 19 prescriptions aimed at curing patients of, or protecting them from, different illnesses and maladies.[16] The last of the above-mentioned individuals, Fernán Álvarez Abarca, was also the author of an almost completely preserved *Praxis medica*, the theoretical contents of which are minimal; this was aimed at the medical practitioner and includes brief instructions on what to do in the case of various illnesses, together with recipes.[17] The collecting of successful remedies was a common practice in learned Latin medical circles.[18] A collection of brief works—*De los pesos de las medicinas* [Of the weights of medicines], *El libro de los olios* [The book of oils], *Tratado de las orinas, de los pulsos e de otras señales* [Treatise on urines, pulses, and other signs]—to be found in manuscript 2262 of the University of Salamanca (ff. 217–263), in which the *Compendium* (ff. 193–216v) is also to be found, may be attributable to Gómez García of Salamanca, according to Beaujouan[19] and Amasuno,[20] although his authorship is not proven.

The recent discovery of new manuscript documentation related to the teaching of medicine in 15th-century Salamanca, together with the study of early printed works from Salamanca which have hitherto not been taken into account, enables us to outline a reconstruction of the arguments and problems that preoccupied the university medical community, and hypothesize that in Salamanca there existed a movement that approached Galenism from the suppositions of a Latin medical form of humanism based on recovering the most representative medical scholars of late 13th and early 14th century Galenism, without disregarding those authors who, in the course of the 15th century, had made interesting contributions to the field of medicine. This chapter aims to offer an approach to these subjects.

METHODS OF LEARNING

The earliest known medical literature likely to have been produced in the sphere of influence of the University of Salamanca is to be found in manuscript 3371 (ff. 84–91) of the Biblioteca Nacional (Madrid). It is a text designed to help the student of medicine remember a series of commonly used concepts, such as that of medicine itself, the elements, complexion (*complexio*), humor, virtue (*virtus*), limb, skin, fat (*pinguedo*), ligament, bone, cartilage, movement of the heart (*motus cordis*), sight (*visus*), generation of the limbs (*generatio membrorum*), and natural things (*res naturales*), among others. The noteworthy aspect is that each definition is accompanied by a reference to the author and work from which it is taken (Isaac Israeli, Avicenna, Aristotle, Johannitius, Hali Abbas, Jean de Saint-Amand, Arnau de Vilanova, Gerard de Solo, Gilbert of Aquila, etc.). This was a literary genre produced in university contexts at the end of the 13th century, which answered the interest that the student community had in a series of medical works and authors which they sought to gain easy access to and to make easier to read and understand. In this way, a considerable volume of information was placed in the hands of the university community by using the scholastic procedure of division and subdivision. The most prominent example can be found in the case of the Parisian master, Jean de Saint-Amand,[21] who, in about 1285, wrote his *Revocativum memorie* to satisfy the curiosity of a university community that was keen to discover the collection of Galen's works which made up the nucleus of the "new Galen", which at that moment was reaching university circles for the first time.[22] It was therefore not fortuitous that the author—"magister Alfonsus", whom it has not been possible to identify—gave his work the title of *Collectiones doctorum in arte medicina ad facilem inventionem capitulorum et memoriam confortandam et recordationem* ("Fragments and summaries of doctors to facilitate locating and remembering subjects"). The manuscript was written in 1433, a date which marked the moment when the Salamanca Faculty of Medicine started to undergo a process of renewal. The work makes constant use of Avicenna (*Canon* and *De viribus cordis*), but also of the above-mentioned writers. Let us consider one example:

> Note—it reminds the reader—that the 'natural things' (*res naturales*) are those that are necessary to maintain health. They are 'things' without which the body as something which can be healed (*sanabile corpus*) cannot exist, as Arnau de Vilanova says in the second chapter of his *Speculum*.[23]

In addition, the interesting aspect of the example which has just been referred to is that it is the first known reference to the great treatise of medical pathology (the *Speculum medicine*) by the *medicus cathalanus* (d. 1311) in the context of the University of Salamanca.

Recently, when studying one of the medical manuscripts included by Beaujouan in his catalogue,[24] I located what appears to be a notebook belonging to a medical student or a young graduate of the Faculty of Medicine at Salamanca in the last third of the 15th century or the first few years of the 16th century. To judge by a manuscript note by the author,[25] the notes were taken in 1504 and prior to this date. They consist of highly varied medical notes, which range from drafts of texts to ask for the degree of bachelor of medicine (*Petitio gradus*), unfortunately undated and without names, to copies of brief medical treatises by Salamanca professors, recipes, lists of simples with their names in Latin and Castilian, and verse descriptions in Castilian of medicinal simples, in the form of aides-memoire. For example, after the copy of the Latin text corresponding to rhubarb root (*rubarbarum radix*) taken from the *Circa instans*, Plateario's famous Salernitan list, the following verse in Castilian is to be found:

> Rhubarb is a medicine/Known by blessed names,/And its properties are ample,/And it cleans the stomach, liver and blood;/For cholera and phlegm it is a benign purgative,/As it safely eliminates them.[26]

The use of verse was a procedure that enabled students to remember the characteristics and instructions for use for the scores of simples available to the medieval physician. It was a technique that enjoyed a certain popularity in academic circles. For example, one can point to the introductory poems with which Jacques Depars (*Jacobus de Partibus*), a professor in Paris, summarized some of the chapters of his commentaries on the *Canon*, which he finished in 1453,[27] and which were widely known in Salamanca, as will be seen below.

THE INTELLECTUAL ENVIRONMENT OF THE FACULTY OF MEDICINE IN SALAMANCA: DISCUSSIONS AND PROBLEMS CONCERNING A RENEWAL OF GALENISM

However, the most interesting aspect of the manuscript in the "Real Academia de la Historia" is the presence of a substantial number of brief observations under the general heading of *Nota quedam medicinalia*.[28] Under this title, the author of these notes collected a wide range of medical doctrines and problems in the form of aphorisms, as well as summaries of doctrines and opinions taken from lessons and texts, apparently to make them easier to learn by heart, in addition to problems of doctrine and medical practice in the form of *quaestiones* and *disputationes*. The latter detail is of particular interest because, for the first time, it enables us to gain an insight into the real intellectual environment of the Faculty of Medicine in Salamanca in the closing years of the 15th century. A paper such as the present one is not the place to offer a full list of these problems, so I will limit myself to pointing out but a few: the problem of the origins of sleep and of the alternating periods of sleep and wakefulness; whether it is possible to base a prognosis on patients' appetite or lack of appetite; the nature of sweat; differing attitudes to the problem of the causes on the part of the physician (*medicus*), the natural philosopher (*philosophus naturalis*), and the metaphysician (*metaphisicus*); the causality of fevers; the situation of the body in what is known as the state of "neutrality", in which the patient's body, according to Galenism, is neither healthy (*sanum*) nor sick (*egrum*), this having been a much-debated subject in Italian academic circles since the 14th century;[29] different problems related to phlebotomy; the nature of apoplexy; the way in which food reached the various limbs depending on their position in the body; the difference between poison and medicine; the way medicines act; different problems connected with the pulse; whether innate heat, the *spiritus*, and blood are one and the same thing, and the way they act; to what extent the physician (*medicus*) is but a helper (*adiutor*) of nature or whether he can reach parts that nature is unable to; problems related to the convenience of bleeding on certain days and with patients of a certain complexion or constitution.

While it is important to discover the questions that concerned the medical academic community in Salamanca, it is even more interesting to discover the intellectual keys with which these same questions were approached. These questions concerned academic Galenism from the late 13th century onward. The answers that Galenism offered in the course of the later Middle Ages were neither unitary nor static. Galenism was sensitive to the different intellectual fashions and influences, as well as to the social factors, which defined the various fields of endeavor in Europe. For example, did the new medical humanism which spread in European academic circles in the 15th century have a bearing on the new interpretation of these questions on the part of the Galenism that was practiced in Salamanca in the late 15th and early 16th centuries? This is a question that it is difficult to answer, although the available evidence allows us to pose the question and outline an initial answer.

In the midst of these *quaestiones*, our student or young graduate noted specific opinions about some of them given by professors with whom he was in contact. For example, he explained the opinion of the "physician to the Queen" as regards the action of medicine (*de actuatione medicinarum*).[30] Our student reflects the way in which medieval professors would expound their opinion on a specific problem. This they did by means of an indirect approach: first of all, they would expound the opinions to be found in what were, from their point of view, the leading works of medical literature, before going on to pronounce which one they judged to be the most solidly based. This technique enables us to discover which scholars dominated university medical discussions at that moment. In this case, the professor of Salamanca (a "physician to the Queen", most probably the third one) adduced three leading (*famosa*) opinions: that of Pietro d'Abano in his *Conciliator*, that of Averroes in his *Colliget* (without forgetting that of Avicenna in his *Canon*), and that of Galen in several of his works. The Salamanca professor decided in favor of that of Galen.[31] Is this preference, an indication of the appearance of the medical humanism of European new Galenism in the circles of academic medicine of Salamanca? This form of humanism sought an intellectual tool to renew Galenism's message in the sources (in this case the works of the Latin Galen). We shall see below how academic Galenism in Salamanca at the turn of the 15th and 16th centuries found (or endeavored to find) other resources to renew itself by means of "going back to the sources".

As regards the question of the way in which medicines acted, another of the theoretical points, although not without practical repercussions, discussed in the university circles of Salamanca's Faculty of Medicine, the problem of the grading of medicines, was put forward; in other words how to establish the different strengths of simples and how to combine them in compound medicines so that the desired effects might be achieved. The problem of differing degrees of strength was not limited to medicines; medieval physicians also considered the possibility of "measuring" the very constitution or complexion of each patient; and that of the nature of the illness. Hence "the importance of knowing how to measure in degrees; in other words to graduate", as "the physician to the Queen" said, according to what was written down in this student's notes.[32] In fact, he noted down the professor of Salamanca's opinion "on the degrees of medicines" (*Materiam de graduatione medicinarum recollectam secundum doctorem Regine*).[33] In the discussion, five problems were raised in the first instance: in the first place, "graduating" (*graduare*) human generation; second, illnesses; third, complexional medicines; fourth, purgative medicines; and fifth, compound medicines.[34] We shall not consider how each of these problems was developed and, in fact, the extant notes do not allow us to analyze them in detail. However, two points that I consider to be of interest should be emphasized.

The first of these is to stress the presence of this type of discussions in the Faculty of Medicine in Salamanca at the end of the 15th century. These very intense debates were initiated in Montpellier in the last decade of the 13th century by Arnau de Vilanova in one of his most original and emblematic works, the *Aphorismi de gradibus*.[35] This work was used and cited by the "physician to the Queen", as is recorded in these notes. It is the second piece of evidence for the survival and continued usage of this approach in late medieval Spanish faculties of medicine. The first person to consider this problem in the Spanish kingdoms was Antoni Ricart (d. 1422), in the opening years of the 15th century, who spent his time between the Faculty of Medicine in Lérida and the *Estudi* of medicine in Barcelona.[36] In addition to dealing with the subject in an original, detailed way, Antoni Ricart echoed the discussions taking place in university circles in Lérida, both from the medical point of view (his own) and from that of natural philosophy [a lost work by the Valencian Nadal Lambrí (*fl.* 1370–1394), which was scorned by Ricart].[37] Almost a century later, the subject emerged in Salamanca, where a "renaissance" of the Galenism of the turn of the 13th and 14th centuries, as embodied by the works of Arnau de Vilanova, Bernard de Gordon, Taddeo Alderotti, and Pietro d'Abano (*Conciliator* and his commentaries on the *Problemata*) among others, is known to have occurred in the last third of the 15th century. As will be seen below, many of the writings of the first two authors were copied or edited by scholars in Salamanca.

The second of the points concerning this discussion "On the degrees of medicine" that I wish to stress is that the professor of Salamanca was concerned that it should be extended to the practical field so that its application to the doctor–patient relationship should not go unnoticed by the student in question. At different moments of the explanation, the student reflected how the professor interrupted his lecture and warned them, "Take into account that when you administer a compound medicine, you should not change one of the components for another..., as some physicians do".[38] Throughout the *quaestio* one can detect a highly practical tone of advice. Indeed, a concern that academic Galenism should be useful to the medical practitioner is predominant, this being a feature of all the known works of Fernán Álvarez Abarca and, generally speaking, one that defined the academic environment of Salamanca in this period.

Arnau de Vilanova's presence in Salamanca was not restricted to only the two highly influential medical works known as the *Aphorismi de gradibus* and his *Speculum medicine*. To them should be added his *Regimen sanitatis ad regem Aragonum*, his *Antidotarius,* his *Experimenta* ("accounts of various medical problems and of methods of treatment that the author has found to be successful in each instance"),[39] and, above all, his *Medicationis parabole*, the aphoristic format and practical nature of which guaranteed that it was widely used. The commentary of Diego Álvarez Chanca (*c*.1450–1515), probably a Salamanca student, on Arnau de Vilanova's *Parabole*, should be placed within this context of recovering Arnau as part of a tradition of Galenism in the Latin West that had to be defended and revived (*revivere*). This commentary appeared in Seville in 1514, but it was almost certainly written more than 10 years before, perhaps even before 1501, as Paniagua has suggested.[40]

Nevertheless, the reconsideration of the problem of the graduation of compound medicines in academic circles in Salamanca meant that the *Aphorismi de gradibus*, together with, logically, al-Kindi's *De gradibus* (*Quia primos*), gained new importance. These two works became authentic intellectual stimuli which reoriented debates of a theoretical nature, although with obvious practical implications, which have hitherto passed unnoticed by medical historians.

In spite of what has been said about Fernán Álvarez Abarca ("the third physician to the Queen"), the leading figure in the renaissance of Arnau's medical works was Francisco Núñez de la Hierba (*c*.1460–1504/1505), the regent of the Vespers Chair in the Faculty of Medicine in Salamanca in 1501, who included the bachelor Fernando del Marmol in his undertaking.[41] The collaboration of these two scholars made it possible to publish al-Kindi's *Liber de gradibus* accompanied by Arnau de Vilanova's *Aphorismi de gradibus*, and his *Medicationis parabole* in 1501.[42] Fernando del Marmol, as he himself explains, carried out the transcription from a *codex unicus* in fairly poor condition belonging to Nuñez de la Hierba, "the letters of which . . . were so worn that it was hardly possible to understand what they said".[43] In his undertaking he applied the requirements of humanistic philology, taking great care to offer an accurate text, over the correction of which both he and Nuñez de la Hierba took great pains.[44] It is, however, still surprising to find, together in the same volume, the area of theoretical medicine (represented by al-Kindi's works and, above all, by Arnau de Vilanova's *Aphorismi*, perhaps the work of the new Galenism developed in Montpellier in the late 13th century in which the mutual dependency between natural philosophy and medicine can be most clearly seen),[45] and the area of practical medicine based on this same new Galenism, the clearest example of which were the same author's *Medicationis parabole*. This may have been fortuitous and imposed by the contents of the manuscript which was used, but may also have been intentionally sought by the editors in order to emphasize the practical aspects of al-Kindi's work, by means of Arnau's interpretation of it; in this way they may have sought to stimulate new reflections on academic Galenism in Salamanca at the opening of the 16th century. The publication of the al-Kindi (*de gradibus*)—Arnau de Vilanova (*Aphorismi*) tandem was considered to be a genuine intellectual stimulus at the same time as a practical proposal by the person who promoted the work, Núñez de la Hierba. In his efforts to find new approaches and paths to Galenism, he caused old forgotten authors to be "reborn" (*reviviscere*) by means of producing their works in a suitably published format, a procedure characteristic of medical humanism. Núñez de la Hierba did not hesitate to reconsider the subject of the degrees of compound medicines (a problem that had not been solved by Galenic therapeutics) through Arnau's interpretation of al-Kindi, at the same time as he defended the novelty of the subject and the validity of the long-deceased al-Kindi's proposals:

> Many have explained . . . the qualities of simples (*erbarum*): for example, this one is warm or this one is cold. But who other than al-Kindi has taught with greater clarity and detail as regards the qualities of compounds (*mixte*), or which degree of the illness they take advantage of, or what the harm they cause is?[46]

In the Prologue that he composed as the introduction to the edition, he did not hide the atmosphere of controversy in existence in Salamanca as regards the subject of the determination of the degrees of compound medicines, both from the point of view of how to determine them in practical terms and from that of the theoretical suppositions required; as has been seen, this is confirmed by the contents of the notes to be found in manuscript 9/443 of the "Real Academia de la Historia" (Madrid) when it provides information about some of the *quaestiones* that Fernán Álvarez Abarca posed in relation to *de graduatione medicinarum* (ff. 65–66v). Offering the scientific community of Salamanca the combined texts of al-Kindi and Arnau de Vilanova was the clearest and most evident way to follow the most daring and recent proposal concerning the subject made on the basis of Galenism. For this reason, it was not by mere chance that he should address his Prologue to "young students" (*ad iuvenes medicine studiosos*):

> Young students, for some time I have been reflecting on the reason why, while volumes by authors old and new are being published all around, however in medicine, the prince of sciences among all others, nothing or very little is written on the degrees of compound medicines. This I attribute to the difficulty of the subject and to the effort that it requires. We can accept that there may be few who are sufficiently intelligent to deal with this subject, but we cannot allow what has been done on the subject, whatever it may be, to remain hidden, or to remain poorly known, without becoming irritated. Who would not criticize those who hide these works, who would not praise those who bring them to light, if in this way the sciences are enriched?
>
> I am convinced . . . , that those who compose something new warrant equal praise, . . . as those who publish what is unknown, who seek out what has been written by others, and do not let it remain hidden for any longer . . .
>
> For that reason, and in my wish to be useful to you, so as not to give the impression that . . . we scorn what has come down to us after having been kept by our ancestors and we lose it, . . . I have managed to make the most ancient al-Kindi, a man of great authority, be reborn . . . I confess that I have not been able to contain the desire, in spite of the efforts and late nights involved, to bring to light this most learned man . . . . From him most scholarly doctrines can be obtained and . . . he will be of great use in order to more than fulfil your desires and also to achieve the goal that we seek sooner.

> Young students, many are the things I could say unto you. However, I will not refrain from saying this: practice, which is a far from insignificant part of medical science, would be highly imperfect and useless without al-Kindi. This cannot be denied by theoreticians, in whose books there are many things that could be neither understood nor even noticed were it not for what has been explained by al-Kindi, whose doctrines introduce us to theoretical problems.[47]

The Salamanca edition was not accompanied by extensive commentaries on the first two works, which were so closely related. At least, they have not come down to us today. How and why did this question become a topical issue in Salamanca a century later than in Lérida, and why via the combination of al-Kindi and Arnau de Vilanova through the text where the latter posed the question in late 13th-century Montpellier? The answer is not known. At present, we can go no further than to state that this question was present in a faculty of medicine whose relationship with the great scholars of Montpellier went back at least to the end of the 14th century. However, it would not be unreasonable to suggest that a group of physicians at the university of Salamanca approached the renewal of Galenism by means of trying to apply philological techniques, a characteristic of the new medical humanism, in order to edit ancient texts correctly. Nevertheless, they applied these techniques (the search for reliable manuscripts, the painstaking correction of texts, and detailed knowledge of the subject matter on the part of the editor) to writers of the university-centered Latin Galenism of the last third of the 13th century, such as Arnold of Vilanova, Bernard of Gordon, Peter of Abano, among others. As is known, these individuals were all key participants in the movement called the "new Galen" which renewed Latin Galenism in the late 13th century.[48] One of the results was the revival of these authors in the late 15th and early 16th centuries, together with the publication of some of their most significant works by the early Castilian printing.

As has already been mentioned, the student who wrote the notes collected in manuscript 9/443 of the Real Academia de la Historia (Madrid) also reflected various medical problems related to the pulse. In this context, he included a summary of the opinion that a certain "doctor de la Parra" (Gonzalo de la Parra, *fl.* 1471–1512)[49] expounded on this subject in his *Tractatus de pulsibus*, which was never printed.[50]

However, perhaps one of the most interesting questions that this student's notes allow us to gain an insight into is what was known as "mal francés" ("French disease") or "mal de vuvas" ("bubonic disease").[51] There can be no doubt that the subject must have been of great interest to both university and non-university medical circles in the area of Castile around Salamanca. Proof of this can be found in the prescriptions for the *morbum gallicum* contained in manuscript 4220 of the Biblioteca Nacional (Madrid), which includes Fernando Álvarez Abarca's *Praxis medica* (ff. 1–68) and the *Recetario* (full of practical advice for young physicians and including an *antidotarium*) brought together by Fernando Fernández de Sepúlveda, a bachelor in medicine and a master apothecary who had received his training at the recently created Estudi General (1499–1500) and Hospital General (1512) of Valencia.[52] According to an anonymous report addressed to the Catholic Monarchs (*c.*1498–1500), Valencia and Zaragoza were the two cities whose rulers were most diligent in facing the professional problems that arose from new approaches to pharmacy.[53]

Until recently, we were only in possession of two works on this new disease that were published in the form of monographic treatises: the *Tractatus de pustulis de saphati nominantur* written by Juan de Fogeda, from Seville, published in Salamanca (1496), and the text by Francisco de Villalobos *Sobre las contagiosas y malditas bubas. Estoria e medecina* (Salamanca, 1498). However, to these should be added the *Tractatus de saphati* written by Doctor de la Hierba (undoubtedly Francisco Núñez de la Yerba), which the author of the notes that are being considered copied in its entirety.[54] This work is in the form of a dialog between an inhabitant of Córdoba (*Cordubensis*) and one of Sagunto (*Saguntinus*), a city in the Kingdom of Valencia. In the body of the work, a solar eclipse (October 10, 1493) is mentioned, and the *explicit* states that the work was finished in Ciudad Rodrigo on St Luke's Day (October 18), 1496. The copier's (our student's) concern for precision enables us to determine that he completed the copy on May 8, 1504 (f. 78v). We are witnessing the medical world's concern about to what was perceived by the majority as a new disease, which marked a new frontier in medical practice in the closing years of the 15th century and the opening ones of the 16th century. It should not be forgotten that not only were the treatises by Juan de Fogeda (1496) and López de Villalobos (1498) published in Salamanca, but also the *Consilium* written by the Valencian Gaspar Torrella on a new form of contagious illness, known in Castilian as "modorrilla" ("drowsiness") (Salamanca, 1505), a work which was not by chance dedicated to Fernán Álvarez Abarca, "the third physician to the Queen", and holder of the Prime Chair in the Salamanca Faculty of Medicine.[55] The above-mentioned details enable us to detect the existence of a network of intellectual relationships within the Iberian Peninsula between Valencia, at that moment an active scientific center, and the University of Salamanca.

The works that have been mentioned are not the only ones which enable us to gain an insight into the world of day-to-day medical teaching in 15th-century Salamanca with the aim of reconstructing, as far as possible, the hitherto almost unknown intellectual atmosphere of this center. A manuscript containing a series of texts by Gentile da Foligno (d. 1348), Pietro Torrigiano (*c.*1270–1350) and Marsilio de Santa Sofía (d. 1405) has also been preserved.[56] This includes extensive fragments of the first and last writers' comments on different parts of the *Canon*; in the case of the second author, it contains fragments of his well-known commentary on Galen's *Tegni*, known by the name of *Plusquam commentum*, one of the most carefully developed expositions of late medieval academic Galenism.[57] What is interesting is that it was a recent Salamanca graduate, Pedro (*Petrus*) Pugo, that copied the more than 102-columned folios. He carried out this task at two moments of his academic life, as he himself explains at two points in the manuscript: during the summer of 1457, when he was undertaking his first year of medical practice after obtaining his bachelor's qualification in medicine,[58] and 4 years later, in September 1461, during his second year, probably after receiving his degree. Our young graduate summarized his world of intellectual concerns related to medicine that were raised by the texts being copied in the form of *quaestiones* in the margins. These annotations, sometimes small synoptic charts, were designed to help him to memorize the huge quantity of information provided by the medical texts that he was copying, which made the task of the medical student the great exercise in memorizing that it still is to this day. For instance, among others, "Whether fat and obesity are of a cold complexion" (*Utrum pinguedo et adeps sunt frigide complexionis*), or what is the mechanism for the generation of human beings (*quid est generatio viventium*). In addition, Pedro Pugo's asides tell us that the requirements of the university constitutions which made it obligatory to fulfill 1 or 2 years' practice after obtaining the degree of bachelor in medicine were being applied in mid-15th century Salamanca.[59]

If the evidence provided by one of the *quaestiones* summarized in our student's notes, mentioned above, and the material supplied by the aide-memoire *Collectiones doctorum* is added to that just considered, we can complete a picture in which the considerable influence of academic Galenism, which originated in Montpellier, Italy and Paris, and revolved around Avicenna's *Canon*, on the Salamanca Faculty of Medicine in the second half of the 15th century is quite noticeable. In practice, when the question (*quaestio*) of the influence of diet on the *regimen* according to the temperament of the bodies of the people being treated is discussed,[60] Galen's opinions found in his various works (*de ingenio sanitatis*, *de regimine sanitatis*, *liber alimentorum*, commentaries on Hippocrates' *Aphorismi*, etc.), together with those of Avicenna (*Canon*) and of Averroes (*Colliget*) are mentioned; but at a certain moment, the author of the *quaestio* starts to bring together and to evaluate the opinions of "more modern" scholars (*moderniores*): the opinions of Taddeo Alderotti (d. 1295), Pietro d'Abano (d.*c*. 1315) in his *Conciliator*, Gerard de Solo (d.*c*. 1360), Pietro Torrigiano (*c*.1270–1350) as stated in his *Plusquam commentum*, but, above all others, of the great corpus of commentarists on the *Canon*, beginning with the commentaries of Dino del Garbo (d. 1327) and Gentile da Foligno (d. 1348) and continuing with those of Ugo de Siena (= Ugo Benzi, 1376–1439) and those of the influential Parisian professor, Jacobus de Partibus (= Jacques Despars, *c*.1380–1458) are paraded before our eyes. In this overview, the reference to Christoforus de Honestis (d. 1392), whose prescriptions and works on remedies can also be seen to have enjoyed great prestige in Salamanca university circles, defining a more complex horizon in the use of *antidotaria*, should also be mentioned.[61]

Bernard de Gordon[62] was to be another of the Montpellier scholars of Arnau de Vilanova's generation who never ceased to be present among those who were being trained in Salamanca in the last third of the 14th century (for instance, Alfonso de Chirino).[63] Those writing in Montpellier at the turn of the 13th and 14th centuries (Arnau de Vilanova and Bernard de Gordon) enjoyed renewed prestige in European university medical circles of the time. The copies of Arnau de Vilanova's writings (both his medical and spiritual works) made by Pier Leoni (d. 1492), physician to Lorenzo de Medici,[64] or those made in 1464 by a German student during his stay in Padua,[65] should also be remembered. However, Avicenna's *Canon*, whether directly or via commentarists, ranging from Gerard de Solo,[66] also a Montpellier professor, to Jacques Despars, from Paris,[67] was the nucleus around which medical intellectual activity coalesced in Salamanca.

The *Canon* was not a work with one single unvarying interpretation in university contexts in the later medieval period. Commentarists used its contents as intellectual stimuli to expound their own opinions and experience or to reconsider problems in the light of the intellectual developments of the moment. It suffices to read the commentaries on Avicenna's text by Gentile da Foligno (d. 1348)[68] or those that Jacques Despars, who was far from unaffected by the humanistic currents of the period between 1430s and 1450s in the Parisian, German (Constance), and Florentine circles which he frequented, dedicated to the same passages a century later.[69] In the mid- and late 15th century, not in vain did the *Canon* continue to be, in university centers throughout Europe, the best compendium of everything a well-trained physician ought to know in order to satisfy the needs of a form of attending patients based on *medicinalis scientia*, as required by the most demanding sector of the market.[70] In fact, in Book I, university professors and with them their students would find the foundations of this medical *scientia*, physiology, and anatomy, as well as the general principles on which to base Galenism's general pathology and therapeutics, as structured by the *Canon*. The second book of the treatise, dedicated to simples, was perhaps of less interest. Avicenna's authority in this field was undeniable, but the 15th century was to witness the production of numerous commentaries on the lists of antidotes and treatises on simples, as well as a substantial accumulation of material. The third book of the *Canon* was without doubt its mainstay. Not in vain was it dedicated to the systematic exposition of all medical pathology in accordance with the traditional approach "from the head to the feet", so highly regarded by medieval man; in addition, this approach to pathology included the wide-ranging field of fevers, a form of illness that affected the whole body and which, in practice, encompassed the greater part of clinical practice as seen by medieval physicians. Any commentary on this book opened up a possibility for masters to explain their clinical experience, although, it must be said, cloaked in the peculiar scholastic form of expression. That part of Book IV also devoted to fevers was also suggestive, although other parts were overwhelmed by the new developments as regards the subjects under discussion that gradually appeared in the 14th and 15th centuries. It should be remembered that it provided the backbone for the collection of illnesses that medieval physicians described as "leprosy", of such great impact on medieval society; this chapter failed to provide sufficient clinical or pathogenic resources to face up to what were known as "new diseases" that physicians, including Castilian ones, described at the turn of the 15th and 16th centuries ["French disease", "buboes", "modorrilla" ("drowsiness"), among others].[71] The entire suggestive chapter on poisons and symptoms resulting from poisonous animal bites was also dealt with in this book.[72] Book IV offered the opportunity to acquire some knowledge of surgical treatment, a field in which the work could not compete with the great treatises of Teodorico Borgognoni, Lanfranc, and Guy de Chauliac, especially the latter, around which revolved the most recent developments in Castilian surgery resulting from the most influential periods of the school of surgery at the Hieronymite monastery of Guadalupe in the second half of the 15th century and the first few years of the following one.[73] All the surgical treatises were translated into Castilian. Book V is entirely devoted to practical pharmacology;

this was the "list of antidotes" that accompanied all medieval medical treatises, the expression of the ultimate level—the climax—of the physician–patient relationship: the moment when the physician wrote a prescription after determining his patient's diagnosis–prognosis.

Such was the intellectual platform upon which Galenism was erected in Salamanca in the second half of the 15th century, the central years of which marked a veritable boom for its Faculty of Medicine, and the opening years of the 16th century. Unfortunately, the commentaries of the Salamanca professors or the *repetitiones* that their substitutes offered during their lengthy absences from the lecture halls are not available to us. This valuable material has unfortunately been lost or has not yet been traced. However, it can be stated that their Galenism was in accordance with that taught in the faculties of medicine in those parts of Europe closest to Castile, where the *Canon* was the cohesive element and an obligatory work of reference.[74] It was, therefore, not by chance that the young Francisco López de Villalobos (*c.*1473–1549), a recent graduate of the Salamanca Faculty of Medicine, with hardly any clinical experience, should set about composing a Castilian verse summary of the *Canon* and should define that a knowledge of this work was a minimum requirement for those who, without any knowledge of Latin, might wish to exercise medicine with a certain degree of propriety and thereby have the possibility "of sampling some elements of medical science".[75]

Even though he had received his training at the Faculty of Medicine in Salamanca, the interesting medical compendia by Alfonso Chirino, written in Castilian, namely *Menor daño de la medicina* [The least harm of medicine] (1406/1411) and *Espejo de medicina* [The speculum of medicine] (1414),[76] had nothing to do with the output of the universities, although at the same time they also reflected a series of points that concerned university-trained physicians, and quite probably those at the institution itself, connected with both theoretical problems (derived from the power of seduction exercised over a proportion of them by Averroes' medical works) and practical ones. Among the latter were those arising from the doctor–patient relationship: medical etiquette and courtesy in the relationship with patients; the effects and efficacy of very complex prescriptions; the medical practices of healers belonging to the Muslim and Jewish minorities; and control of professional activity, among others. Neither did the *Sumario de la medicina* [Summary of medicine] (1498) by López de Villalobos, which included his work *Sobre... las bubas*, [On... buboes], also written in Castilian, have anything to do with the production of the university, as it was not a work written by a member of this academic institution.[77] Medical intellectual works started to be produced at both Salamanca and Valladolid in the 16th century. Unfortunately, not a single work which includes, not even in the form of notes, commentaries on the different works which must have served as a basis for medical teaching, from the texts of the *Articella* to the *Canon* itself, via the works of Galen, has been preserved; nor do we possess any of the discussions that took place in the academic context to mark the awarding of degrees. Only by means of the manuscripts that have been considered can we catch a glimpse of the university's medical intellectual world.

THE LATE 15TH CENTURY REVIVAL

It is highly significant that the earliest evidence that we have concerning the presence in the University of Salamanca library not only of medical works, but rather of the *libri naturales*, which shook the European academic world in the 13th century, should date from the second half of the 15th century.[78] To be precise, on June 9, 1468, the University took the trouble to acquire a series of books, which it evidently did not possess:

> Similarly, they ordered that any books which might be necessary for the library of said centre of study, both texts and commentaries, should be purchased with the funds of the University . . . and that the books that were necessary for it to see and to know should be purchased.[79]

As from the 1450s, whether by means of donations [for instance those made by the masters Juan of Segovia (*c.* 1457) and Pascual Ruys (d. 1479)] or of purchase, the University of Salamanca came to possess works on *quaestiones naturales* by 14th and 15th-century authors (John Marbres, John Buridan, John Versor, and Blasius Parmensis), in addition to Aristotle's and Albert Magnus' works on natural philosophy (from physics to embryology), together with a series of medical works. The latter included Salernitan medicine, the collection of works by Hippocrates and Galen that were in use and known in the course of the 12th and 13th centuries, plus the developments of scholastic medicine dating from the turn of the 13th and 14th centuries. An indication of the weakness of Castilian university institutions was the period of a 150 years that the University of Salamanca took to acquire the tools of intellectual work characteristic of scholastic medical institutions of the last third of the 13th century and the first decade of the 14th century. This is likewise indicated by the belated presence in the university library (which effectively began to operate as from 1465)[80] of the medical literature of Salerno (dating to the 11th and 12th centuries), exemplified by the *Viaticum* translated by Constantine the African, of Hippocrates' works, and those of the "new Galen", represented in the "bank of medicine" by the *De regimine acutorum* (probably with Galenic commentaries) and by the important work by Galen entitled *De ingenio sanitatis* (= *De methodo medendi*). This was also the moment when works by Bernard de Gordon, Arnau de Vilanova, Jean de Saint-Amand, and most probably by Pietro d'Abano, whose *Conciliator* is admiringly referred to by Francisco López de Villalobos in the *Prohemium* to his *Sumario de la medicina* (1498) and, which, as has been seen, was frequently used for teaching purposes, became available in the library. A similar time lag can also be detected as regards the arrival of Aristotle's *libri naturales* and Albert Magnus' works, commonly used volumes in European university circles in the second half of the 13th century and also in the Castilian cathedral chapters of Burgo de Osma and Toledo in the same period.[81]

This does not mean that the important corpus of natural philosophy constituted by Aristotle's *libri naturales* and the commentaries by Albert Magnus or Thomas Aquinas were unknown in intellectual circles in Salamanca. I have already mentioned the existence of the Aristotelian *corpus recentius* in the library of the College of

San Bartolomé, donated at the end of the first third of 15th century by Diego de Anaya, founder of the institution. One can also refer, for example, to the work of the Dominican Lope de Barrientos (1382–1469), who, from the friary of San Esteban (Salamanca) and later from his episcopal see in Cuenca, put forward a wide-reaching program to spread natural questions from an Aristotelian viewpoint. In his unfinished work *Clavis sapientie*,[82] unfortunately still an unedited manuscript with interesting medical contents, such as the way medicine was dealt with by natural philosophers, not with the aim of practicing themselves, but since the health-illness question, as a natural problem, was not alien to them.[83] This work was a true encyclopedia continuing the tradition of this late medieval literary genre, which sought to place the contents of the *libri naturales*, in the broadest sense that this term had at the time, in the hands of an educated public (with a knowledge of Latin), by summarizing the contents of a by now extremely abundant and complex form of literature in alphabetical order.[84] By means of a knowledge of the meaning of terms, all of which were converted into genuine technical expressions (e.g., *alimentum, animalia, complexio, corpus, cor, eclipsis, elementum, fisonomia, gustus, imaginatio, lux, luna, motus, natura, naturalia, odor, passio, phisis, planetis, pestis, res, sensus, sphera celestis, tactus, vita, de formatione vocis*, among others), Lope de Barrientos aimed to provide a tool to provide the reader with an introduction to the contemporary world of knowledge, of which the natural things formed a part. This is why he called his work *Clavis sapientie*, in other words "The key to knowledge". The final aim of this knowledge (*sapientia*) was no longer just knowledge of God, but also knowledge of the natural world created by Him, of which human health and illness, considered as something characteristic of human nature, formed part.

It should likewise be remembered that it was the king's consultations that led to Lope de Barrientos' brief works of natural philosophy, which were nothing other than a popularization in Castilian of Aristotle's *Parva naturalia*. In this context, I refer to his texts entitled *Tractado del dormir y despertar y del soñar* [Treatise on sleep, waking, and dreaming], *Tractado de la divinança e sus espeçies, que son las espeçies de la arte mágica* [Treatise on divination and its various kinds of magical art] (in which he dealt with, for example, the problem of medical prognosis), and *Del caso y fortuna* [Of the case and fortune].[85]

It should not be forgotten that Lope de Barrientos' work was carried out in the context of learned courtly circles, such as those of the Castilian courts of Enrique III (1390–1406) and Juan II (1406–1454),[86] and those of some aristocrats who were very sensitive to the natural questions and to the Aristotelian *libri naturales*, such as the Marquis of Santillana (1398–1458)[87] or Juan Alonso Pimentel, first Count of Benavente (d. 1420),[88] or Rodrigo de Mendoza, Marquis of Cenete (1470–1523) and son of Cardinal Pedro González de Mendoza (1428–1495). Rodrigo de Mendoza's collection of Aristotelian manuscripts included texts of many of the natural works. His library reflects the interests of both Rodrigo and his father.[89] We know that the Castilian court often came into direct contact with Italian humanism and with Italian university doctors, and a part of the Castilian nobility is also known to have been highly impressed by Italian humanism.[90] Moreover, we should not forget that the rationalist Jews who had strong links with the court—amongst them the scholar Meyr Alguades,[91] one of the physicians of King Enrique III—became so influenced by Aristotelism that Alguades designed a program of translation into Hebrew from the Latin of some of the Aristotelian works, particularly the *Ethics*, the *Economics*, and the *Politics*. Meyr Alguades himself only completed the translation of the *Ethics*.[92] The close intellectual relationship of Alguades with the intellectualized Aragonese Jews was decisive. His translation was in response to the requests of Benveniste ibn Lavi, one of the most influential members of the Aragonese Jewish community.[93] Jewish scholars' use of Latin to gain access to Greek natural philosophy should not surprise us. As I mentioned in the introduction to this article, this was a consequence of the intellectual power of attraction of natural philosophy and scholastic medicine for rational Jewish scholars.[94]

Two worlds—Italian humanism and the intellectual Jews of the Crown of Aragon—appear as the stimuli of the Aristotelian renaissance in Castile. This does not imply that a local newly recovered Castilian Aristotelian tradition did not also act as a stimulus. The first evidence of this tradition can be identified in Santiago de Compostela, between 1222 and 1230, in the midst of the archiepiscopal library and the Franciscan and Dominican monasteries.[95]

An illustration of the interest aroused in natural questions among an educated Castilian-speaking public lying outside intellectual circles of academic origin can be found in the Castilian translation of Aristotle's *De animalibus*, probably written around the same date as Lope de Barrientos' work; this was a considerable intellectual feat, and it is unfortunate that the name of the author remains unknown.[96]

Nevertheless, in this context, two details must be pointed out: in the first place, the fact that, in the mid-15th century, problems that had been expounded in Castile itself with the same intellectual tools in the mid- and late 13th century (take, for example, Juan Gil of Zamora's *Historia naturalis* of *c*.1280)[97] were being raised, commented on and made known; secondly, the limitations of intellectual curiosity that Catholic orthodoxy imposed on natural philosophers such as Lope de Barrientos, as the 15th century advanced, in comparison with the approach of brethren of the same order in the early and mid-13th century. Aristotelian rationalism led to a loss of intellectual freshness and made it difficult to deal with subjects such as alchemy, necromancy, and magic, which were totally rejected. Only those matters that were compatible with the Catholic orthodoxy of the time were retained. For instance, for Barrientos, the existence and influence of devils was as real as the devastating effects of plagues. Only in this way can the sarcasm which Lope de Barrientos displayed toward those who cultivated "çiençias" ("sciences") such as necromancy or alchemy, on the occasion of Enrique de Villena's death in poverty (1434), be explained.[98] It is not without interest that Barrientos should have played a leading role in expurgating and burning the latter's library after his death. King Juan II ordered that this intellectual aristocrat's library should be burnt following Lope de Barrientos' direct advice, an incident that Barrientos himself describes.[99]

Let us return to the medical world of 15th-century Salamanca. By means of the manuscripts that have been mentioned—Pedro Pugo's copy of medical works, the notes taken by the anonymous student and the aide-memoire known as the *Collectiones doctorum*—we have extracted information about the "modern" scholars whose opinions were explained to students or whose works they were recommended to consult and/or read. Unfortunately, no inventory of the works contained in the library belonging to any of the professors of Salamanca of the period in question has survived. The introduction of new subjects in the academic medical world of Salamanca at the turn of the 15th and 16th centuries (new diseases, the problem of the degrees in compound medicines, among others) has already been commented on above. In this context, we must mention two developments of differing importance, which cannot be discussed within the context of this article. First, the growing importance of astronomy–astrology as from the 1460s, when Alfonso X's astronomical tables were reintroduced in the Faculty of Arts of Salamanca.[100] This was to have major repercussions for the renovation of medical Galenism in late 15th century Castile, although the leading role in this movement was to be played by a physician who did not belong to academic circles, but rather to the court of Queen Isabella.[101] The second development was the production of the first work dealing with children's illnesses to be printed in Europe, that by the Italian Paolo Bagellardo de Flumine (1410/1420–1490/1492), a professor in Padua, by the new printing press of Salamanca. His *Libellus de egritudinem infantium*, inspired by Rhazes' *Libellus de morbis infantium*,[102] translated by Gerard de Cremona, with his own personal clinical observations, was printed in Padua in 1472.[103] Almost 50 years later, in about 1515, it was printed in Salamanca.[104]

RESTRICTIONS ON INTELLECTUAL ACTIVITY

In order to start up a program of medical studies, time and money are required, in addition to experience and knowledge on the part of those who are to carry out the project. It cannot be denied that the professors known to have been in Salamanca possessed both experience and knowledge; proof of this is that they were held in esteem for their medical knowledge and, from what is known about certain of them, they enjoyed considerable professional prestige. As far as money was concerned, not only is it known that salaries were not particularly high, but also that there were problems involved in receiving them, the closing of the university at different moments in the later Middle Ages not being unrelated to financial difficulties. I consider that such economic problems, together with personal greed, were closely connected with the enthusiasm, which many of Salamanca's professors of medicine displayed for associating themselves with ecclesiastical benefices that enabled them to dispose of the income thus produced. The case of Fernando Díaz of Toledo, Vespers professor between 1409 and 1413, who simultaneously held the posts of archdeacon of Niebla and canon of Toledo, was not an exception.[105]

The availability of time, however, was a real limiting factor; it can be stated that all the ordinary professors of medicine, both those attached to the Prime chair and those associated with the Vespers chair, lived subject to the whims of private clients (an inescapable feature of medical practice at the time), which resulted in frequent absences from the city. Their main clients were the members of the royal family as well as of other noble families. The itinerant nature of the court accentuated and stimulated such absences from their university posts, which brought about low academic and intellectual performance. This was even more serious if the increase in student numbers as from the mid-15th century is taken into account. Such absences annoyed both students and the university authorities, since they led to a decline in prestige and an unnecessary expense for the university's depleted coffers. All the measures introduced by the university authorities in order to avoid such absences and to encourage the continued presence of ordinary professors (such as the automatic loss of their chair) were doomed to failure because of the wishes of the members of the royal and noble households to dispose of the services of those physicians who embodied the prestige of university medical learning.[106] We cannot fail to mention the absenteeism of the ordinary professors of medicine as one of the reasons for their low intellectual performance, which is demonstrated by the lack of academic medical literature. Even though this factor was also present in other European academic centers, elsewhere professors succeeded in combining private practice with the production of intellectual works. Why, then, did the professors of Salamanca produce so few works? Was López de Villalobos right when he mentioned in 1514, well into middle age and well-situated at court, to one of the holders of a chair, Fernando Álvarez Abarca (third physician to the queen), who was also a royal physician and who is not known to have written any academic works, the lack of enthusiasm for producing academic studies on the part of individuals who might well have done so in view of their talent, their apparent lack of interest in intellectual matters, and, finally, university physicians' thirst for money alone?:

> Most famous doctor, he said, some days ago you had a conversation with me in which you expressed yourself to be indignant with the unforgivable silence of physicians, owing to their laziness and their reprehensible lack of literary activity, since, (in a place) where so many sources of wisdom could burst forth and so many doctrines could flourish, you note that teaching withers and dries up and that understanding falls silent and they remain without nurture because of the apathy of highly learned persons, and you regret that they are only concerned with arguing and legal disputes about pragmatic matters (rather than intellectual ones) and obtaining money.[107]

The presence of works by the great medical authorities, of both Greek and Arab origin, as well as those by authors trained in scholastic medicine at Montpellier, Paris or the various Italian centers active at the turn of the 13th and 14th centuries, in both the private libraries of the members of Salamanca University and in that of the university itself in the 14th and 15th centuries, demonstrates that Salamanca was more a center of consumption of *scientia medica* than a center of production. Nevertheless, this does not mean that movements related to the renewal of Galenism, a form of Galenism focusing upon medical practice, directed by university physicians who were far from indifferent to clinical developments and new intellectual currents such as medical humanism and astrology, were not present at its heart.

NOTES

*This is a revised and shortened English version of an article previously published in Spanish in *Dynamis* 20 (2000). English version by Dr Philip Banks to whom I am indebted.

[1] H. Rashdall, *The Universities of Europe in the Middle Ages*. A new edition... edited by F. M. Powicke and A. B. Emden, 3 Vols. (London: Oxford University Press, 1936) [Repr. 1969]; J. Ijsewijn and J. Paquet (eds.), *The Universities in the Late Middle Ages* (Leuven: Leuven University Press, 1978); H. De Riddder-Symoens (ed.), *A History of the University in Europe*. Vol I: *Universities in the Middle Ages* (Cambridge: Cambridge University Press, 1992); O. Pedersen, *The First Universities. 'Studium Generale' and the Origins of University Education in Europe*. English transl. by R. North (Cambridge: Cambridge University Press, 1997). The last work focuses on the faculties of *Artes*.

[2] V. L. Bullough, *Achievement, Professionalization, and the University"*, in Ilsewijn and Paquet (eds.) (note 1), pp. 497–510.

[3] See N. Siraisi, "The Faculty of Medicine", in H. De Ridder-Symoens, (ed.) (note 1), pp. 360–387.

[4] L. García-Ballester, L. Ferre, and E. Feliu, "Jewish Appreciation of Fourteenth-Century Scholastic Medicine", *Osiris* 6:85–117 (1990).

[5] P. U. González de la Calle and A. Huarte y Echenique, *Constituciones y Bulas complementarias dadas a la Universidad de Salamanca por el Pontífice Benedicto XIII (Pedro de Luna)* (Zaragoza, 1932), p. 42. Reproduced by V. Beltrán de Heredia, *Bulario de la Universidad de Salamanca (1219–1549)*, 3 Vols. (Salamanca: Universidad de Salamanca, 1966–1967), II, num. 4444. Quoted by M. Amasuno, *La Escuela de Medicina del Estudio salamantino* (Salamanca: Universidad de Salamanca, 1990), p. 51. Benedict XIII's statutes for the university (1411) were published by H. Denifle, *Urkunden zur Geschichte der mittelalterlichen Universitäten. Die päpstlichen Documente für die Universität Salamanca* (Freiburg i. Br.: Herder, 1889 [Archiv für Literatur- und Kirchen- Geschichte des Mittelalters, Bd. V]).

[6] Rashdall (note 1), II, pp. 64–65.

[7] L. García-Ballester, "Medical Science in Thirteenth-Century Castile: Problems and Prospects", *Bulletin of the History of Medicine* 61:183–201 (1987); G. Beaujouan, *La science en Espagne au XIVe et XVe siècles* (Paris: Conférence du Palais de la Découverte, 1967) [Repr. In *Science médiévale d'Espagne et d'alentour* (Aldershot: Ashgate-Variorum, 1992)].

[8] García-Ballester (note 7).

[9] E. Esperabé Arteaga, *Historia de la Universidad de Salamanca*, 2 vols. (Salamanca: 1914–1917), I, p. 22.

[10] Beltrán de Heredia (note 5), I, num. 24, pp. 330–331.

[11] It appears in the address of a document of Pope Urban V (1363), in which the bachelor of medicine Lorenzo Juan (*Laurentiis Joaniis*) is mentioned, it being added that "he had taught medicine for several years in the *Estudio* of Salamanca". "Item Laurentio Joannis, ..., baccalario in medicina, qui de eadem in studio Salamanticense per aliquos an[n]os legit". Beltrán de Heredia, (note 5), III, num. 1367, pp. 307–308.

[12] This was the degree obtained by Angelo de Costefort, physician to Charles II of Navarre; in 1362, the latter awarded Angelo a substantial sum of money to cover the considerable expenses involved in obtaining a master's degree in medicine at Salamanca, a degree which he was in possession of by 1363. V. Beltrán de Heredia, *Cartulario de la Universidad de Salamanca (1218–1600)*, 6 Vols. (Salamanca: Universidad de Salamanca, 1970–1973), I, num. 59, p. 639.

[13] Esperabé Arteaga (note 9), I, pp. 41–42.

[14] See Amasuno (note 5), pp. 48–125.

[15] See D. Jacquart, "La question disputé dans les facultés de médicine", in B. C. Bazàn, J. W. Wippel, G. Fransen, and D. Jacquart (eds.), *Les questions disputées et les questions quodlivétiques dans les facultés de théologie, de droit et de médecine* (Turhout: Brepols, 1985), pp. 279–315.

[16] R. Sancho de San Román, *Tres escritos sobre pestilencia del Renacimiento español. Fernando Álvarez. Diego Álvarez Chanca. Licenciado Fores* (Salamanca: Universidad de Salamanca, 1979). The *Compendio* and the writings by both Diego de Torres and Gómez García de Salamanca have been edited by M. Amasuno, *El 'Compendio de medicina' del doctor Gómez de Salamanca* (Salamanca: Universidad de Salamanca, 1971); *Idem*, *Un texto médico-astrológico del siglo XV: 'Eclipse sel sol', del licenciado Diego de Torres* (Salamanca: Universidad de Salamanca, 1972). The short work by Gomez García de Salamanca, in *Idem*, *Medicina castellano-leonesa bajomedieval* (Valladolid: Universidad de Valladolid, 1991), pp. 36–84 [Acta Historico-Medica Vallisoletana XXXII].

[17] Biblioteca Nacional (henceforth BN), Madrid MS 4220, ff. 1–68. The *Praxis medica* is incomplete, lacking the initial part. It follows the arrangement established by Bernard of Gordon's *Lilium medicine* (also known as *Praxis*): in the first place fevers are explained, followed by the remaining illnesses arranged "from the head to the feet", an arrangement very familiar to medieval practitioners.

[18] M. R. McVaugh, "Two Montpellier recipe collections", *Manuscripta* 20:175–180 (1976).

[19] G. Beaujouan, *Manuscrits scientifiques médiévaux de l'Université de Salamanque et de ses 'Colegios Mayores'* (Bordeaux: Féret & Fils, 1962), pp. 135–137.

[20] Amasuno (note 16), p. 106, accepts this authorship with a considerable degree of caution.

[21] On this Parisian master, see W. O. Schalick III, *Add One Part Pharmacy to One Part Surgery and One Part Medicine: Jean de Saint-Amand and the Development of Medical Pharmacology in Thirteenth-Century Paris* (D. Phil., Johns Hopkins University, 1997).

[22] Jean de Saint Amand's aim was clear: "So that students may quickly locate what they are so earnestly seeking in Galen's books (*ut scolares qui saepius in libri Galeni quaerendo . . . et citius inveniant*)", in O. Paderstein(ed.), *Revocativum memorie*, (Berlin: Prohemio, 1892), p. 10. See L. García-Ballester, "Arnau de Vilanova (*c.*1240–1311) y la reforma de los estudios médicos en Montpellier (1309): El Hipócrates latino y la introducción del nuevo Galeno", *Dynamis* 2:97–158 (1982), in pp. 105–107. A shorter and revised version in, *Idem*, "The 'New Galen': A Challenge to Latin Galenism in Thirteenth-Century Montpellier", in K.-D. Fischer, D. Nickel, and P. Porter (eds.), *Text and Tradition. Studies in Ancient Medicine and Its Transmission Presented to Jutta Kollesch* (Leiden, Boston, Köln: Brill, 1998), pp. 55–83.

[23] "Nota quod res naturales sunt ille quod ad esse sanitatis sunt necessarie secundum rationem sui generis. Et ideo dicitur res sine quibus sanabile corpus esse non potest. Arnaldus in *Speculo*, capitulo secundo", BN, Madrid MS 3371, f. 90va.

[24] Real Academia de la Historia (thereafter RAH), Madrid MS 9/443. See G. Beaujouan, "Manuscripts médicaux du moyen age conservés en Espagne", *Mélanges de la Casa de Velázquez* 8:161–221 (1972) [*Science médiévale d'Espagne et d'alentour* (Aldershot: Ashgate-Variorum, 1992)].

[25] RAH, Madrid MS 9/443, f. 78v.

[26] "El ruybarbo es tal medicina/de las que se llama por nombre benditas,/et sus propiedades son muy infinitas,/et stómago et hígado et la sangre afina;/a cólera et phlegma es purga benigna/sacándolas mucho con seguridad". RAH, Madrid MS 9/443, f. 104.

[27] D. Jacquart, *La médecine médiévale dans le cadre parisien: XIVe-Xve siècle* (Paris: Fayard, 1998), p. 498.

[28] RAH, Madrid MS 9/443, ff. 58v–101.

[29] See T. Joutsivuo, *Scholastic Tradition and Humanist Innovation. The Concept of Neutrum in Renaissance Medicine* (Helsinki: Bookstore Tiedekirja, 1999); T. Pesenti, "The Teaching of the *Tegni* in Italian Universities in the Second Half of the Fourteenth Century", *Dynamis* 20:159–208 (2000).

[30] The date suggests that this must have been Fernán Álvarez Abarca (*c.*1456–1526), also known as "the third physician to the Queen" and the holder of the Prime chair at the Faculty of Medicine from 1497/1498 onward. He was a brother of Gabriel Álvarez Abarca (d.*c.* 1496/1497), who was also a holder of the Prime chair at the Faculty of Medicine, known as "the second physician

to the Queen", whom he was to succeed in the chair, and son of Fernán Álvarez Malla (d. in 1469), "the first physician to the Queen", and likewise Prime chairholder at the Faculty of Medicine. See Amasuno (note 5), p. 99 onward. Amasuno does not consider the subject we deal with in this article.

[31] RAH, Madrid MS 9/443,ff. 63–64v.

[32] "In medicinis necesse est nobis scire graduare", *Ibid.* f. 65.

[33] *Ibid.*, ff. 65–66v.

[34] "Primum est graduare humanam generationem nam hoc multum iuvat in conservatione et curatione. Secundum est graduare egritudines.... Tertium est graduare medicinas complexionales. Quartum graduare medicinas purgativas. Quintum graduare medicinas compositas", *Ibid.* f. 65.

[35] See the excellent Introduction by M. R. McVaugh to his edition of Arnald of Vilanova's *Aphorismi de gradibus*, in *Arnaldi de Villanova Opera Medica Omnia* (henceforth *AVOMO*), 2nd ed. (Barcelona: Universidad de Barcelona, 1992), II, pp. 1–136.

[36] See J.-M. Dureau-Lapeyssonnie, "L'oeuvre d'Antoine Ricart, médecin catalan du XVe siècle", in G. Beaujouan et al., *Médecine humaine et vétérinaire à la fin du moyen age* (Genève-Paris: Droz, 1966), pp. 171–364.

[37] See L. García-Ballester, *La medicina a la València medieval. Medicina i societat en un pais medieval mediterrani* (València: Institució Valenciana d'Estudis i Investigació, 1988), p. 93.

[38] "Nota consilium ipsius doctoris (i.e., doctoris Regine) et est quod in medicina composita ab aliquis ne tollas aliquam medicinarum ipsas componentium sicut sunt quidam medicorum...", RAH, Madrid MS 9/443, f. 65v.

[39] See G. Beaujouan, *Manuscrits scientifiques médiévaux de l'Université de Salamanque et de ses 'Colegios Mayores'* (Bordeaux: Féret & Fils, 1962), pp. 110–111. Part of them were edited and commented on by M. R. McVaugh, "The *Experimenta* of Arnald of Villanova", *Journal of Medieval and Renaissance Studies* 1:107–118 (1971), in p. 107.

[40] J. A. Paniagua, *El doctor Chanca y su obra médica* (Madrid: Ediciones Cultura Hispánica, 1977).

[41] Hardly anything is known of the lives of these two physicians of the Salamanca university school. As regards Fernando del Marmol we have no information whatsoever, with the exception of his collaboration with Núñez de la Hierba. As for the latter, he is known to have been born in Salamanca in about 1460, and to have studied Arts and Medicine in that city's university, obtaining the maximum academic degree (*Magister*) in both disciplines in 1487. He must have been connected with the Faculty of Arts to judge by the edition of Pomponius Mela's *Cosmographia* that he prepared (Salamanca: 1498). He does not seem to have held a chair at the Faculty of Medicine. He disappears without trace from Salamanca academic circles in June 1504. For the few known biographical details and his relationship with natural philosophy, see C. Flórez Miguel, P. García Castillo, and R. Albares, *La ciencia de la tierra. Cosmografía y cosmógrafos salmantinos del Renacimiento* (Salamanca: Cajade Ahorros y Monte de Piedad, 1990), pp. 443 onward.

[42] *Alkindus de gradibus medicinarum compositarum. Parabole magistri Arnaldi de Villanova. Aphorismi de graduatione medicinarum compositarum cum commento eiusdem* (Salamanca: Juan de Porras, 1501); J. A. Paniagua drew attention to this edition. See J. A. Paniagua, "Las ediciones renacentistas de *Medicationis parabole*", in *Medicina e Historia* (Madrid: Universidad Complutense, 1980), pp. 27–43.

[43] "Erant enim littere tui exemplaris apud nos unici, quod mihi transcribendum tradidisti, ita caduce, ut quod sibi vellent, vix intelligere possem", *Alkindus de gradibus medicinarum* (note 34), f. 32v (Letter of the bachelor Fernando del Marmol to Doctor de la Hierba).

[44] "Quod, ni tua suffultus essem ope, ad finem minime perduxissem... Meo labore ergo et industria effectum est ut omnibus hec opera utilissima imprimerentur. Sed tua eminentissima doctrina tantum valuit, ut sine mendis in manus omnium venirent, et que pannosa situque obsita hactenus latebant, castigatissima undecumque promulgarentur", *Ibid.* (note 34), f. 32v.

45 Introduction by McVaugh to his edition of Arnold of Vilanova's *Aphorismi de gradibus* (note 34), pp. 89 onward.

46 "Qualitates enim erbarum aliarumque rerum multis verbis multi alii ostendere. Puta hec calida, hec frigida est. Est complurime mixte quam vim habeant, cui gradui egritudinis prosint, cui moceant, quis melius, quis clarius, et copiosus distincte nostro Alchindo docuit". Franciscus Nunnez de la Hierba. Prologus, *Alchindus de gradibus medicinarum compositarum* (note 34), f. 1v.

47 "Francisci Nunnez de la Hierba, in medicina doctoris, cathedram vespertinam medicine Academia salmanticensi as presens regentis, ad iuvenes medicine studiosos Prologus./Diu multumque cogitabam studiosa iuventus quid nam esset in causa cum passim et veterum et novorum auctorum omnimodo doctrine varia volummina imprimerentur, in medicina scientia relinquarum facile principe aut nihil, aut parva quedam de gradibus medicinarum compositarum scriberentur. Quod sane operis difficultati, ac inmenso labori imputavi existimans non mediocri acumine ingenii ad huius rei cognitionem opus esse. Hoc ferendum est, pauci fortassis reperiuntur, quorum ingenium huic materie parsit. Sed quis non irascatur, hec eadem (quantulacumque sint) de hac re elucubrata delitescere, et vix ad nos pervenire. Quis non succenseat occultantibus, quis non laudet promulgantes, quod locupletent scientias./Ego autem viri doctissimi mihi persuasi in pari laudis meritorum fastigio, et qui edunt condita, et a aliis scripta scrutantur... /Qua de causa vobis gratificari cupiens; vestris commodis inservire anhelans, et ne quid a maioribus nostris servatum ad nos pervenit id oscitantes dissutis undique malis amittere videremur, et negligentia tanto amulumento obstaret, usuique communi officeret, effeci, ut antiquissimus Alchindus vir quiedemmagne auctoritatis iterum revivisceret.... /Non me potui continere (fateor) quin quibusvis laboribus et vigiliis comptentis hunc doctissimum virum in lucem ederem. Callebam enim ex eo uberrimam doctrinam elici posse, et hunc penitus enucleatum maximo adiumento vobis fore, et ad vestrum desiderium libentius expexplendum, et ad sanctissimum propositum facilius obtinendum... /Plura in hanc sententiam dicerem iuvenes eruditissimi, nisi vobis ist hec nota putarem. Illud tamen unum non pretermittam practicam, que non ignobilis scientie medicine pars habetur, omnino imperfectam, que non ignobilis scientie medicine pars habetur, omnino imperfectam, ac veluti mancam sine Alchindus esse. Quod nec intelligi, nec percipi sine huius prestantissimi viri doctrina possint, ut eorum sit inditium". Nuñez de la Hierba. Prologus. *Alchindus de gradibus medicinarum compositarum* (nore 34), f. 1v.

48 On this renewed scientific movement of Galenism, see the articles by García-Ballester (note 22).

49 Beaujouan (note 24 s.v.) suggests Juan de la Parra, but he would seem to have been a somewhat later scholar, more closely related to courtly circles than academic ones. See the extensive work on this individual by N. Alonso Cortés, "Dos médicos de los Reyes Católicos", *Hispania* 11:607–657 (1951).

50 Gonzalo de la Parra was considered by his colleagues to be a good clinical practitioner. Reference to the way in which he was praised by Fernando Fernández de Sepúlveda, doctor in medicine and apothecary, who practiced medicine in the city of Salamanca, can be found in the latter's *Manipulus medicinarum in quo continentur omnes medicine tam simplices quam composite secundum quod in usu apud doctores habentur. Utilis medicis necnon Aromatarijs nuper editus* (Salamanca, 1523), f. 26ra.

51 See J. Arrizabalaga, J. Henderson, and R. French, *The Great Pox. The French Disease in Renaissance Europe* (New Haven-London: Yale University Press, 1999).

52 BN, Madrid MS 4220. The *Praxis medica* is only incompletely preserved (ff. 1–68). Fernando Fernández de Sepúlveda (b.*c*. 1485) was trained as a physician in both Salamanca and Valencia, where he received particular instruction in pharmacy at the recently created unified hospital (1512). On his return to Salamanca, prior to 1520, he exercised both professions and maintained a close relationship with university medical circles. He wrote a practical treatise on medicine and pharmacy (*Manipulus medicinarum*, note 49), with a large number of autobiographical details.

A collection of recipes (*Recetario*), attributed to him, including advice for young physicians who are starting to undertake medical practice, in addition to an outline of an *antidotarium* (ff. 69–136v; 139–162v), can be found in MS 4220; in particular, the author emphasized that those who were embarking on a medical career should possess a solid background in pharmacy. The contents of this manuscript deserve further examination. See, L. García Ballester, "La çiençia y el ofiçio de la boticaría", in *Historia de la Ciencia y de la Técnica en la Corona de Castilla*, Vol. I (Valladolid: Junta de Castilla y León, 2002) pp. 865–915.

[53] "Item. Que los boticarios sean examinados por los fysicos y boticarios que oviere en los lugares donde vivieren, como se haze en Valencia y Çaragoça y otros lugares bien regidos", Archivo General de Simancas (thereafter AGS), Diversos de Castilla, leg. 1, doc. 55 (*c.*1498). Reproduced by I. de la Villa, *Los médicos y la medicina en la época de los Reyes Católicos. Comentarios a unas Ordenanzas del siglo XV reproducidas del Archivo de Simancas* (Valladolid: Talleres Tipográficos Cuesta, 1939), p. 45.

[54] RAH, Madrid MS 9/443, ff. 69–78v (*Tractatus de saphati*).

[55] See J. Arrizabalaga, "El 'Consilium de modorrilla' (Roma y Salamanca, 1505): una aportación nosográfica de Gaspar Torrella", *Dynamis* 5–6:59–94 (1985–1986).

[56] BN, Madrid MS 12366, ff. 48–157va.

[57] See N. G. Siraisi, *Avicenna in Renaissance Italy. The 'Canon' and Medical Teaching in Italian Universities after 1500* (Princeton, New Jersey: Princeton University Press, 1987), p. 108.

[58] BN, Madrid MS 12366, f. 131vb.

[59] "Ego Petrus Pugo in medicina bachallarius incepi scribere istud librum in [*illegible*] dum ibi practicabam in estate et primo anno mee practice .12. die mensis junij anno a nativitate 1457 et finivi eum secunda die augusti eiusdem annij", *ibid.,* f. 131vb.

[60] RAH, Madrid MS 9/443, ff. 60–62v.

[61] An author frequently referred to by Fernández de Sepúlveda in his *Manipulus medicinarum* (Salamanca, 1523) (note 49), and by Bernardino de Laredo in his work *Modus faciendi* (Seville, 1527).

[62] On this physician, see L. E. Demaitre, *Doctor Bernard de Gordon: Professor and Practitioner* (Toronto: Pontifical Institute of Mediaeval Studies, 1980).

[63] See the biography of this Castilian physician by M. Amasuno, *Alfonso Chirino, un médico de monarcas castellanos* (Valladolid: Junta de Castilla y León, 1993).

[64] See L. García-Ballester and E. Sánchez Salor, Introduction to *Arnaldi de Villanova Commentum supra tractatum Galieni de malicia complexionis diverse cum textu Galieni* (Barcelona: Universidad de Barcelona, 1985), p. 142 [*AVOMO*, XV].

[65] See the *explicit* of theVatican manuscript Vat. pal. lat. 1098: "Scriptum per me Johannem Schureissen in famosissimo studio Padue tunc temporis illic medicine scolaris vel studens anno domini 1464". L. Schuba, *Die medizinischen Handschriften der Codices Palatini Latini in der Vatikanischen Bibliothek* (Wiesbaden: Dr Ludwig Reichert Verlag, 1981).

[66] His *Commentarius super Canonem Avicenne* concentrating on the part of book IV dealing with fevers. See D. Jacquart, *Supplément* to E. Wickersheimer, *Dictionnaire biographique desmédecins en France au moyen age* (Genève: Droz, 1979), pp. 85–86. *Suppl.*, pp. 85–86.

[67] See Jacquart (note 26).

[68] As an example we can refer to his commentaries and *quaestiones* on mineral waters and their therapeutic effects. See L. García-Ballester, "Sobre el origen de los tratados de baños (*de balneis*) como género literario en la medicina medieval. A propósito del poema médico *Nomina et virtutes balneorum Puteoli et Baiarum* de Pedro de Éboli (*c.*1160–1220) y la *Tabula super balneis Puteoli*, atribuida a Arnau de Vilanova (m. 1311), contenidos en el MS 860 de la Biblioteca Universitaria de Valencia", *Cronos* 1:7–50 (1998).

[69] Jacquart (note 26), pp. 204–227.

[70] See the above-mentioned book by Siraisi (note 56), p. 7.

[71] See Arrizabalaga et al. (note 50), and Arrizabalaga (note 54).

[72] It should be remembered that Juan Gil (d.*c.* 1320), a Franciscan from Zamora (Kingdom of Castile), made use of it in his encyclopedia (*Historia naturalis*) and in his treatise *Contra venena et animalia venenosa* ("On poisonous animals"). This scientific encyclopedia has been edited by A. Domínguez and L. García-Ballester, *Johannis Aegidii Zamorensis Historia naturalis. Introducción, edición crítica, traducción castellana e índices*, 3 Vols. (Salamanca: Junta de Castilla y León, 1994). The treatise *Contra venena* was edited by M. de Castro y Castro, "Johannes Aegidii Zamorensis OFM. 'Contra venena et animalia venenosa' ", *Archivos Ibero-Americanos* 36:3–116 (1976).
[73] G. Beaujouan, "La bibliothèque et l'ecole médicale du Monastère de Guadalupe a l'aube de la Renaissance", in G. Beaujouan et al. (note 35), pp. 367–468.
[74] See the above-mentioned book by Siraisi (note 56).
[75] "huius (prescriptions) vero notitiam adimplere nequeunt, si vestigia aliqua medicine alfacere non urgeantur", *Sumario de la medicina (1498). Proemium, facsimil* edition by L. S. Granjel et al. (Salamanca: Universidad de Salamanca, 1998), p. 46. Both translation and punctuation are mine.
[76] A. Chirino, in A. González Palencia and L. Contreras Pozas (eds.) *Menor daño de la medicina. Espejo de medicina por Alonso Chirino, con un estudio preliminar acerca del autor* (Madrid: Real Academia de Medicina, 1945) [Biblioteca clásica de la medicina española, XIV]; *Idem, Menor daño de la medicina de Alonso de Chirino*. Edición crítica y glosario por M. T. Herrera (Salamanca: Universidad de Salamanca, 1973) [Acta Salmanticensia, Filosofía y Letras 75].
[77] See note 74 above.
[78] Among the first works to enter the library of the College of San Bartolomé, founded in 1413–1418 by Diego de Anaya, former bishop of Salamanca (1392) and Archbishop of Seville (1417–1437), were Aristotle's *libri naturales* in William of Moerbecke's Latin translation from the Greek (*corpus recentius*). These were donated by Diego de Anaya himself in the early 1430s; he must have acquired them through the *scriptoria* or dealers in manuscripts in Paris or Italy. See Beaujouan (note 19), pp. 17–22.
[79] "Item mandaron comprar qualesquier libros que fuesen menester para la libreria del dicho estudio, asy testos como lecturas, de los dineros de la Universidad . . . e fazer comprar los libros que fuesen para ella menester e verlos e cognoscerlos", Archivo de la Universidad, Salamanca, reg. 1, f. 126. See Beaujouan (note 19), p. 2.
[80] Beaujouan (note 19), p. 2.
[81] See L. García-Ballester, "El papel de las instituciones de consumo y difusión de ciencia médica en la Castilla del siglo XIII: el monasterio, la catedral y la universidad", *Dynamis* 4:33–63 (1984); R. Gonzálvez Ruiz, *Hombres y libros de Toledo, 1086–1300* (Madrid: Fundación Ramón Areces, 1997) [Monumenta Ecclesiae Toletanae Historica. Series V: Studia I].
[82] BN, Madrid MS 1795, ff. 26–160va.
[83] See L. García-Ballester, "*Artifex factivus sanitatis*: Health and Medical Care in Medieval Latin Galenism", in D. Bates (ed.), *Knowledge and the Scholarly Medical Traditions* (Cambridge: Cambridge University Press, 1995), pp. 127–150; *Idem*, "The Construction of a New Form of Learning and Practicing Medicine in Medieval Latin Europe", *Science in Context* 8:75–102 (1995).
[84] BN, Madrid MS 1795, f. 26vb.
[85] All these works were edited by L. G. A. Getino, *Vida y obra de Fray Lope de Barrientos*, Vol. I (Salamanca: 1927) [Anales Salmantinos]. See also A. Martínez Casado, *Lope de Barrientos un intelectual de la corte de Juan II* (Salamanca: Editorial San Esteban, 1994), a work with a traditional historiographical approach. The *Tractado de la divinança* has been re-edited together with a full introduction by P. Cuenca Muñoz, *El 'Tractado de la divinança' de Lope de Barrientos. La magia medieval en la visión de un obispo de Cuenca* (Cuenca: Ayuntamiento de Cuenca, 1994). Lope de Barrientos' work continued the intellectual tradition of the members of the mendicant orders in the first half of the 15th century, both in Europe as a whole and in Castile (in this case through the Dominicans), characterized by their attention to natural

questions. This tradition had been started in Castile by Juan Fernández (*fl.* 1225) in the then recently founded Dominican house of Bonaval in Santiago de Compostela, by studying the contents of Aristotle's works on natural philosophy and Arab commentaries on them, and by expressing his interest in mathematical questions (Euclid) and astronomy and astrology (al-Farghani's *Elementa astronomica*); this was continued by the group that translated various medical works from Arabic into Latin in the Dominican *Studium* in Murcia in the 1270s. See L. García-Ballester, "Nature and Science in Thirteenth-Century Castile. The origins of a Tradition: The Franciscan and Dominican *Studia*", in L. García-Ballester (ed.), *Medicine in a Multicultural Society: Christian, Jewish and Muslim practitioners in the Spanish Kingdoms*, 1222–1610 (Aldershot: Ashgate-Variorum, 2001, II); *Idem*, García-Ballester (note 7).

[86] It should be remembered that Diego de Anaya, who was particularly concerned about higher education and in possession of a rich library with a high proportion of scientific works, was tutor to Enrique III. See Beaujouan (note 19). Juan III appointed Lope de Barrientos as tutor to the *infante* Enrique, the future Enrique IV (1454–1474).

[87] He was the owner of the manuscript with the only Spanish translation of Aristotle's *De animalibus*, see M. Schiff, *La Bibliothèque du Marquis de Santillane* (Paris: Librairie Émile Bouillon, 1905), pp. 34–36.

[88] J. H. Elsdom, *The Library of the Counts of Benavente* (Ann Arbor, 1962), pp. 23–27.

[89] F. J. Sánchez Cantón, *La biblioteca del Marqués de Cenete iniciada por el Cardenal Mendoza (1470–1523)* (Madrid: CSIC, 1942).

[90] See A. R. D. Pagden, "The Diffusion of Aristotle's Moral Philosophy in Spain, ca. 1400–ca. 1600", *Traditio*, 31:287–313 (1975). King Enrique III (d. 1406), in his last illness, called upon the services of the Italian physician Pietro da Tossignano (d. 1407), a prestigious professor at the Faculty of Medicine in Bolonia. See Amasuno (note 62), pp. 68–71.

[91] M. Steinschneider, *Die hebraeischen Übersetzungen des Mittelalters und die Juden als Dolmetscher* (Berlin: 1893) (reimpr. Graz: Akademische Druck- u. Verlagsanstalt, 1956), p. 210, n. 26; the reader will find interesting remarks about Meyr Alguades in Y. Baer, *History of the Jews in Christian Spain*, 2 Vols. (Philadelphia, 1978). I employ the terms "rationalist" (and its contrary "antirationalist") for the sake of convenience, although they are overly simplistic. See I. Twersky, "Rabbi Abraham Ben David of Posquières: His Attitude to and Acquaintance with Secular Learning", *Proceedings of the American Academy for Jewish Research* 26:161–192 (1957), on pp. 164, 184–185; and Ch. Mopsik, *Lettre sur la sainteté: Le secret de la rélation entre l'homme et la femme dans la Cabale* (Lagrasse: Editions Verdier, 1985), pp. 32–34, 48–49.

[92] See the important paper by L. V. Berman, "The Latin-into-Hebrew Translation of the Nichomachean Ethics", in *Jerusalem Studies in Jewish Thought* 7:147–169 (1988) (Shlomo Pines Jubilee Volume on the Occasion of his Eightieth Birthday) (in Hebrew). Berman's paper includes the Hebrew edition of Alguades' Prologue. I am very grateful to Eduard Feliu (Barcelona), who gave me access to this fascinating Prologue.

[93] See the above-mentioned article by Berman (note 91). On Benveniste ibn Lavi and his important role in the Aragonese Jewish community, see Baer (note 90).

[94] García-Ballester et al. (note 4), and the literature cited therein.

[95] See García-Ballester (note 84).

[96] BN, Madrid MS 10198, ff. 1–93. As mentioned above this manuscript belongs to library of the Marquis of Santillana (see note 86) and is still unedited. J. Martínez Gázquez (Autonomous University of Barcelona) and myself are at present engaged on this task.

[97] Edited by A. Domínguez and L. García Ballester (note 71).

[98] "Este don Enrrique fue muy grant sabio en todas çiençias, en espeçial en la Theología e Nigromançia, e aun fue grant alquimista. Y con todo esto vino en tan grant menester, al tiempo que fallesçió non se falló en su cámara con qué le pudiesen enterrar. Y fue cosa de Nuestro Señor, porque las gentes conoscan quánto aprovechan las semejantes çiençias". (This Don Enrique was very knowledgeable about all the sciences, especially Theology and Necromancy,

and he was also a great alchemist. And in spite of this, he was so needy that when he died there was not enough to be found in his chamber for him to be buried.) J. De Mata Carriazo (ed.), *Refundición de la Crónica del Halconero por el obispo don Lope de Barrientos* (Madrid: Espasa-Calpe, 1946), p. 170.

[99] "Y después que él fallesció, el Rey mandó traer a su cámara todos los libros que este don Enrrique tenía en Yniesta, e mandó a fray Lope de Barrientos, maestro del Prínçipe, que catase si avía algunos dellos de çiençia defendida. E el maestro católos, e falló bien çinquenta volú[me]nes de libros de malas artes. E dio por consejo al Rey que los mandase quemar. El Rey dio cargo dello al dicho maestro, e él púsolo luego en esecuçión, e todos ellos fueron quemados". (And after he died, the king ordered that all the books that this don Enrique had in Iniesta should be brought to his chamber, and he ordered Brother Lope de Barrientos, tutor to the prince, to determine whether any of them were of the forbidden sciences. And the master examined them and found at least 50 volumes of books of the evil arts. And he advised the King that he should order them to be burnt. The King ordered this master to do so, and he thus carried it out, and they were all burnt.) See J. De Mata Carriazo, (ed.), *Refundición* (note 95), p. 171.

[100] See J. Chabás, "Astronomy in Salamanca in the Mid-Fifteenth Century: The *Tabulae resolutae*", *Journal for the History of Astronomy* 29:167–175 (1998).

[101] I refer to the important book by J. Gutiérrez de Toledo (*fl.* 1491–1515), *De computatione dierum creticorum* (Toledo: Antonio Téllez, 1495). See my forthcoming book, *La medicina en una sociedad multicultural. Las ciencias de la salud en la Corona de Castilla, siglos XIII a XVI* (Barcelona: Ediciones Península, 2001).

[102] *Abubetri Rhazae Liber ad regem Mansorem* (Basel: in officina Henrichi Petri, 1544), pp. 509–517 (Repr. Brussels: 1973).

[103] For Paolo Bagellardo and his work on pediatrics, see K. Sudhoff, *Erstlingeder pädiatrischen Literatur* (Leipzig: 1925); A. Simili, "I trattati di pediatria di paolo Bagellardo da Fiume, di Iaccopo Tronconi, di Leonello de Vittori da Faenza", *Episteme* 8:375–397 (1974); A. Peiper, *Quellen zur Geschichte der Kinderheilkunde* (Bern-Stuttgart: Verlag Hans Huber, 1966), pp. 45–47.

[104] See F. J. Norton, *A Descriptive Catalogue of Printing in Spain and Portugal, 1501–1520* (London, New York, Melbourne: Cambridge University Press, 1978), num. 575; L. Ruiz Fidalgo, *La imprenta en Salamanca (1501–1600)*, 3 Vols. (Madrid: Ollero & Ramos, 1994), num. 108.

[105] Beaujouan (note 72), pp. 387, 398.9, 430–434; Amasuno (note 5), pp. 60–61, 147.

[106] The following letter sent to the academic authorities of Salamanca University by both Queen Isabella of Castile and by King Ferdinand of Aragon is quite significant. In it the Catholic Monarchs order the academic authorities not to apply academic discipline to professor Diego de Torres, who was frequently absentee from his post: "El Rey e la Reina. Rector a maestrescuela e consiliarios e doctores de la Universidad del Estudio de Salamanca. Nos hobimos enviado mandar al licenciado de Torres [Diego] que fuese a curar a la condesa de tendilla, que estaba mal dispuesta de su persona, el cual asimismo habemos agora enviado mandar que tenga cargo de curar de la salud de la condesa de Cifuentes, que nos dizen que está mal. Por ende nos vos mandamos e encargamos que en tanto que el dicho licenciado estoviese curando de la dicha condesa de Cifuentes, no entendais de fazer mudanza alguna en la cátedra que el dicho licenciado tiene en ese Estudio, porque estando las dichas condesas en buena disposición de salud, luego se irá el dicho licenciado a residir en su cátedra. De Madrid a 28 de septiembre de 94 años" (AGS. Libros de Cámara, lib. primero, f. 146). Reproduced by Beltrán de Heredia (note 11), II, num. 216, p. 149. Another similar case was that of Juan Fernández, Prime professor at the Faculty of Medicine, who was in the service of the young Juan II and the regent queen Catalina of Lancaster. The royal house ordered the academic authorities not to apply the Constitutions of the Salamanca Studium and deprive the professor of his chair, despite his absence of more than 6 months, as he had been serving the royal household. See Beltran de Heredia (note 5),

II, num. 508, pp. 75–76. See also Amasuno (note 5), pp. 64–65. These two examples pose an interesting question as regards the science-power relationship in the Kingdom of Castile, an almost unstudied subject in Spain for the period in question, although this is not the moment to do so.

[107] "Cum mecum, celeberrime doctor, superioribus diebus colloquium haberes indignabundus de medicorum [in]curie detestando silentio, ocioque vituperando in re litteraria, nam ubi tot tantique sapientie fontes scaturire possent atque multiformis doctrina florescere, ibi disciplinas marcidas et arescentes conspexeris, et intellectus abmutescentes, et non irriguos, peritissimorum hominum desidia, defleres, quia rerum mechanicarum contentiones et iurgia dumtaxat, aurumque perpetrandum in cura esset". F. López de Villalobos, *Congressiones: vel duodecim principiorum liber nuper editus* (Salamanca: ex expensis v. Viri Laurentii de Liom de deis, 1514). Dedication letter.

JOSÉ M. LÓPEZ PIÑERO

THE FACULTY OF MEDICINE OF VALENCIA: ITS POSITION IN RENAISSANCE EUROPE

MEDICAL EDUCATION DURING THE RENAISSANCE

In order to situate historically, the development of medical education during the Renaissance, one must remember that at that time, medicine was the only scientific type of occupation which had been professionalized. Qualifications had begun to be regulated in the 12th century, and in the 13th century so had the education to be received by those wishing to qualify as physicians. In the regulations promulgated by Frederick II in 1240 for the kingdom of Sicily, universities were already the institutions where such studies were taught. This model of medical teaching, practiced at the major late medieval universities, led by Bologna, Padua, and Montpellier, was limited to physicians strictly speaking. Surgeons were excluded from university education, partly as a result of the negative appraisal of manual work and techniques, deemed to be tasks of a lower category, a concept derived from the writings of Plato and Aristotle. The training received by surgeons continued to consist basically of a craftsman apprenticeship with its own teaching institutions in certain countries from the 14th century onward.

The regulations governing the university education of physicians were consolidated and developed during the Renaissance. Generally speaking, it was necessary to study first in the arts faculty and graduate as a Bachelor of Arts or with an equivalent qualification. This was followed by an average of 3 or 4 years in the faculty of medicine culminating in the Bachelor of Medicine which was normally required to be a practicing physician. However, the only requirement in order to receive the title of licenciate or doctor consisted of merely defending a thesis on a specific subject. The didactic method was scholastic, based on "lectio" that consisted of texts by classic authorities being read and interpreted, and any difficult or obscure passages giving rise to problems ("quaestiones") were then discussed ("disputatio"). Until the time of the hegemony of the humanist movement which published the original texts in Greek—their Latin versions being translated directly from the medical works of Classical times—the texts used in "lectiones" were the Latin versions retranslated from the Arabic of some of these ancient works and of others by Arab authors from the late Middle Ages. Direct, polished translations by humanist physicians were introduced into certain universities in the mid-16th century and into the majority several decades later. This was not the only change which occurred in university medical education in Renaissance times. One group of advanced universities also introduced

M. Feingold and V. Navarro-Brotóns (eds.), Universities and Science in the Early Modern Period, 65–82.

some innovations which caused the scholastic teaching method to be superseded. This renovation was influenced considerably by the position of the faculty of medicine in each university, generally subordinate to that of the faculties of theology and canon and civil law. Faculties of medicine were given preferential treatment by the authorities at only a few universities. A classic example is the encouragement given by the Venetian Senate to the faculty of medicine at the University of Padua which made a decisive contribution to it becoming the most important in Europe at that time. Other similar cases include the University of Valencia whose faculty of medicine was the most prestigious and best equipped of the Hispanic kingdoms thanks to the support received from the municipal urban oligarchy upon which it depended.

The most outstanding innovation concerned how anatomy was taught. The practice of dissecting human corpses had begun at the University of Bologna in the early 14th century, spreading later to the Universities of Padua, Montpellier, Lerida, and others mainly in Italy, France, and Spain. Usually however there would be one autopsy at the most during the winter months, and the autopsy would moreover be designed to exemplify Galenist doctrines. As Andreas Vesalius (1515–1564) was to say later in the preface to his *De humanis corporis fabrica* (1543), the professor remained seated in the chair "reciting information about facts he had never seen personally but had learnt by memory from books", whilst someone else, "incapable of explaining the dissection to the students", carried out the autopsy, "destroying what he was supposed to show".[1] Vesalius himself, after being appointed professor of surgery in 1537 at the University of Padua, revolutionized how anatomy was taught by coming down from his chair and approaching the dissection table where he carried out the autopsies personally and showed the parts of the corpse to his students. This innovation was quickly adopted in the Universities of Bologna, Rome, Pisa, Pavia, Ferrara, and Naples which together with the University of Padua employed the most prominent anatomy professors at that time including Realdo Colombo (d. 1558), Gabriele Fallopio (d. 1562), Giulio Cesare Aranzio (1530–1589), and Girolamo Fabrizi d'Aquapendente (1553–1619). This method was also adopted at an early date at the University of Valencia following its introduction in 1547 by Pedro Jimeno (d. 1551), a Vesalius disciple as was his successor, Luis Collado (d. 1589), who converted Valencia into the heart of the Spanish Vesalian movement, responsible for establishing chairs in anatomy at the Universities of Alcalá (1549) and Salamanca (1551). This new anatomy was warmly received in France at the University of Montpellier by Guillaume Rondelet (1507–1566) although the chair was not established until 1593. The University of Paris on the other hand arose as the major center opposed to this reform, and was the base from which Jacques du Bois (Sylvius, 1478–1555), a former professor of Vesalius, launched his scathing criticism. In the German-speaking world, the University of Basle was the first participant in the Vesalian movement (1570) thanks to its professor Felix Platter (1536–1614). In England, the Tomlins Readership in anatomy of the University of Oxford was not set up until 1624.

Another important innovation was that related to the teaching of materia medica. Mention is often made of chairs in botany being established, but in fact these were chairs of "simples", encompassing medicines derived mainly from plants and also animal and mineral elements. The first of these was created at the University of Padua (1533) and as in the case of the chairs of anatomy, others were set up at various Italian universities and at universities in countries more directly influenced by the Italian Renaissance Schools of Medicine. These were sometimes "chairs of anatomy and simples", teaching anatomy in the autumn and winter and materia medica during the spring and summer, although such chairs were split into two separate ones at a later date. Their functions were not merely of a scholastic nature, particularly because students were directly involved in handling and collecting the plants on outings, to be herborized in different areas close to the cities where the universities were found. The utility of medicinal herb gardens was obvious, and the Venetian Senate set up the one at the University of Padua in 1545. The one belonging to the Faculty of Medicine at the University of Pisa was created at virtually the same time, under the management of such a prominent figure as Luca Ghini (d. 1556), followed by those at Valencia (1567), Bologna (1568), Leipzig (1579), Leiden (1587), Paris (1591), Montpellier (1598), etc. It must be remembered that these were not botanical gardens in the strict sense of the word, the aim was simply to collect as many medicinal plants as possible to enable students of simples and physicians to become familiar with them. This did not prevent some of them making a considerable contribution to botanical research.

The third noteworthy Renaissance innovation was the shift into the foreground of clinical observations and environmentalist-oriented public health studies. Whilst not manifestly breaking away from Galenism, the Hippocratic Collection was adopted as a model, particularly its group of clinical histories concerning "the constitution" or environmental circumstances in *Epidemics* and the treatise *of airs, waters, and places.* This "Hippocratic" Galenism developed from the mid-16th century onward mainly in Italy, Spain, and France.

Surgeons continued to be excluded from university medical education throughout the Renaissance in most parts of Europe. The conflict between surgeons and physicians in certain places is exemplified by that between the Fraternity of Saint Côme, in Paris and the United Company of Barber-Surgeons of London. In Italy on the other hand, several universities had chairs of surgery, the first of which was established in the late 13th century in Bologna. The most prestigious chair of surgery in the 16th century was that of Padua, where Vesalius started innovating the teaching of anatomy, and which was occupied at a later date by Fabrizzi d'Acquapendente. Surgery had also been taught in Montpellier in association with the University there, since the Late Middle Ages. The Italian influence led to several Spanish universities establishing chairs of surgery, the first being that of Valencia (1502).[2]

THE ORGANIZATION OF THE TEACHING OF MEDICINE AT THE UNIVERSITY OF VALENCIA IN THE 16TH CENTURY

The University of Valencia was a typical municipal university sustained financially and governed directly by the local bourgeois oligarchy. It gave preferential treatment to the development of the teaching of medicine and relegated the teaching of theology and law to the background, unlike what was happening in other major universities at that time. In addition, the Senate of Medicine acted as a professional guild until the privileges of the Kingdom of Valencia were abolished in the early 18th century.

When the structure of the new university was being planned in 1499, it was decided not only that it should have a chair of medicine but also that it should incorporate the School of Surgery that had been operating in the city since 1462. However, when the first professors were appointed in 1501, another chair was added, whose purpose was to teach anatomy and simple medicines. The University of Valencia was consequently home to the first Hispanic chairs of both surgery and also anatomy and simples. The aim of the first chair became the systematic exposition of the basic doctrines of medicine.

In the early decades of the 16th century, standards in the teaching of medicine were very low and strictly confined to traditional theories. This situation changed radically however from the 40s onward due to the activities of a large renewal-oriented group that managed to impose Renaissance trends taking up the classics and converted the University of Valencia into an avant-garde center as regards morphology and materia medica. Anatomy and simples were taught, in keeping with the new trends, on the basis of dissecting human corpses and herborizations, practices which were adopted subsequently in other Spanish universities. In 1560, this teaching was divided into two separate chairs, one of anatomy and the other of simples or "herbs". Meanwhile in 1548, a chair of practice had been established to be followed later by a chair of Hippocrates (1567) and another of special practice (1574). The method employed by the two chairs of practice followed Galenist theories: the first taught the general basics of diagnosis and therapeutical guidelines, whilst the second covered the application of same to different illnesses. The chair of Hippocrates was in line with the approach mentioned earlier derived from Hippocratic writings which emphasized the importance of clinical observation and an environmentalist approach to the study of disease, whilst not actually breaking away from Galenism. In 1590, yet another chair was created entitled *De remediis morborum secretis,* which existed for just one academic year despite being of great historical importance for, as we shall see, it was the only one devoted to chemical medicaments in 16th century Europe in keeping with the ideas inherent in the Paracelsan movement. As a result, the Faculty had eight chairs plus two minor chairs or regencies created in 1584: a remarkable number in a period when most major universities had only two or three chairs of medicine. Professors were appointed annually by municipal "juries", there being no competitive examinations or other procedures involved.

Teaching did not take place in the main building of the university but in the General Hospital, the outcome of the unification in 1512 of the city's welfare institutions. The facilities available for teaching, apart from those of the hospital itself, consisted of a classroom and also, from the second half of the century onward, an anatomy amphitheater. A small botanical garden was established in 1567, although it was not a permanent fixture like those of the 18th century but a modest medicinal garden to complement practical classes based largely on herborizations in the country.[3]

THE RENOVATION OF ANATOMY: THE VESALIAN MOVEMENT

The dissection of human corpses as a regular practice began in Valencia under a royal privilege granted in 1478 to the School of Surgery founded, as mentioned above, 16 years earlier. This explains partly why a chair dedicated to the teaching of anatomy and simples had been in existence at the University of Valencia since 1502 along the lines of the Italian model, and was the first of its kind in the Hispanic kingdoms and one of the first in Europe. In the initial decades of its existence this chair was occupied by obscure figures who did little more than analyze Galen's *De usu partium* and conduct the occasional dissection in the cold winter season.

This is the background against which certain innovations began to appear prior to the Vesalian reformation. One such innovation reveals the approach to anatomy adopted by Miguel Jerónimo Ledesma, who led the introduction of medical humanism into the University of Valencia in the 1540s. In the debate about blood-letting in the treatment of pleurisy or "aching sides", that typified the clash between the followers of arabized Galenism and the supporters of the humanist, Ledesma assumed the same stance as Vesalius: he resorted to anatomical data. His work on this matter (1546) includes a morphological summary that, despite closing with a reference to the Galenist treatise *De anatomicis administrationibus*, does devote a great deal of attention to the description of the flow of blood through the veins in the thorax based on the observations made whilst dissecting human corpses. This summary is illustrated by two diagrams, one of which is by Vesalius, an author upon whom praised is heaped.[4]

In 1545 and 1546, the chair of anatomy and simples—occupied until then as mentioned earlier by insignificant professors—was held by Pedro Jaime Esteve, one of the most brilliant followers of the humanist school. His education took place in Paris and also in Montpellier where he studied under Rondelet. Esteve was interested in botany, a subject on which he wrote several works we will mention later. He also paid particular attention to anatomy, along lines totally in keeping with his education in the two French cities. Like Ledesma, he believed the Galenic treatise *De anatomicis administrationibus*, hardly disseminated at all in the Late Middle Ages, to be the most important morphological text, but he emphasized that the basic source of anatomical knowledge was the dissection of human corpses, an activity he practiced frequently.

Esteve's attitude to Vesalius is in keeping with his mentality. Although he does not quote Sylvius, the leader of the reaction against the Vesalian reformation who had been his teacher in Paris, his influence in some of Esteve's statements is obvious, although at the same time, the education received under Rondelet does offset this. Basically he believes that "suspecting that Galen never dissected a human body is ridiculous, and daring to assert it is absolute madness".[5] In this way, when studying the first two cervical vertebrae and their role in how the head moves, he attacks Vesalius for his criticism of Galen in a fashion worthy of Sylvius: "I know not what demon in Galen has incited that Vesalius, surely a man born to slander".[6] He was not however completely against the man, unlike the Parisian professor, adopting a completely different tone on another occasion: "Vesalius—a very wise but rather sarcastic man—has on several occasions accused Galen of using incorrect vocabulary when describing the joints".[7] On another instance, he even manifestly recognized the importance of the contribution made to describing the veins of the thorax correctly: "We must applaud Vesalius' contribution to this matter enthusiastically, being the first to raise this question. The way we are means we are not ashamed to publicly recognise those who have made our progress possible".[8]

This ambivalent attitude toward Vesalius, typical of a French-educated physician such as Esteve, gave way at the University of Valencia to one of unconditional support when Pedro Jimeno came to occupy the chair of anatomy and simples in 1547. Educated in Italy and having been taught by Vesalius himself, Jimeno carried out activities that constituted the cornerstone of the Valencian School of Anatomy and the Spanish Vesalian movement.

Born in Valencia in the third lustrum of the 16th century, Jimeno studied arts in the university there and possibly also medicine. In the 1540–1543 triennial, he attended Vesalius' lessons on anatomy in Padua, a turning point in his scientific career for, from this moment onward he became one of his earliest and faithful followers. After returning to Valencia, he was appointed to the chair of anatomy and simples in 1547, a post he held for 2 years. He was appointed to another chair in 1549 but at the end of the academic year he moved to Alcalá and was the first person to occupy the recently created chair of anatomy in the University. He died there shortly after.

Within the confines of barely one decade, Jimeno carried out remarkably fertile scientific activities. Firstly, he converted the University of Valencia into one of the first in Europe where anatomy was taught in accordance with Vesalius' ideas. He was of the opinion that the best method was for the dissection of human corpses to be carried out and explained by the same professor. He considered verbal expositions on the other hand to be extremely limited since anatomy was usually "ardua atque difficillima dictu, fieri longe facillima".[9] Jimeno also incorporated other considerations typical of the Vesalian teaching reform such as the use of a complete skeleton in his lessons instead of isolated bones as had been usual until then.

Another of Jimeno's contributions was the publication of the first text on anatomy after *De human corporis fabrica* (1543) by his master, which incorporated the new Vesalian morphology completely, with the addition of the results of his own research. The work was printed in Valencia under the title *Dialogus de re medica, compendiaria ratione, praeter quaedam alia, universam anatomem humani corporis prerstringens* in 1549, when the author had already held the chair of anatomy and simples for 2 years. It opens with a long nuncupative epistle addressed to Pedro Lozano, physician to the duchess of Calabria, then the main patron of Renaissance humanism in Valencia. Jimeno was a staunch follower of said mentality, not only from the medical point of view but also because of his admiration of Erasmus and Luis Vives. He called the former, "an incomparable man, veritable restorer of the entire literary republic, patron and ideal sponsor of all scholars". Likewise he penned enthusiastic references to "our very famous and distinguished Juan Luis Vives, a truly brilliant leading figure of Spain and student of this illustrious city".[10]

Apart from one appendix devoted to dietary problems, Jimeno's book is devoted to the exposition of the anatomy of the human body. Its structure is that of the Renaissance dialog genre. The questions posed by "Gaspar, a citizen" are answered by "Andrés, a physician", obviously the Valencian anatomist's recognition of his master. The support drawn from Vesalius' *Fabrica* is constant. "Indeed", declared C. O'Malley, "*Dialogus* is a succinct summary of this work with some sentences quoted literally".[11] Jimeno, however, did not limit himself to merely assimilating Vesalius' work passively. This enthusiastic and expert cultivator of the dissection of human corpses was able to employ said work not only as a teaching tool, but also as a research method. It was this that led to his discovery of the stapes, the third ossicle in the middle ear, of which he published the first printed description.

Furthermore, Jimeno was also aware of the importance that the new Vesalian anatomy would have in a great variety of medical problems. His third major contribution was precisely his role as a catalyst of the influx of Vesalian anatomy upon medicine as a whole. Apart from the development in this respect of the Valencian School itself, the teaching of Jimeno in Alcalá was decisive in the orientation of Francisco Valles and Francisco Díaz, two key figures in Castilian medicine at that time. Valles, the author of the most important and influential work in the medicine of Renaissance Europe with a Hippocratic outlook, declared when taking care of his lessons on the locations of illnesses (1559): "In previous courses I behaved in such a way that I did not dare deal with the works of any small part unless I myself had observed their complete formation and had demonstrated it in the sight of my students with the skill and help of my good friend Jimeno, who had come from Valencia to Alcalá to explain the art of dissection, in which he was highly skilled, and who died not long after whilst working here. He did his very best to enable me to practice a great deal and teach my students".[12] Díaz' work *Tratado de todas las enfermedades de los riñones, vexiga y carnosidades de la verga y urina* (1588) was based on morphology and is generally considered to be the basic text of modern urology, in which he refers to "Ximeno, excellent Valencian doctor and the first to begin with great elegance and skill to cut and do anatomy in the city of Valencia, where medicine and anatomy are so brilliant at the present time, and likewise the history [description] of herbs and other curiosities related to this faculty. And I hide not my boast that I have spent some time in this city and havc had thc most expert doctor Collado and doctor Ximeno as masters".[13]

When Jimeno left Valencia in the summer of 1550, Luis Collado was appointed to replace him in his university post. Collado was to become the key figure in the consolidation of the Valencian School of Anatomy and its solid support for the Vesalian reform.

Collado was also born in Valencia, probably at a somewhat later date than Jimeno. He studied arts and medicine in the university there and, in circumstances we are unaware of, trained as an anatomist at Vesalius' side: "he was my only master in the knowledge of anatomy (I confess openly) and whatever my skill in dissection may be worth, it is to him and no other that I owe it".[14] One of his main influences in Valencia itself was Miguel Jerónimo Ledesma, whose posthumous work he was responsible for publishing in 1547. He was consequently an ardent follower of the humanist school, and as intransigent with Avicennism and medieval tradition as Jimeno, Esteve, or Ledesma himself. After teaching surgery in 1546 and 1548, he alternately held the chair of anatomy and simples, and those of "principles" and "practices" between 1550 and 1574, in accordance with the rote system then in force at the University of Valencia. In 1574 he took up the new chair of special practice, established at his request which he held for 10 academic years without interruption. A key figure in medicine in Valencia at that time, he enjoyed great prestige and influence and was a "protophysician and visitor of the Kingdom" from 1576 until his death in 1589.

Collado's main contribution to anatomy is the volume entitled *Cl. Galeni Pergameni liber de ossibus . . . Enarrationibus illustratus,* printed in Valencia in 1555. The work contains three texts of different lengths: a review of said Galenic work, a description of the cranial sinuses and orifices, and a "letter to the reader" in which he justifies his defense of Vesalius from the attacks of Sylvius.

The historical importance of this book by Collado was not fully understood until its immediate antecedent was taken into account: the attack launched upon Vesalius in previous years in two books by Sylvius [Jacques Dubois], his former teacher in Paris. In 1549, Sylvius had published a revised edition of Balamio's translation of *Liber de ossibus*, accompanied by a review intended to defend Galen from the criticism made of his osteology in *Fabrica*. His main argument was that this book by Galen "dealt solely with human bones", despite the fact that a "slanderer" insisted that it concerned the bones of monkeys. Two years later an even more scathing attack was launched in the famous *Vaesani cuiusdam calumniarum in Hippocratis Galenique rem anatomicam depulsio* (1551), the title of which includes the play on words "Vesalius-vaesanus [mad]", that Sylvius had already used in his 1549 commentary.

Collado's whole book and not only his "letter to the reader" is a defense of Vesalius against the attacks of his former master. Like Sylvius, he uses Balamio's translation, beginning with the statement that the Parisian professor had falsified the Galen text by means of serious errors and omissions. His intention is mainly to underline the opposing standpoints of "Vesalius, eminent restorer of anatomy" and "Jacob Sylvius, remarkable imitator of Galen". Collado's decision in favor of what he refers to as "philosophical freedom"—a remarkable expression for a mid-16th century author—is categorical.[15] Consequently, his commentary consists of an osteological exposition along the lines of *Fabrica* and his own experience of dissection. His main argument in favor of Vesalius is his observations in the many dissections he had conducted himself. Despite his admiration of Vesalius, he does not hesitate to challenge his opinion when they contradict the data gathered from his own observations, as occurred for example when dealing with the etmoid bone, sphenoid suture, and the holes found at the root of the incisors. He also aimed to complete Vesalius' description, including in this respect that of the stapes which he claims to have discovered years earlier in conjunction with his student Cosme Medina, who at that time held the chair of anatomy at the University of Salamanca.

Collado's praise of Vesalius is amongst the most clever to be received by the great Flemish anatomist in his lifetime. *Apologia* (1555) by Renatus Henerus, a physician from Lindau, is usually quoted as being the first manifest defense of Vesalius against the onslaught by Sylvius. The book by Collado published in the same year must be grouped together with the one by Henerus from this point of view. It is however of far greater importance, for it mirrors one of the first European schools to cultivate the teaching and research of anatomy along the line laid down by Vesalius. The third of the texts it contains entitled *Ossium capitis foraminum, et sinum ad tyrones brevis descriptio*, is in fact one of Collado's lessons, published at the request of his students.

The influence of Collado, upheld over a longer period than that of Jimeno, was equal to or greater than the latter's. During the years when he was the greatest figure in medicine at the University of Valencia, the teaching of anatomy progressed there to a considerable extent. In 1560 as we know, the chair of anatomy and simples was divided into two separate chairs, with the subjects of each expanding to a full academic year. Later, in 1567, the teaching of anatomy was organized over 2 years, with the practice of obligatory dissections being stringently supervised.

The Valencia School of Anatomy was the hub of the Vesalian movement in Spain. In the Crown of Aragon, the organization of its teaching served as an immediate model for the Universities of Saragossa and Barcelona, although it was not until the following century that their chairs of anatomy came to encompass any notable activities in the realms of practice. In the Crown of Castile as we have already seen, Jimeno was the first to occupy the chair at Alcalá. Following his premature death, he was succeeded by Pedro Marcos de Ayala, of Valencia as were virtually all those who occupied the Alcalá chair in the last third of the 16th century and the early years of the 17th century. The chair of anatomy at Salamanca was created with the consensus of the body of professors in September 1551, with Cosme de Medina, mentioned earlier as a pupil of Collado who had worked with him on the description of the ossicles in the middle ear being appointed to this position. He held it for 10 academic years and was responsible for the section of anatomy in the famous statutes of the University of Salamanca of 1561, that contain the rules governing the teaching of morphology, directly inspired by the Valencian model which were possibly the most minute and exacting rules promulgated in 16th century Europe. The Valencian anatomists saw the formation of outstanding figures who introduced new morphological proposals in different fields of medicine and other realms of culture. Their influence on the works of Francisco Valles and Francisco Díaz has been mentioned earlier. Similar comments may be made of other foremost figures such as Juan Tomás Porcell who, in the 1564 epidemic, carried out the first systematized autopsies of plague victims in Saragossa, a veritable milestone in the origins of modern pathological anatomy, and the famous goldsmith Juan de Arfe, whose scientific works include one of the most important treatises of artistic anatomy of Renaissance times.[16]

THE CHAIR OF SIMPLES

The chair known commonly as "herbs", was as we already know actually dedicated to "simples", in other words, materia medica. As occurred with anatomy, its teaching was reoriented in 1545, when the chair was held by Pedro Jaime Esteve. The works by Esteve include an edition of Nicander's *Theriaca*, with a Latin translation and in-depth commentaries, and the *Diccionario de las yerbas y plantas medicinales que se hallan en el Reino de Valencia.* In his comments upon the text by Nicander, which is a didactic poem about toxicology concerned mainly with animal poisons, Esteve went beyond this subject and discussed animals and plants, with comments on the locations in Valencia where they could be found and the common names they were known by there.[17] His *Diccionario* was disseminated like many other books at that time in handwritten copies, none of which unfortunately have been located. However, Gaspar Escolano provides in his *Décadas* (1610) a summary of it, mentioning some of the plants it includes, with a list of the common names in Castilian or Valencian of 120 species, some accompanied by details of their places of origin, medicinal or dietary purposes, properties, etc.[18] The importance of Esteve's contribution is mainly that the date when it was written (between 1545 and 1556) makes it one of the earliest attempts in Europe to gather evidence about regional flora.

The foremost figure to hold the chair of simples in the 16th century was Juan Plaza, born in Valencia toward 1525 and educated as a physician in his University. He occupied the chair from 1567 to 1583, the period in which a botanical garden was established: the first of the university type in Spain and one of the first in Europe. We have already said that it was not a garden on a par with those of the 18th century, but a modest facility designed to complement teaching methods based largely on herborizations conducted in different areas of the Valencian territory.

Plaza maintained close scientific connections with the great Flemish naturalist Charles de l'Escluse (Clusius), whose contributions to botany were disseminated across Europe mainly in the form of his famous work *Rariorum aliquot stirpium per Hispanias observatarum Historia* (1576). They include exact descriptions of species of local flora, most of which Plaza, together with Clusius and their students and followers, baptized with names incorporating the adjective "Valencian", hence "Chrysantemum Valentinum" (*Chamomilla suaveolens* (Pursh) Rydb.), "Hemerocallis Valentina" (*Pancratium maritimum* L.), "Hippoglossum Valentinum" (*Globularia alypum* L.), "Linaria Valentina" (*Linaria tryphilla* (L.) Mill.), "Scammonea Valentina" (*Cynanchun acutum* L.), etc. In some instances this adjective has been retained in modern nomenclature as for example "Polygala Valentina" (*Coronilla valentina* L.).[19]

Although Plaza's contributions dealt mainly with Valencian flora, he also contributed to the study of certain exotic species, particularly from America. The first mention of an acclimatized avocado pear in Europe (*Persea americana* Mill.) was the specimen presented by Plaza in Valencia to Clusius in 1563. It was in full bloom which enabled the Flemish botanist to become familiar with the characteristics of the fruit—which he expounded upon in his book—thanks to his Valencian colleague. Similar circumstances occurred with the maguey or agave (*Agave* sp.). "Primum mihi hanc plantam demostravit clarissimus vir D. Joannes Plaça, medicus et professor Valentinus", states Clusius in his work, going on to say that "the people of Valencia call it '*fil y agulla*'(thread and needle) because of the sharp tips of its leaves and its fiber that can be used as thread".[20] It is interesting to note that this common Valencian name was upheld amongst the denominations used for the species of this genus by central European botanists for almost a century. Caspar Bauhin for example, when discussing the maguey in his *Pinax* or *Theatrum botanicum* (1623), which constituted a milestone in modern taxonomy and botanical nomenclature, mentioned the synonym "*Fill y agulla*, id est, filum et acus, Hispanis".[21]

Of the works that Plaza left as manuscripts and were thought to be lost, we have managed to locate a copy of his *Practica generalis*, the analysis of which, currently under way, will contribute to the knowledge of his work whilst occupying the chair of special practice, which he held, following Collado, from 1584 until his death in 1603. *Officina medicamentorum*, the first Valencian pharmacopoeia, published in 1601 by the Association of Apothecaries, not only received Plaza's seal of approval but also featured several of his magistral formulae, including one for betony syrup that was very popular. The effects, overestimated at that time, of the *Stachys officinalis* Trev. it contained, were associated with those of other considerably more effective plants such as *Salvia officinalis* L., *Ruta graveolens* L., and *Satureja fruticosa* Beg. or savory which was virtually a Valencian endemism.

Jaime Honorato Pomar succeeded Plaza in the chair of simples. Born in the city of Valencia itself in about 1550, he studied medicine at the university there under professors such as Luis Collado and Plaza himself. After obtaining his title of doctor in October 1573, he remained in Valencia as a practicing physician in direct contact with university medical circles. He occupied the chair of anatomy for 5 years (1574–1578) and in 1584 was appointed to the chair of simples. In a similar fashion as had occurred when Plaza was appointed, Pomar insisted on a practical approach to teaching. Documentary evidence shows that during the three lustra when he held the chair, the organization of didactic herborizations was developed and extended. The detailed rules that appeared in 1611 in the byelaws of the University of Valencia shortly after his death, may be considered to be the remnants of the rules stemming from his teaching activities.

Pomar's fame led to his becoming advisor to Philip II on natural history matters. He initially carried out this role from his chair in Valencia, but in April 1598 the monarch created a position of royal botanist for him in Madrid and he consequently moved to the Court where he lived until his death in early 1606. As a sign of his appreciation, Philip II gave him a splendid pictorial codex with more than 200 paintings of plants and animals that constituted one of the most important natural history atlases of that period.[22]

The plants and animals portrayed in the drawings in the Pomar Codex are divided unequally between the Old World and America. Those from the Old World can be divided in turn into the species that originated in Western Europe and the Mediterranean area and the exotic or "pilgrim" species, i.e., those that originated in the Near East and more distant areas of Asia and Africa. The codex also contains paintings of 7 animals and 25 plants from America, several of which were discovered during the first scientific expedition to the New World (1571–1577) that Philip II commissioned Francisco Hernández to lead. Pomar was the author of the texts that indicate the names of these plants and animals, mainly in Latin and/or Castilian Spanish, although there are also almost 20 names in Valencian, two in Italian and six in Amerindian languages.

Most of the animal names are those from Pliny's denominations and those of plants from Dioscurides', although both include names taken from different classic authors and Renaissance naturalists such as Clusius, Mattioli, and Hernández himself. His Valencian origins are revealed not only in the use of Valencian words and the presence of Valencianized Castilian Spanish words, but also in the use of certain classic terms such as for example "sphacheon" in reference to the "poisonous spider of streams", in keeping with Nicander's *Theriaca* published by Pedro Jaime Esteve.[23]

INCREASING PROMINENCE OF CLINICAL OBSERVATIONS AND ENVIRONMENTAL STUDIES IN PUBLIC HEALTH

In addition to his important labor in the realms of anatomy, Luis Collado was a follower of the school of thought known as Hippocratic Galenism, a school reflected in his book *Ex Hippocratis et Galeni monumentis Isagoge* (1561) and in his many manuscripts on pathological and clinical subjects, particularly the one entitled *Observationes in praxi*, a study of the diseases that predominated in Valencia in the years 1571 and 1572 from the Hippocratic environmentalist approach.

The contribution made by Miguel Juan Pascual had even greater influence on bringing clinical observations and environmentalist studies in public health into the foreground. Born in Castellon in about 1505, Pascual studied medicine in at the Universities of Alcalá and Montpellier, following which he was appointed professor at the University of Valencia in 1548, a post he held until his death in 1561. His treatise on practical medicine *Morborum internorum ... curatio* (1555), which studied diseases following the then usual order of "a capite ad calces", followed by a second part dedicated to fevers, is based not only on clinical observations but also on ecological conditions. The author analyses the so-called "predominant diseases" in Valencia in certain years in accordance with the largely climate-orientated interpretation found in the Hippocratic environmentalist tradition. Far from adhering to dry and abstract scholastic expositions, he takes care of the specific social circumstances in which different ailments were treated. He criticizes the empirical practices of the populace and quacks harshly and likewise the irresponsible prescriptions of academic professionals.

This book by Miguel Juan Pascual was held in very high esteem in several European countries, which explains its 11 editions in just over one century. From the seventh edition onward (1579) it was printed together with the *Scholia* penned by Pedro Pablo Pereda, a physician from the town now known as Jativa. His chapter on syphilis was moreover reproduced in the editions included from 1566 onward in the collection of texts on venereology by the Italian Luigi Luvigini.

As an appendix to his treatise on medical practice, Miguel Juan Pascual published a brief treatise entitled *Medica disputatio. An cannabis et aqua in qua mollitur possint aërem inficere* that made its appearance in the 1555 edition. This is one of the first texts printed in Europe about urban pollution from the public health standpoint. He wrote it at the request of the Valencian inquisitors to whom he was physician, driven by the comment of some of his colleagues that the reasons for the "outbreak of serious fevers" that had occurred the previous autumn in Valencia and the surrounding region were due to the stench of the pools where hemp was steeped. He adduced the opinions of Galen and other classic authors and contradicted, being a faithful follower of Hippocratic Galenism, those of Avicenna. His baseline is however that of experience, "on which one must rely above all". His conclusion is that "the source of these ailments cannot be attributed to the pools in which hemp is steeped", being of the opinion that, "these are no cause for concern unlike the other waters around the region; the area closest to the sea is the most unhealthy together with those in which the royal palace and all the houses between the road to Sagunto and the sea are is situated ... if the smell of hemp is deemed to be unpleasant, how much more so is that of the beasts and worms swarming over Valencia; if the smell of hemp is unwelcome, far more so is that of human excrements, the evacuation of which we cannot do without and which is more widespread because of the many sewers that are always open and exude terrible odours".[24]

The ecological approach of the Hippocratic environmentalist tradition adopted by followers of Hippocratic Galenism such as Luis Collado and Miguel Juan Pascual moved beyond the astrological interpretations of earlier physicians with a humanist mentality such as Pedro Jaime Esteve and Miguel Servet. In practice, this led to astrology being excluded from academic medicine. One opinion that attracted many adherents was one that manifestly discredited astrology, relegating it to the popular genre of lunar calendars that associated astrological forecasts on health with those on farming and other activities, such as the one by the Valencian Jerónimo Cortés (1596), that was to continue to be reprinted over the following three centuries. This is the context in which the response of academic circles in its defense must be situated, which includes the book *Apología en defensa de la astrología contra algunos médicos que dicen mal della* (1599) published by the Valencian physician Manuel Ledesma.

THE TEACHING OF SURGERY

One of the foremost works in the realms of surgery was that of Juan Calvo, a follower of Collado, who lived for a while in Montpellier and returned to live in Valencia until his death in 1599. The first edition of his *Cirugia universal y particular del cuerpo humano* appeared in 1580, with 10 subsequent reprints in Castilian Spanish and two in French. He wrote it with teaching in mind after teaching surgery for 12 years in Valencia. His lessons became very famous and were attended by the surgeons themselves of different Hispanic kingdoms and also physicians seeking to complete their training.

Of the great texts on surgery in 16th century Spain, *Cirugia* by Juan Calvo is the one that best matches the structure and style of a didactic treatise. This format has certain obvious limitations particularly the bookish tone emphasized even further by the author's inclination toward scholastic approaches, but it does nevertheless offer certain advantages, the most important being the systematic and orderly nature of the exposition. Consequently the descriptions of the operating stages are more exact and more "modern" in style than those found in the other Spanish works of that period. One example is the methodical and in-depth description of the cranial trepanation technique, a matter in which he was a staunch interventionist, although he declared himself to be eclectic as regards the use of trepans or periosteotomes, tailoring his choice to the circumstances of each case. In the debate on healing wounds, he also maintained an intermediate attitude between the "dry" or first intention method defended by Bartolomé Hidalgo de Agüero and the "wet" method proposed by Juan Fragoso. He conceived of several useful techniques including those concerning the surgical treatment of varicose veins and fistulas, and the removal of malignant tumors.

Calvo's work includes a *Tratado de anotomía* of certain length in which he quotes his teacher Luis Collado and other post-Vesalian authors, particularly Fallopio, on several occasions. This book also includes a *Tratado de morbo gálico*, the most in-depth study of the symptoms and treatment of syphilis of all those published by surgeons in that period.[25]

THE BREAKAWAY FROM TRADITIONAL MEDICAL KNOWLEDGE: PARACELSISM

In the 16th century, the only school of thought that constituted an absolute and manifest breakaway from Galenism was that of the iatrochemistry derived from Paracelsism. The works by Paracelsus himself were disseminated little in his own lifetime and the two decades following his death in 1541. It was only after the 1560s that virtually all of them were published in both the original German and in Latin translations. The publication of these works together with disenchantment with the results of humanist Galenism led to what Thorndike baptized as the "Paracelsan Revival" which manifested itself in two forms: the Paracelsan movement in the strict sense of the word and the response of academic medicine to the doctrines of Paracelsus, which ranged from discrediting it vehemently to assimilating a variation of his contributions.[26]

The "Paracelsan Revival" manifested itself at the University of Valencia in a remarkable occurrence: the creation of a chair of chemical medicine. Lorenzo Cózar, one of the most outstanding medical figures in the city at that time, was appointed to the position. He had occupied the chair of surgery and in 1589 he was named "protophysician of the City and of the Kingdom of Valencia". In the same year he published his book *Dialogus veros medicinae fontes indicans*, in which he denounced the lackings inherent in traditional Galenic medicine, particularly from a therapeutical viewpoint and proposed a new foundation based on the principles of Paracelsism.

Cózar declares that he began by learning "the art of making chemical medicaments from the careful observation of many experiments and carefully reading the works of new experts in said art". He had overcome the traditional concept of alchemy and used the Paracelsan term "the art of separating": "Alchemy is not used here in the sense of the active transmutation of metals.... On the contrary, by alchemy we mean that part of the art of separating that reveals hidden properties, separates the pure from the impure and manifests the countless differences to be found in waters, oils and balsams, powders, and salts".[27] He underlined the importance of chemistry for physicians from both practical and theoretical viewpoints. He considered it to be a technique making it possible to obtain chemical medicaments that were more effective than traditional medicaments; and also, a method of investigating nature on the basis of theories different from the classic ones. He rejected the latter completely, announcing in his *Dialogus* another work of his that was never published: "The phenomena of this chemical art are remarkable but in order to understand their causes, does the universal philosophy of Plato and Aristotle suffice? Of course not, as expounded in the work on the elements that I have written".[28]

Cózar's program was condemned in Valencian medical circles. He mentioned this on numerous occasions and attempted to redress the situation by experimental demonstrations. "The most obvious chemical phenomena go unnoticed by non-experts, particularly by those physicians who dislike the practice of any philosophy and deny the utility of and need for this extremely important art. Consequently, it would be advisable first of all to convince them and beg them to at least observe the operations conducted by good chemists. The spirit of some is so degenerate however, that they would rather look away and cover their ears for fear that, being convinced of the truth, they might be obliged to admit that they were wrong".[29]

Despite the opposition of the local supporters of Galenism, Cózar managed to convince the University of Valencia to create a chair of chemical medicaments. Although operative for only one academic year (1591–1592), probably due to the death of its holder, this chair—in which Cózar taught the preparation of said remedies and how to administer them—was an exceptional case from a European viewpoint of the incorporation of the Paracelsan movement into an academic institution.[30]

THE POSITION OF THE FACULTY IN LATER PERIODS

And finally, we would like to offer a short overview of the main stages in the later development of the Faculty of Medicine at the University of Valencia.

The approach to teaching came to focus increasingly on practicals during the 17th century: the facilities of the anatomy amphitheater and the medicinal garden for example were enlarged whilst dissections and herborization outings became more frequent. However, the influence of the neoscholastic ideology of the Counter-Reformation converted the Faculty, for the most part of the century, into a nucleus of intransigent Galenism, opposed to the innovations of the Scientific Revolution.

This situation was completely reversed from the 80s onward when it became one of the most important centers in the *novator* movement, which clashed headlong with traditional medicine and its methodological and epistemological fundamentals. It is to this movement that we owe not only the complete assimilation of the new bases of anatomy and physiology and those of the iatrochemical and iatromechanical systems, but also certain original contributions of note. The most outstanding were those by Crisóstomo Martínez, the author of one of the finest atlases of anatomy in the history of morphological illustration, and also one of the first to cultivate microscopic research.

The members of the *novator* movement enabled the Valencia Faculty of Medicine to continue developing during the 18th century at the same rate as the most advanced

European institutions, a process which culminated with the reorganization of 1787, which implemented an avant-garde study programme on trends in Enlightenment.

This process ground to a halt however when Spanish scientific activity in general collapsed between 1808 and 1834 as the result of the War of Independence against Napoleon and the subsequent absolutist repression and economic ruin that occurred during the reign of Ferdinand VII.

It was only after the uphill struggle to recovery in the central decades of the 19th century that the Faculty was at last able to become, in the last two decades of that century, an institution forming an integral part of the new medicine based on experimental laboratory research. It was in this context for example that Santiago Ramón y Cajal undertook his great neurohistological work and that the world's very first cholera vaccination was applied during the epidemic of 1885.

This blossoming development was however the result of circumstances and came to an abrupt halt due to several adverse factors including for example the pronounced centralism in the organization of Spanish universities. In the century now drawing to a close, the Faculty of Medicine at the University of Valencia has undergone a series of pronounced ups and downs but has not at any time moved beyond the condition of a provincial academic institution.[31]

NOTES

[1] A. Vesalius, *De humani corporis fabrica* . . . (Basileae: J. Oporinus, 1543), "Praefatio", a3.

[2] *Cf.* C. D. O'Malley (ed.), *The History of Medical Education* (Berkeley: University of California Press, 1970); T. Ogawa (ed.), *History of Medical Education. Proceedings of the 6th International Symposium on the Comparative History of Medicine* (Tokyo: Saikon, 1983); J. M. López Piñero, *Ciencia y técnica en la sociedad española de los siglos XVI y XVII* (Barcelona: Labor, 1979).

[3] *Cf.* M. Peset Reig, M. F. Mamcebo, M. Martinez Gomis, and P. García Trobat, *Historia de las universidades valencianas,* 2 Vols. (Alicante: Inst. Juan Gil Albert, 1993. A. Felipo Orts, *La Universidad de Valencia durante el siglo XVI (1499–1611)* (Valencia: Departamento de Historia Moderna, 1993); J. M. López Piñero, *La Facultad de Medicina de la Universidad de Valencia. Aproximación a su historia* (Valencia: Facultad de Medicina, 1980); J. M. López Piñero, *Clásicos médicos valencianos del siglo XVI* (Valencia: Conselleria de Sanitat i Consum, 1990).

[4] M. J. Ledesma, *De pleuritide commentariolus* (Valentiae: Per Joannem Mey, 1546), f. 25v–6r.

[5] P. J. Esteve, *Hippocratis Coi . . . Epodemion Liber Secundus . . . Latinitate donats, et fusissimis commentariis illustratus* (Valentiae: Apud Ioanem Mey, 1551), f. 147v.

[6] *Ibid.*, f. 86r.

[7] *Ibid.*, f. 83v.

[8] *Ibid.*, f. 148r.

[9] P. Jimeno, *Dialogus de re medica, compendiaria ratione, praeter quaedam alia, universam anatomem humani corporem perstringens* (Valentiae: per Ioannem Mey, 1549), f. 41v.

[10] *Ibid.*, f. 1v.

[11] C. D. O'Malley, "Pedro Jimeno: Valencian Anatomist of the Mid-Sixteenth Century", in *Science, Medicine and Society. Essays to Honor Walter Pagel* (London: Heinemann), p. 70.

[12] F. Valles, *Cl. Galeni Pergameni de Locis Patientibus Libri Sex, cum Scholiis, Lugduni, Per Claudium Pontanum,* 1559, p. 5.

[13] F. Díaz, *Tratado de todas las enfermedades de los riñones, vexiga, y carnosidades de la verga y urina* (Madrid: Francisco Sánchez, 1588), f. 19v.
[14] L. Collado, *Cl. Galeni Pergameni Liber de Ossibus . . . , enarrationibus illustratus* (Valentiae, Ex Typographia Ioannis Mey, 1555), "Ludovicus Colladus medicus letoris", s. f.
[15] *Ibid.*, A5r.
[16] J. M. López Piñero, "The Vesalian Movement in Sixteenth Century Spain", *Journal of the History of Biology* 12:45–81 (1979); J. M. López Piñero, "La disección anatómica y la reforma vesaliana en la España del siglo XVI", in *Medicina moderna y sociedad española (Siglos XVI–XIX)* (Valencia: Cátedra de Historia de la Medicina, 1976), pp. 61–130; J. M. López Piñero and L. García Ballester, *Antología de la escuela anatómica valenciana del siglo XVI* (Valencia: Cátedra e Instituto de Historia de la Medicina, 1962); J. M. López Piñero, F. Jerez Moliner, A. Martinez Almagro, *Clásicos morfológicos valencianos, del Renacimiento al siglo XIX* (Valencia: Morphos Ediciones, 1997) [*Archivo Español de Morfología, 2,* número extraordinario]; C. D. O'Malley, "Pedro Jimeno: Valencian Anatomist of the Mid-Sixteenth Century", en *Science, Medicine and Society. Essays to Honor Walter Pagel* (London: Heinemann), pp. 69–72; J. B. Peset Vidal, *Recuerdo apologético de Luis Collado* (Valencia: Instituto Médico Valenciano, 1878); P. Casanova Ciurana, *El Doctor Luis Collado, catedrático del siglo XVI* (Valencia: Imp. de F. Domenech, 1895); J. M. López Piñero and F. Calero, *Las "Controversias" (1556) de Francisco Valles y la medicina renacentista* (Madrid: C.S.I.C., 1988); J. M. López Piñero and M. L. Terrada, "La obra de Juan Tomás Porcell (1565) y los orígenes de la anatomía patológica moderna", *Medicina Española* 52: 237–250 (1965).
[17] P. J. Esteve, *Nicandri Colophoni poetae et medici antiquissimi clarissimique Theriaca, Petri Iacobi Steve medico valentino interprete et enarratore* (Valentiae: Joannes Mey Flandrus, 1552). J. M. López Piñero, *La edicion grecolatina de la* Theriaca *de Nicandro, con adiciones botánicas y zoológicas, par Pedro Jaime Esteve (1552)*, Valencia, Bibliogilic, 2005.
[18] G. Escolano, *Década Primera de la Historia de la Insigne y Coronada Ciudad y Reyno de Valencia* (Valencia: Pedro Patricio Mey, 1610), col. 687.
[19] C. Clusius, *Rariorum aliquot stirpium per Hispanias observatarum Historia . . .* (Antverpiae: Ex Officina Christophori Plantini, 1576), pp. 179–181, 181–183, 196–198, 220–221, 225–226, 254, 287–289, 350, 368, 391, 422, 427, 468, 479, 484.
[20] C. Clusius, *op. cit.*, p. 444.
[21] C. Bauhin, *Pinax. Theatri botanici . . . sive index in Theophrasti, Dioscoridis, Plinii et Botanicorum qui a seculo scripserunt opera* (Basileae: Sumptibus et Typis Ludovici Regis, 1623), p. 286.
[22] *El Atlas de Historia Natural donado por Felipe II a Jaime Honorato Pomar,* Edición facsímil y estudio introductorio por José Mª López Piñero, 2 Vols. (Valencia: Vicent García Eds., 1990); J. M. López Piñero, *El Códice de Jaume Honorat Pomar (c. 1550–1606): Plantas y animales del Viejo Mundo y de América* (Valencia: Ajuntament de València, 2000).
[23] *Cf.* F. Martí Grajales, *El doctor Juan Plaza. Estudio biográfico* (Valencia: Imp. M. Alufre, 1893); V. Peset Cervera, *Noticia histórica del catedrático valenciano de materia médica Dr. Juan Plaza* (Valencia: Imp. F. Domenech, 1895); J. M. López Piñero, *Atlas y diccionario histórico de las plantas medianzles*, Valencia, Faxémil, 2005; J. M. López Piñero, *El Códice Pomar (ca. 1590) el interés de Felipe II por la Historia natural y la expedición Hernández a América* (Valencia: Instituto de Estudios Documentales e Históricos sobre la Ciencia, 1991); J. M. López Piñero, "The Pomar Codex (*ca.* 1590): Plants and Animals of the Old World and from Hernández Expedition to America", *Nuncius* 7:35–52 (1992). J. M. López Piñero and M. L. López Terrada "Las plantas americanas en la relación de Clusius con los naturalistas españoles", in *La influencia española en la introducción en Europa de las plantas americanas (1493–1623)* (Valencia: Instituto de Estudios Documentales e Históricos sobre la Ciencia, 1992), pp. 66–103.
[24] M. J. Pascual, *Medica disputatio. An cannabis et aqua in qua mollitur possint aërem inficere.* Appendix to:*Morborum internorum . . . curatio brevi methodo comprehensa* (Valentiae: Typis

Ioannes Mey Flandri, 1555). *Cf.* V. Guillén Marco, *Recuerdo apologético del Dr. Miguel Juan Pascual, catedrático de medicina en el Estudio General de Valencia, en el siglo XVI* (Valencia: Instituto Médico Valenciano, 1908); J. M. López Piñero, "La tradición del ambientalismo hipocrático. El informe de Miguel Juan Pascual sobre la contaminación (1555)", in *Los orígenes en España de los estudios sobre la salud pública* (Madrid: Ministerio de Sanidad y Consumo, 1989), pp. 21–24, 69–74.

[25] *Cf.* J. Aguilar Lara, *Recuerdo apologético de Juan Calvo* (Valencia: Instituto Médico Valenciano, 1879); R. San Román Gómez, "La obra quirúrgica de Juan Calvo", *La Medicina Contemporánea* 79:71–104 (1961); J. L. Fresquet Febrer, *La "Cirugía Universal y Particular" (1580), de Juan Calvo. Análisis de texto y estudio de las referencias* (Valencia: tesis de licenciatura, 1979); J. L. Fresquet Febrer, "El tratado de anatomía de Juan Calvo. Contribución al estudio de la morfología posvesaliana española", en *Estudios dedicados a Juan Peset Aleixandre*, Vol. II (Valencia: Universidad de Valencia, 1982), pp. 17–28.

[26] L. Thorndike, "The Paracelsan Revival", in *History of Magic and Experimental Science*, Vol. V (New York: Columbia University Press, 1959), pp. 617–651.

[27] L. Cozar, *Dialogus veros medicinae fontes indicans* (Valentiae: P. P. Mey, 1589), f. 13v.

[28] *Ibid.*, f. 20v.

[29] *Ibid.*, f. 15v.

[30] *Cf.* J. M. López Piñero, "Paracelsus and his Work in 16th and 17th Century Spain", *Clio Medica* 8:113–141 (1973); J. M. López Piñero, *El "Dialogus" (1589) del paracelsista Llorenç Coçar y la cátedra de medicamentos químicos de la Universidad de Valencia* (Valencia: Cátedra e Instituto de Historia de la Medicina, 1977); J. Pardo Tomás, "Llorenç Coçar y la Inquisición valenciana", in *Homenatge al Doctor Sebastià Garcia Martínez*, Vol. I (València: Generalitat Valenciana, 1988), pp. 363–373.

[31] J. M. López Piñero (ed.), *Historia de la medicina valenciana*, 3 Vols. (Valencia: Vicent García Eds., 1988–1992); J. M. López Piñero et al., *Bibliographia Medica Hispanica, 1475–1950,* 8 Vols. (Valencia: Instituto de Estudios Documentales e Históricos sobre la Ciencia, 1987–1996); J. M. López Piñero, T. F. Glick, V. Navarro Brotóns, and E. Portela Marco (eds.), *Diccionario histórico de la ciencia moderna en España,* 2 Vols. (Barcelona: Península, 1983).

VÍCTOR NAVARRO-BROTÓNS

THE CULTIVATION OF ASTRONOMY IN SPANISH UNIVERSITIES IN THE LATTER HALF OF THE 16TH CENTURY

The four universities known to have taught mathematics in the 16th century in what is now known as Spain, were Salamanca, Valencia, Alcalá, and, at the end of the century, Seville. In addition to being taught at university, astronomy was also taught in other institutions such as for example, the Casa de la Contratación of Seville, the so-called Mathematics Academy of Madrid, certain naval academies and, towards the end of the century, certain Jesuit schools.

At the University of Valencia, following the official foundation of the *Estudi* or center of learning in 1500, a chair of mathematics was set up in 1503, although we have no documentary evidence of the subjects taught there in the early decades of that century. The first person to occupy the chair was Tomás Durán, a Dominican from Salamanca who published (Valencia, 1503) Bradwardine's *Arithmetic* and *Geometry* and also Pecham's *Perspectiva*, together with *Questiones super perspectivam* by Henricus or Heinrich of Hesse (or of Langenstein). None of Durán's successors before the mid-16th century are known to have published and we have been able to find almost no trace of their activities. Despite the considerable interest in astrology of some teachers from the faculty of medicine, and the fact that they published books on this subject, in the early years of the university, the teaching of mathematics must have concentrated basically on preparing pupils to study natural philosophy and logic, mainly under the influence of the University of Paris. It must however be said that the commentaries on Aristotle's treatise *De caelo* and the one by Juan de Celaya of Valencia, also included descriptions of planetary motion in keeping with Ptolemaic models situating eccentrics and epicycles in concentric spherical shells concentric around the earth.[1]

From the year 1540 onwards, it was compulsory for students of medicine to have an arts degree, as was already the case for students of law and theology. This must have increased the interest in the study of astronomy and astrology. According to certain documents of that time, in the years 1540–1550 the study of mathematics included arithmetic, geometry, geometrical optics, music, judicial astrology, and cosmography (which included astronomy and geography). The upsurge in humanism led to greater interest in astronomy and astrology amongst humanist physicians, who took them as a basis for the interpretation of Hippocratic texts. The chair of astronomy was occupied in the 1555–1556 academic year by the noteworthy physician and humanist Pedro Jaime Esteve. His commentaries on the second book of *Epidemics* in

M. Feingold and V. Navarro-Brotóns (eds.), Universities and Science in the Early Modern Period, 83–98.

the Hippocratic Collection highlighted the importance of astronomy for physicians: "medicus astronomiae ignarus non est sectator hipocratis".[2] In the introduction to this book, Esteve demonstrated his knowledge of astronomy by his references to Ptolemy, Regiomontanus, Jacobus Ziegler, Gemma Frisius, and other authors. He also provided a description of the stars of the zodiac and the northern hemisphere, and the variations that had occurred in their positions since the time of Ptolemy due to the precession of the equinoxes, and reported his own observations in this respect. He deemed all this to be essential in order to study the climate of different regions and to forecast changes in the air. It has been said that Esteve also drew up a series of ephemeredes for the years 1488–1600.[3]

In general, very little documentary evidence survives of the subjects taught by astronomy teachers prior to the time when this chair was held by Jerónimo Muñoz in the mid-sixties. Mention must be made of the treatise *De Sphaera* (1553) by Baltasar Manuel Bou, professor of astronomy in 1559–1562. In this book, Bou described himself as pupil of the Italian astronomer and astrologer Luca Gaurico, so we can assume that he was educated in these subjects in Italy. Bou's book is a free version of *Sphaera* of Sacrobosco, structured essentially on the framework that author established. The book contains two prefaces by the physicians Pedro Jaime Esteve and Antonio José Villafranca. It is interesting to note that although Bou set forth the classic theory of the elements, when discussing the sphere of fire he acknowledged that many authors were doubted its existence. He went on to say that any discussion of such questions was a matter for natural philosophy and not for astronomy. As regards the theory of zones, Bou points out that lands thought to be inhabitable by the ancients were shown by experience to be habitable. The treatise abounds with references to Renaissance astronomers such as Regiomontanus, Gemma Frisius, Petrus Apianus as well as classical astronomers, in keeping with the preferences of the period. It also features tables to calculate the rising and setting of planets, the position of the sun and its height above the horizon of Valencia.

As regards mathematical subjects, the constitutions of 1561 mention only the teaching of astronomy, and list the following topics: the sphere, the *theorica* of the planets, tables, and the use of astrolabes. The statutes of 1555, however, had already established the need to teach certain principles of geometry (to enable Aristotle's *Analytica Priora* and *Posteriora* to be understood) and also arithmetic and geometry (in order to grasp natural philosophy) and this practice was probably maintained.[4] It was in all likelihood in relation to this that Pedro Juan Monzó, a mathematics professor in 1562–1564, published his *Elementa Arithmeticae, ac geometricae* in 1569 (certain authors date earlier editions in 1559 and 1566). This succinct work on the theory of numbers and proportions, contained topics of geometry that Monzó deemed necessary in order to understand Aristotelian natural philosophy and dialectics. Monzó's influence on logic has been regarded as crucial—he contributed to refocusing the study of this subject using a humanist approach that rejected terminist logic.[5] He did not however seem to be interested in astronomy.

The teaching of mathematics and related subjects in the *Estudio General* in Valencia reached considerable heights during the period when the chair was occupied by Jerónimo Muñoz, one of the most outstanding scientists of 16th century Spain. Muñoz began his studies in Valencia, where he graduated a Bachelor of Arts, and continued them in different locations in Europe. His own comments tell us that he was a student of Oronce Finé and Gemma Frisius.[6] He lived in Italy for sometime and taught Hebrew at the University of Ancona. According to a testimony of that time, Jews who came to listen to him believed him to be Jewish or that at least that he had been brought up by Jews, because of his fluent Hebrew.[7] The hypothesis that he was in fact a converted Jew cannot be ruled out. Following his return to Valencia, he was appointed to the chair of Hebrew in 1563 and in 1564 he combined this chair with that of mathematics, a position he held until 1578, the year he moved to the University of Salamanca.

In Valencia, in his chair of mathematics, Muñoz taught arithmetic, geometry and trigonometry, geometrical optics, astronomy, geography, and astrology. Although Muñoz published very few works, holograph documents or copies made by students of his in all these subjects are still to be found in several libraries across Europe.[8]

I do not intend to analyze the teachings of Muñoz here, having already done so in other papers. My aim here is to highlight, and comment on, some of the most noteworthy features of his astronomical teaching.

Muñoz became very well known in Spain as a mathematician, geographer, Hellenist, and Hebraist. His fame elsewhere in Europe was due mainly to his study of the supernova of 1572, disseminated in his *Libro del nuevo cometa* (Valencia, 1573) which he wrote in response to Philip II's request for his opinion on the phenomenon. This book was translated into French by Guy Lefèvre de la Boderie, a pupil of Guillaume Postel who collaborated with the Polyglot Bible of Antwerp as a Hebraist.[9] He also became known because of the detailed descriptions of his results and conclusions that were made by such prominent authors as Cornelius Gemma and Thaddaeus Hagecius. Muñoz also corresponded with Hagecius and Bartholomaeus Reisacherus of Vienna, another of the authors to deal with the supernova. Hagecius also furnished Tycho Brahe with letters he and Reisacherus had received from Muñoz about the supernova, which Brahe copied and used in his discussion of the works by Muñoz in the *Astronomiae Instauratae Progimnasmata*.[10]

Muñoz went on to pinpoint with remarkable precision the position of the nova in relation to the stars in Cassiopeia, and also its equatorial and ecliptical coordinates. He also attempted to measure its parallax and determined that it was completely imperceptible, indicating that the star was situated much further beyond the moon. In short, Muñoz concluded that the nova was a comet of a superlunary nature and origin like most of those comets which last a long time, although its appearance and behavior did not comply with those described in literature on this topic, being very similar to fixed stars, to the extent that it was "more like a star than a comet" in appearance. His reason for classifying this phenomenon as a comet stemmed from his wish to interpret the origin of the nova in terms of natural causes based on astrological tradition. Many authors who classified it as a star and consequently accepted its superlunary origin, nevertheless interpreted it as outside the regular, ordered course of nature; like miracles, produced not by *potencia dei ordinata* but by *potencia dei absoluta*. This was how it was interpreted by leading mathematicians and astronomers such as Cornelius Gemma, Thomas Digges, Tycho Brahe, and Thaddaeus Hagecius. Muñoz, committed to interpret the formation and apparition of the nova in terms of natural causes, was one of those best able to draw out its cosmological implications, showing for example that the evidence of its observed appearance and behavior could not be reconciled with the established doctrine of the incorruptibility—that is, the changelessness—of the celestial region.[11]

Hence, in this work on the nova, Muñoz comes across as an astronomer who feels perfectly entitled to extract conclusions of a cosmological nature from his observations. He evidently did not feel bound by the scholastic norm of *metabasis*, forbidding the crossing of boundaries between one *scientia* and another (the accusation from which Osiander had sought to defend Copernicus). He proudly declared indeed that God had granted him "unfettered and well-disposed ingeniousness suitable for understanding any subject", and that he had been forced by natural reasons and geometric proofs to accept that corruption and fires did exist in the heavens.[12]

Studies of the manuscripts of Muñoz and statements made by his students have enabled us to demonstrate that when he observed the supernova, he already had certain well defined ideas on cosmology that were patently anti-Aristotelian in important details and similar to the doctrines of the Stoic tradition. They also showed that Muñoz used to discuss these cosmological matters at the Universities of Valencia and Salamanca.

The most comprehensive exposition of Muñoz' cosmological ideas is to be found in his *Commentaries* on the second book of Pliny's *Natural History*.[13] Briefly, Muñoz held that the entire universe, from the earth located at its center to its limits, was full of air, which in addition, impregnated all the things in the world and served as a link between them. Muñoz consequently rejected the existence of the sphere of fire thought to separate the two regions: the sublunary and the celestial. He also rejected any other abrupt discontinuity in the heavens that celestial orbs or spheres would constitute. According to Muñoz, the cosmos has no exact dimensions but it is finite and ends where the air, after becoming increasingly rarer, can become no thinner. The outer limit has no defined shape and beyond it, there is possibly a huge nothingness. The planets are propelled by their own force or nature through the cosmic air like fish through the sea or birds through the air around the earth and are not dragged by orbs. Stars move in the same fashion, not dragged along embedded in a sphere. The heavens are corruptible and the composition of the planets and stars contains elements and qualities similar to those found on earth, albeit in a purer form. Comets are formed in the heavens and are therefore celestial bodies.

Muñoz' *Commentaries* adheres strictly to the canons of humanism, and is an attempt to interpret Pliny's text—by analyzing his sources and discussing his correct and mistaken premises—whilst setting forth his own ideas. In the usual humanist fashion, he brought into play a very wide range of quotations by authors of classical times, including poets, historians, geographers, mathematicians, and philosophers. However, this was not merely a literary or theoretical device; it also served to place philosophical assertions on the same level of opinion, their worth depending on the validity of their arguments. In addition, and despite the fact that Muñoz insisted on the distinction between the truths of faith and the truths of reason, he did not fail to point out that the cosmology he proposed was more in line with Christian theology, and that in general, reason must be compatible with the faith. Mention must be made in this respect of his criticism of Aristotle, Theophrastus, and other authors who stated that the world was eternal: a criticism which Muñoz made from the standpoint of the "true faith", but which is at the same time in keeping with his belief that the cosmos is surrounded by a vacuum since it has to burn:[14] "since those things which burn do turn into a more subtle substance, they must occupy a larger space; consequently, the place of the whole world in flames shall therefore be larger".

One of the main ideas of the cosmologies put forward by Muñoz is that of cosmic air, or in the words of Pliny, "the spirit known by both Greek authors and our own by the same term, 'air,' a vital something able to penetrate everything and which is in everything".[15] In his commentary on the text by Pliny, Muñoz quotes the famous verses by Virgil, "since the creation of the world, a single, inner spirit has nourished the heavens, the Earth and the watery plains and the luminous globe of the Moon and the titanic stars".[16] Along with Virgil, and also with regards to the air, he quotes two other poets of stoic influence, the famous authors of didactic poems on astronomy and astrology: Aratus and Manilius. The author to whom he pays the greatest attention, however, is Hippocrates, whose statements allow him to claim that his opinion of the substance of the sky "is not new but very ancient, albeit somewhat obscured by Aristotelian commentaries".[17] Both here and elsewhere, Muñoz resorts to the credentials of truth and genuineness stemming from the most ancient or "original" things: another topic dear to humanists and related to the cyclical conception of the history of human culture. In a different context however, Muñoz points out the progressive nature of knowledge,[18] and wrote a letter to his friend Reisacherus to the effect that as regards matters that can be tested, no authority to be taken on trust: not even Ptolemy, Alfonso, Regiomontanus or Copernicus.[19]

In addition, Muñoz defends astrology from its detractors. His cosmological ideas are closely linked to his astrological convictions, according to which the stars influence the earth by their light, heat and hidden processes, and it is the air, saturating the entire universe, which transmits these influences.

As regards the motion of the planets, since they are not dragged by spheres but move freely though the air, Muñoz considered that the only plausible explanation for the two basic visible movements—the one from east to west that they all have in common and the one each has from west to east—was that in reality they have only one movement, from east to west. In the case of the planets, this movement would be rather slower that that of fixed stars the closer the planet is to the earth the more slowly it moves, since the air becomes thicker in the same proportion and would consequently offers more resistance. Furthermore, the movement of the stars is not regular: they follow irregular paths with variable axis referred to by Muñoz as *spirae* or spirals.[20]

These commentaries on Pliny, the holograph of which is kept in Copenhagen University, are dated 1568, apparently the date when Muñoz presented them at the

University of Valencia in some extraordinary lessons given by some of the most outstanding professors, to which public figures resident in or visiting the city were invited.[21] In the foreword, in typical humanist style designed for the *captatio benevolentia* of the listeners, Muñoz introduces himself as a theologian and teacher of the Old Testament and warns that his mission is firstly to demonstrate "which of Pliny's ideas go against the Christian religion and which others are in line with it",[22] but, also, at different points in the text, in order to settle certain questions on the form and nature of the heavens or the nature of comets, he does declare the certainty of mathematics as opposed to the merely probable reasons offered by philosophers.[23]

In this way, Muñoz made skilful use of his dual condition as theologian and mathematician-astronomer to legitimate his criticism of Aristotelian cosmology and to set forth his own alternative ideas. In the treatise he prepared for his classes in Valencia as an introduction to astronomy and geography, Muñoz set forth a classification of the mathematical disciplines in line with Aristotelian tradition. In addition, he declined to explain the theory of the elements, saying that such questions belonged to the realm of physics. Nevertheless, when explaining that the Earth was motionless in the center of the world, he commented upon and criticised Copernicus' theory. And although he did make an effort to employ strictly astronomical arguments and accepted Copernicus' *dictum* that mathematics should be judged by mathematicians, he nevertheless mentioned in passing some of the cosmological problems arising from the concept of a planetary earth. Muñoz laid particular emphasis on the cosmic chaos that would be caused should the sun be in the center of the world and the earth in the fourth heaven, since it would be neither heavy nor of an elemental nature but celestial.[24]

Muñoz's most extensive and ambitious work on astronomy was his translation of and commentaries on Theon of Alexandria's *Commentaries* on Ptolemy's *Almagest*, which he began in Valencia in about 1568 and finished in Salamanca in 1582, although he continued to add notes and data to his comments until at least 1589.[25] In this work, in his comments and comprehensive additions to Theon's text, Muñoz reviewed many aspects of Ptolemaic astronomy, comparing them with the observations, techniques, and calculations of other classic, medieval, and renaissance astronomers including Copernicus whom he quoted often. He also provided his own tables and numerous observations made in Valencia and Salamanca, and described in detail a variety of observation instruments for astronomy and their respective advantages. Like its model, Ptolemy's *Almagest*, this work is to a large extent a highly technical treatise on mathematical astronomy and Muñoz takes great pains to clarify the most difficult sections for the sake of beginners (*ad tirones*), as he himself points out. However, he also includes broad-based discussions of cosmological matters, ranging from those concerning the position of the earth in the world, with a discussion of the heliocentric theory, to the nature of comets, and in this respect, he sets forth certain ideas similar to those in his *Commentaries on Pliny*.

On the heliocentric theory, Muñoz gave a broad outline of Copernicus' system together with an illustration based on the diagram in Chapter 10 of Book I of *De revolutionibus*.[26] Muñóz's criticisms of Copernicus' system are, as in his introductory treatise on astronomy and geography, basically astronomical and are designed to show that this system is incompatible with the phenomena. But Muñoz failed to take sufficiently into account all the movements that Copernicus attributed to the Earth and his complex models for the motion of in longitude of the Earth. Nor did he take into account that the diagram was no more than a simplification of Copernicus' theory, intended to show the general organization of the heavenly bodies derived from that theory. In short, Muñoz was not particularly interested in detailed analysis of Copernicus' theory, in contrast with his meticulous discussion, elsewhere in the book, of the Polish astronomer's calculations. In fact, Muñoz approached this question completely convinced from the outset that the heliocentric theory was false: his cosmological ideas implied the earth at rest in the centre of the cosmos, surrounded by an immense ocean of air, whose life depended on the influences received from the sun, the moon, the planets, and the others stars around it.

After rejecting Copernicus' model, Muñoz set forth his own cosmological ideas, identical to those described in the *Commentaries on Pliny*. As far as comets were concerned, here Muñoz considered all of them without exception to be heavenly bodies that appeared in the sky due to the concentration of planetary rays, referring to his treatise on the nova. As regards the sky, he stated that it had the four primary qualities: hot, cold, dry, and wet, and also that its thinness and rarity made the hypotheses of heavenly bodies invented by astrologers and philosophers impossible. For, as Muñoz commented, how could such orbs continue in existence over time? Instead, he stated that the planets moved thanks to the force supplied by their own nature, making their way across the sky like fish through the sea or birds in the air; they could not have two opposed motions, one from west to east as a result of their own motion and one from east to west with the motion of the universe, but could only be assigned a single motion. He went on to describe planetary motion again on the basis of the theory of delay (of the motion of the planets in comparison to that of fixed stars) and affirmed, rather unconvincingly, that this theory was equivalent to the Ptolemaic models he used in his analysis of astronomy in the rest of the book.[27]

Another noteworthy aspect of Muñoz' teachings is the attention he devoted to the applications of astronomy, in particular to geography and cartography and to the art of navigation. In both his introductory treatise of *Astronomy and Geography* and his additions to Theon's *Commentary*, he paid particular attention to how geographic coordinates are established reviewing all known methods and commenting on the instruments commonly used for this purpose. Muñoz was an expert geographer; he had established the latitudes of certain locations on the Peninsula with remarkable precision. He also estimated the longitudes of several places, though less successfully. He began the geodesic triangulation of the territory of Valencia, employing the method described by Gemma Frisius, and explained this procedure in his classes using actual examples. The first known map of the Kingdom of Valencia, included by Abraham Ortelio in his famous *Theatrum Orbium Terrarum*, is based to a considerable extent on the work of Jerónimo Muñoz.[28]

Although Muñoz was one of the best paid professors at the University of Valencia, his salary was considerably lower than those paid at universities in Castile. The prestige of the University of Salamanca, and its greater proximity to the seat of royal power, was probably also a factor in Muñoz' decision to accept the offer made to him by this university and his move there in 1578.

The chair of mathematics and astronomy in Salamanca, also known as the chair of astrology, had been occupied until 1576 by Hernando de Aguilera. Hernando de Aguilera was responsible together with his brother Juan for the works of Copernicus being incorporated into the statutes of 1561 as a text that could be used if students voted for it, as an alternative to Ptolemy or one of his commentarists: Geber or Regiomontanus.[29] This fact, highly unusual in Europe at that time, gives rise to the question of whether the work of Copernicus, including his heliocentric theory, was in fact taught. The books in which the visits of the rector to the chair are recorded show that Aguilera used to teach the Euclid' *Elements* (books I to VI), the sphere, parts of the *Almagest*, theories (models) of planets, and the *Tables* of Alfonso X, the astrolabe and its use, cosmography according to Petrus Apianus and Gemma Frisius, and astrology according to Alcabitius. The name of Copernicus does not figure in these records. I am of the opinion, however, that the possibility cannot be ruled out that on certain occasions Hernando de Aguilera followed Copernicus' *De revolutionibus* or did at least comment on aspects of the work when expounding subjects related to the sphere or planetary theories. The information contained in the books of visits is, in any case, not complete. In the 1562/1563 and 1563/1564 academic years for example, there is no reference to the chair of astrology.[30] The most usual interpretation, in keeping with the most widespread attitude in Europe towards the work of Copernicus, is that the interest of the Aguilera brothers in *De revolutionibus* was more concerned with the models, data and tables than the cosmological ideas it contained. To date however, documentary evidence has not enabled this question to be answered nor this hypothesis to be confirmed.

The Salamanca chair became vacant in 1576, probably due to the death of its occupant, Hernando de Aguilera. The body of professors at Salamanca was particularly keen to find a highly competent person to occupy the post, for the lack of someone in this position had been felt most acutely when the Pope had asked this university for its opinion on the reform of the calendar.[31]

The subject matter taught by Muñoz in Salamanca was very similar to what he had taught in Valencia. In astronomy, he explained the sphere or introduction to astronomy, models or theories of planets, tables and instruments, and astrology, geography, cartography, and the art of navigation.[32] Given the similarity in the subjects to be taught, he probably used the same texts as in Valencia.

From the late 15th century onwards, the University of Salamanca was a center of study and debate about cosmographic matters—due to a considerable extent to the influence of scientific humanism led by Nebrija and also the increased importance of cosmography attained in the enterprises of sea-faring expansion, the control of the empire, and the construction of the State. The teaching of cosmography already formed part of the statutes of 1538, which included an earlier project launched by the humanist Fernan Pérez de Oliva, and in the 1594 statutes, the pertinent regulations were made even more specific. Some philosophy teachers however, also included cosmographic matters in their exposition of Aristotelian natural philosophy, reflecting recent geographic discoveries.[33]

The need to train skilled cosmographers was one of the main reasons that led Philip II to establish the Mathematics Academy of Madrid in about 1582, which consisted basically of a mathematics and cosmography chair, the contents of which were very similar to those of the chair at the University of Salamanca.[34] It was with a similar desire to train cosmographers in about 1590, and in response to the wishes of Philip II to intensify the teaching of mathematics-related subjects, that the teaching of these subjects was enlarged in the University of Salamanca where a division (partido) or appointment for an associate teacher of mathematics was set up. In the letter from Philip II granting the economic assistance requested by the University, the king emphasized how necessary it was to "to train well-prepared and skilful persons able both to teach the subject at university and to be available at seaports or elsewhere, for they are very necessary and navigation depends upon them".[35] The division was assigned to one of Muñoz' pupils, Gabriel Serrano, until 1592. That year, after the death of Muñoz, Serrano himself came into possession of the chair when Diego Pérez de Mesa, another one of Muñoz's pupils who had been awarded the post in a competitive exam (oposición), resigned.[36]

Serrano held this position until his death in 1598. In 1593 the physician Antonio Nuñez Zamora—also a follower of Muñoz, who was later to replace Serrano in the chair—was appointed to the division. Both Serrano and Núñez were competent astronomers and faithful followers of the teachings of their master.[37]

In 1594 new statutes were drafted for the university. The texts or tables recommended for astronomy were those by Ptolemy, Peurbach, the *Tables* by Regiomontanus (of the "first motive") or Reinhold, Clavius, the *Alfonsine Tables*, and the work of Copernicus, and it was stipulated that Ptolemy's *Almagest* should be studied in combination with Copernicus' *De revolutionibus*, always beginning with the former. This meant that the choice between Ptolemy and Copernicus was no longer subject to the vote of the listeners. Furthermore, astrology continued to be taught and included the study of the comets, and it was also compulsory to study Ptolemy's *Geography*, Apianus' *Cosmografía*, cartography, the use of the different types of astrolabe, the Radius Astronomicus (or Jacob Staff), the art of navigation and the military art (artillery and fortification).[38] These statutes of the chair of mathematics must have been drafted by Muñoz' students for they were strictly in line with his teaching. Serrano, originally from Castalla (Alicante) publish nothing but left manuscripts on astronomy and astrology. He corresponded with Clavius, professor at the Collegio Romano and the most outstanding Jesuit mathematician of his time. Clavius was also one of the principal protagonists of the calendar reform. One letter from Serrano to Clavius, dated 1598, praised his master Muñoz "not only well-versed in mathematics but also in languages, the sacred scriptures and all the other sciences, his knowledge of which is so outstanding that he can rightly be compared with the sages of ancient times". He also told Clavius that he was using books by Clavius to teach mathematics and astronomy, and asked about the progress his works were making.[39] Meanwhile, Núñez Zamora, in his treatise on comets and the nova of 1604, upheld Muñoz' affirmation that comets were begotten in the heavens and that the heavens were corruptible. He also based his statements on the authority of Clavius whose treatise on *Sphaera* accepted the interpretation of the supernova of 1572 as a celestial phenomenon.[40] To judge from the books of visits, neither Serrano nor Núñez expounded Copernicus' *De revolutionibus* closely.[41] These authors in all likelihood followed the practice of their master Muñoz, expounding and commenting on Ptolemaic astronomy and comparing its quantitative results with the data and tables of other authors including Copernicus and Erasmus Reinhold.

It must nevertheless be pointed out that although Núñez Zamora defended the demonstrative nature of astronomy, the wording of his statements is cautious and eclectic and employs the traditional Scholastic technique of questions, referring to a variety of carefully chosen authorities. All this suggests that the academic environment restricted the exposition of free ideas in public to a considerable extent.

The University of Alcalá had a chair of mathematics and astronomy since its foundation but even today, very little is known about the subjects taught there. Pedro Esquivel taught at this university in the mid-16th century until leaving the chair in 1559 when he was appointed Palace mathematician and chaplain by Philip II. As a technician, Esquivel carried out several assignments to improve inland waterways in Castile and Aragon and in 1566 the king commissioned him to conduct a topographical description of Spain. Esquivel used of the geodesic triangulation method described by Gemma Frisius, and produced truly remarkable cartography.[42]

As far as can be gathered, Gabriel Serrano was also a professor at Alcalá, before moving to Salamanca.[43] In about 1586 the chair of mathematics and astronomy at Alcalá de Henares was occupied by Diego Pérez de Mesa.[44] Born in Ronda in 1563, Pérez de Mesa studied arts at the University of Salamanca (1577–1581), and graduated as a master. It was at this university that he took the courses taught by Jerónimo Muñoz in his chair of astronomy and mathematics.[45] He also began to study theology but we do not know whether he graduated from this faculty. In 1591, as we mentioned earlier, he competed for the chair at Salamanca left vacant by Jerónimo Muñoz. He was awarded the chair but did not take possession of it, deciding to remain in Alcalá where he negotiated and obtained a salary increase. In 1595, Pérez de Mesa, apparently upon the king's commission or order, moved to Seville to take up the chair created by the Town Council, at the request of the Courts (Cortes) in Madrid, in collaboration with the University and the *Casa de la Contratación* in Seville. The creation of this chair therefore stems from the same concern of Philip II, mentioned earlier, for the training of well-prepared cosmographers, pilots, and skilled technicians.[46]

We are not aware of the subjects taught by Pérez de Mesa at the University of Alcalá although they may be supposed to be similar, at least as regards astronomy, to what he taught in Seville where he gave also classes of arithmetic and algebra, practical geometry, astrology, and their applications in medicine and navigation, giving all classes in Spanish.[47]

In his *Comentarios de Sphera*, written for the classes he gave in Seville, Pérez de Mesa defines the purpose of cosmography and indicates that this subject is a "science almost mixed with philosophy and therefore it resolves many most wonderful questions of philosophy", such as whether or not there is a sphere of fire in the concavity of the moon, whether it is possible that the Earth moves, whether the stars move "by themselves or together with spheres, being fixed upon them"and"whether the substance of the sky is quintessential and incorruptible". In the first part of the work he discusses all these questions in detail.[48]

Evidently, like his master Jerónimo Muñoz, Pérez de Mesa considered that astronomers were perfectly entitled to make statements about natural philosophy; he devoted the first part of his commentary to a discussion of matters concerning cosmology itself. He denied that there was a sphere of fire in the concavity of the moon and quoted Copernicus and Cardan among other authors, in support. He also denied the existence of and need for celestial spheres, as well as the incorruptibility of the heavens; in support of this last point he mentioned the observations of the supernova made in 1572 by Muñoz. He devoted an entire chapter (Chapter 6) to the motion of the earth, although he referred only to its rotary motion. For Pérez de Mesa, the answer to this question could not be one of absolute certainty but rather of possibility. Consequently he admitted that the earth was "likely" to move (rotate), but nonetheless reached the final conclusion that "the earth was more likely to be at rest than in motion". He mentioned the traditional arguments of Aristotle and Ptolemy together with the evidence of the scriptures as bases for his conclusion. However, he also stated the solutions put forward by Copernicus to make the motion of the Earth physically plausible.[49] As regards the motion of the stars, he also followed Muñoz and the theory of spiral motion "which seems furthermore to be true".

He also paid particular attention to defending that the centre of the Earth, of gravity and of water are the same, in terms similar to those used by Copernicus, and criticised Paulus Venetus and Melchor Cano and, in general, those who had declared that the area of the earth's surface covered by water was ten times greater that the area of dry land.[50]

By way of conclusion, it must be said that astronomy was cultivated in Spanish universities in relation to medicine, humanism, natural philosophy, and cosmography. In the second half of the 16th century, astronomy was taught at the Universities of Valencia, Salamanca, Alcalá, and Seville by the noteworthy astronomer Jerónimo Muñoz and his followers. As well as an astronomer and mathematician, Muñoz was a geographer and topographer, Hebraist and Hellenist, and was also a follower and sustainer of humanist projects. Despite the presence of a certain demarcation between the subjects of astronomy and natural philosophy, Muñoz and his followers expounded and debated upon cosmological matters and considered themselves perfectly entitled to do so. Finally, it has been pointed out that the great importance of cosmography in the reign of Philip II, particularly in response to the needs of geographers, cartographers, and professors of the art of navigation, was a driving force behind the study of this subject at university, together with the study of astronomy. In order to explain the subsequent decline in the 17th century of the interest in astronomy and mathematics-related subjects in general in Spain, one must take into account all the factors behind its development, as well as its crisis or disappearance. It must nonetheless be emphasized that this relative decline should not be confused with a total dearth of cultivation. This, however, is to be discussed in future papers.

NOTES

[1] As Edward Grant, *Planets, Stars, and Orbs. The Medieval Cosmos, 1200–1687* (New York: Cambridge University Press, 1993), demonstrated (1993), pp. 271 ff., scholastic authors in the late middle ages generally acknowledged that Ptolemaic models based on eccentrics were of greater predictive value than the concentric spheres used by Aristotle, and accepted the existence of eccentric orbs. Celaya followed this tradition. In addition, although Celaya did not teach Aristotelian physics in Valencia, his influence and prestige in the city were quite considerable; after returning to Paris he was appointed permanent rector of the University of Valencia. For Celaya's physics and cosmology, see W. A. Wallace, *Prelude to Galileo. Essays in Medieval and Sixteenth-Century Sources of Galileo's Thought* (Dordrecht: Reidel, 1981); V. Navarro-Brotons, "Juan de Celaya", in J. M. L. Piñero, T. F. Glick, E. P. Marco, and V. N. Brotóns (eds.), *Diccionario Histórico De La Ciencia Moderna En España*, 2 Vols (Barcelona: Península, 1983), Vol. I, pp. 203–206; José María López Piñero y Víctor Navarro Brotons, *Història de la ciència al País Valencià* (València: Ed. Alfons el Magnànim, 1995), pp. 83–93, and the references mentioned in these works.

[2] P. J. Esteve, *Hipocrates Coi Medicorum Omnium Principis Epidemion Liber Secundus*, Valencia, 1551, fols. 4r ff. For Esteve as a physician, see J. M. López Piñero, "Pedro Jaime Esteve", *Diccionario Histórico*, Vol. I, pp. 312–314.

[3] V. Ximeno, *Escritores Del Reyno De Valencia*, 2 Vols (Valencia: J. E. Dolz, 1749), Vol. I, p. 112, mentions a *Libro de las Ephemerides*, "known commonly as those of Esteve".

[4] For a recent edition of the by-laws, see *Bulas, Constituciones y Estatutos De La Universidad De Valencia*, 2 Vols (Valencia: Universidad de Valencia, 1999).

[5] J. G. Salvadores, "La enseñanza de la metafísica en la Universidad de Valencia durante el siglo XVI", *Analecta Sacra Tarraconensis*, 45:137–172 (1972).

[6] In his introductory treatise to astronomy and geography *Astrologicarum et Geographicarum Institutionum Libri Sex* (copy in the Biblioteca Apostolica Vaticana, Ms. VL 6.997), Muñoz refers to Gemma Frisius as "institutor noster" (54v) and to as Finé "preceptor noster" (68v). An edition and Spanish translations of this manuscript, in *Jerónimo Muñoz: Introducción a la Astronomía y la Geografía,* V. Navarro (ed.), translation by V. Navarro, A. Pastor, E. Pastor, V. Salavert (Valencia, Consell Valenciá de Cultura, 2003). For Muñoz's manuscripts, see also Víctor Navarro Brotóns, Enrique Rodríguez Galdeano, *Matemáticas, Cosmología y Humanismo en la España del siglo XVI. Los* Comentarios al Segundo Libro de la Historia Natural de Plinio *de Jerónimo Muñoz* (Valencia: Instituto de Estudios Documentales e Históricos sobre la Ciencia, 1998).

[7] See P. A. Morlá, *Emporium Utriusque Iuris Quaestionum,…* (Valencia: Alvaro Franco and Diego de la Torre, 1599) in the *Epistola nuncupatoria.*

[8] See Navarro, Rodríguez, *Matemáticas, Cosmología Y Humanismo.*

[9] J. Muñoz, *Traicté Du Nouveau Comete* (Paris, 1574). On Lefèvre de la Boderie, see F. Secret, *L'Esoterisme de Guy Le Févre de la Boderie* (Genève: Droz, 1960). For the relationship between Lefèvre and Postel and his contribution to the Polyglot Bible of Antwerp, see B. Rekers, *Arias Montano* (London: The Warburg Institute, 1972).

[10] For Cornelius Gemma's discussion about Muñoz' study of the nova, see C. Gemma, *De Naturae Divinis Characterismis; seu Raris et Admirandis Spectaculis in Universo, Libri II* (Amberes: C. Plantin, 1575), Vol. II, pp. 267–274. Muñoz' correspondence with Hagecius and Reisacherus is housed in the Oesterreichische National Bibliothek, Cod. 10.868, n°66 and Cod. 10.689, n°41, fols. 1r–6v. J. L. E. Dreyer published these letters in his edition of *Tychonis Brahe. Opera Omnia*, 15 Vols (Copenhague: Gyldendaliana, 1913–1919; new ed. facsimile, Amsterdam, 1972), Vol. VII, 395–403. Jerónimo Muñoz, *Libro del Nuevo Cometa (Valencia, Pedro De Huete, 1573). Littera ad Bartholomaeum Reisacherum (1574). Summa del Prognostico del Cometa (Valencia, Juan Navarro, 1578)*. Introduction: "The Astronomical Work of Jerónimo Muñoz", Appendices and Anthology by Víctor Navarro Brotóns (Valencia:

Hispaniae Scientia, 1981) includes a transcription and translation into Spanish and English accompanied by a facsimile edition of the letter to Reisacherus, according to the copy of Cod.10.689 mentioned earlier. This copy was apparently made by Tycho Brahe, see Navarro, Rodríguez, *Matemáticas, Cosmología y Humamismo*, pp. 207–208.

[11] See V. N. Brotóns, "The Astronomical Work of Jerónimo Muñoz", pp. 11–111; see also A. Ingegno, *Cosmologia e Filosofia nel pensiero di Giordano Bruno* (Fireze: La Nuova Italia Editrice, 1978), p. 1 ff.; Michel-Pierre Lerner, *Le Monde des Sphères*, 2 Vols (Paris: Les Belles Lettres, 1996–1997), Vol. II, p. 21 ff., and Miguel Angel Granada, "Cálculos cronológicos, novedades cosmológicas y expectativas escatológicas en la Europa del siglo XVI", *Rinascimento*, 2ª ser., 37:357–435 (1998). In his study of the works and opinions about the nova of different authors from the Protestant region, Charlotte Methuen, *Kepler's Tübingen. Stimulus to a Theological Mathematics* (Aldershot: Ashgate, 1997), and *id.*, " This Comet or New Star": Theology and the Interpretation of the Nova of 1572, *Perspectives on Science*, 5:499–516 (1999), pointed out that *Providentia Specialis* and *Providentia Generalis* would be a more suitable way of characterizing them, so that "that the essential difference between the events of special providence and those of general providence could be used to legitimate observations which contradicted accepted physics. The decision that the underlying explanatory system must be revised thus required a theological shift as well as contradiction by observation" (Methuen, "This Comet . . . "). For example, Philip Apianus, accepted that comets could be heavenly bodies, and therefore "he needs to invoke special providence only when explaining why the nova appears at this particular time", since, according to this astronomer "the comet or star has been created as a warning by Almighty God" (Methuen, *Kepler's Tübingen*, p. 508).

[12] Muñoz, *Libro del Nuevo Cometa*, A3r.

[13] The holograph manuscript is housed at the Arnamagnaeanske Institute, Copenhagen, AM 8812 4°, fols. 1–47. Published and translated into Spanish by Navarro, Rodríguez, *Matemáticas, Cosmología Y Humanismo*.

[14] According to St Peter, the Stoics, Heraclitus and Hippasus Metapontinus, quoted by Muñoz in relation to the world on fire. See *Commentaria Plinii . . .* , p. 292, ed. Navarro, Rodríguez, *Matematicas, Cosmologia y Humanismo*.

[15] Pliny, HN 2,10.

[16] Verg. *Eneid.* 6, 725, quoted by Muñoz, *Commentaria Plinii . . .* , ed. Navarro, Rodríguez in *Matemáticas, Cosmología Y Humanismo*, pp. 382–384. In connection with this, in his *Comentarios a Alcabitius*, fol. 113r (Madrid, Bibl. Nacional, Ms. 9287)—a text on astrology also related to his classes, Muñoz compares cosmic air with the spirit issuing forth from the heart to revive the body. For this manuscript, see Navarro, Rodríguez, *Matemáticas, Cosmología y Humanismo*.

[17] Muñoz, *Commentaria Plinii . . .* , p. 386, ed. Navarro, Rodríguez.

[18] At the beginning of his treatise *De Planispherii Parallelogrammi Inventione* (copy in Biblioteca Apostolica Vaticana, Ms.VL 6.997, fols. 1r–71v; another copy in Bayerische Staatsbibliothek, Clm 10.674, fols. 278r–336v) Muñoz states that both sciences and arts, like rivers, have obscure origins, are fed by many affluents and grow continuously until they finally flow into the sea.

[19] For this letter, see footnote 10 above.

[20] On the theory of delay, Muñoz cites Martianus Capella, and in his *Commentaries on Theon*, fols. 36v–37r, says that according to Capella the "ancients" peripatetics supported such a theory. Ptolemy (*Almagest*, 1.8, H28 and Theon, *Com.*, 1.8, 439 10 ff., ed. Rome, 100, ed. Halma, 35–36, ed. Basel), discuss the theory rejecting it because besides the movement toward the East, the planets have also a movement of latitude. Is in relation to such criticism, that Muñoz integrate the movement of latitude in a resultant movement according to *spirae*. As real paths of planets, spiral lines can be traced back to Plato. Moreover, the idea of the delay was incorporated by Alpetragius in his astronomical theories and had a certain diffusion in the XVIth century and even in the XVIIth century in relation to the desire to avoid assigning contrary motions to

one and the same celestial body. See Grant, *Planets, Stars, and Orbs* (cit. note 1), 563 ff. On Alpetragius, see Bernard R. Goldstein, *Al-Bitruji: On the Principles of Astronomy*, 2 Vols (New Haven: Yale University Press, 1971).

[21] Folio 1r says: "Anno 1568 mensis julii XV die incepit Hieronymus Munnos Plinii secundum librum explicare".

[22] Muñoz, *Commentaria Plinii . . .*, p. 278, ed. Navarro, Rodríguez.

[23] On pp. 341 ff., *Commentaria Plinii . . .* ed. Navarro, Rodríguez, Muñoz discusses the shape of the world and the motion of planets and remits the reader to his *Commentaries on Theon*, insisting that his arguments are "solid demonstrations and not merely verbal discussions, . . . (demonstrations) that are ignored by those not familiar with mathematics". On pp. 554–555, with regards to the shooting stars and comets he says, of the former, that Aristotle's explanation would seem to be correct but as regards the location and substance of comets, "I follow not Aristotle but mathematicians who expound things more precisely than philosophers".

[24] Muñoz, *Astrologicam et Geographicarum . . .*, 14r ff. The astronomical arguments that Muñoz puts forward are based on those laid down by Ptolemy and developed by Theon in his *Commentaries on Almagest*, a work that Muñoz knew well having commented and translated it into Latin. Muñoz, like Ptolemy, dealt with the two questions separately: whether or not the earth is in the center of the heavens and whether the earth is in motion and moves away from the center or rotates on its axis. But no mention is made of the fact that Ptolemy and Theon had acknowledged that if the earth rotated on its axis and the heavenly sphere remained motionless, there would be no variation in daily phenomena, from an astronomical viewpoint.

[25] The manuscript ends with the words *Hieronymus Munnos . . . Translation Commmentariorum Theonis Alexandrini in Magnam Constructionem CL. Ptolemaei . . .* The holograph is housed at the National Library in Naples, Ms. VIII, fols. 21r–300r. See Navarro, Rodríguez, *Matemáticas, Cosmología y Humanismo*.

[26] Muñoz, *Theonis Alexandrini . . . .*, 34v ff. The diagram on 35r.

[27] Muñoz, *Theonis Alexandrini . . .*, 35v ff.

[28] See Navarro, Rodríguez, *Matemáticas, Cosmología Y Humanismo*.

[29] On Juan and Hernando de Aguilera, see Eugenio Bustos Tovar, "La introducción de las teorías de Copérnico en la Universidad de Salamanca", *Revista de la Academia de Ciencias Exactas, Físicas y Naturales*, 67:, 235–252 (1973), and the articles on these authors by V. Navarro in *Diccionario Histórico* (cit. note 1), Vol. 1, pp. 28–30. On the statutes of Salamanca University see, in addition to Bustos, Víctor Navarro Brotóns (1995), "The reception of Copernicus's Work in Sixteenth-Century Spain: The Case of Diego de Zúñiga", *Isis* 86: 52–78 (1995); *id.* "El Renacimiento científico (siglo XVI) y la Enseñanza de las Disciplinas Matemáticas en Las Universidades de Valencia y Salamanca". In: *Doctores y Escolares. II Congreso Internacional sobre las Universidades Hispánicas* (Valencia, 1995), 2 Vols. (Valencia: Universidad de Valencia, 1998), Vol. I, 141–159.

[30] The purpose of the inspections (*visitas*) stipulated in the 1561 statutes was to supervise the conduct of the professors occupying chairs. Five times a year, the rector would visit each faculty, alone or accompanied by the senior professor, and question two students. The students' comments about the punctuality, rigor, topics taught by the professors, and the effectiveness of their expositions were noted in the *libros de visitas* (books of visits). For *libros de visitas*, see Manuel Fernández Álvarez, *Copérnico y su Huella en la Salamanca del Barroco* (Salamanca: Universidad de Salamanca, 1974) and Navarro, "El Renacimiento Científico".

[31] See Vicente Beltran de Heredia, *Cartulario De la Universidad de Salamanca*, 6Vols (Salamanca: Universidad de Salamanca, 1970–1973), Vol. IV, doc. 1628, p. 323.

[32] As can be deduced from the books of visits. See Fernández, "Copérnico y su Huella", and Navarro, "El Renacimiento Científico".

[33] See Cirilo Flórez Miguel, Pablo García Castillo y Roberto Albares Albares, *La Ciencia de la Tierra. Cosmografía y Cosmógrafos Salmantinos del Renacimeinto* (Salamanca: Caja de Ahorros, 1990); W. G. L. Randles, "Classical Models of World Geography and Their

Transformation Following the Discovery of America", in W. Haase y and M. Reinhold (eds.), *The Classical Tradition and the Americas* (Berlin-New York, Walter de Gruyter, 1994), Vol. I, pp. 6–76; Navarro, "Te reception of Copernicus' theory", *id.* "La Cosmografía en la época de los descubrimientos". In: *CongresoHispano-portugués:Las relaciones entre Portugal y España en la época de los descubrimientos y la expansión colonial*, Ana María Carabias Torres, (ed.) (Salamanca: Universidad de Salamanca, 1994), pp. 195–207. For the assertions of Pérez de Oliva, see *Estatutos de la Universidad de Salamanca, 1529: Mandato de Pérez de Oliva, Rector*, ed. José Luís Fuertes (Salamanca: Univ, de Salamanca, 1984).

[34] For the Mathematics Academy of Madrid, see María Isabel Vicente Maroto y Mariano Esteban Piñeiro, M. (1991), *Aspectos De La Ciencia Aplicada En La España Del Siglo De Oro*, Valladolid, Junta de Castilla y León.

[35] Letter from Philip II, March 26, 1593, in Esteban Esperabé de Arteaga (1914–1917), *Historia pragmática e interna de la Universidad de Salamanca*, 2 Vols (Salamanca, 1914–1917), Vol. I, pp. 608–609; also published in Bustos, "La Introducción de las teorías de Copérnico", p. 239, note 12.

[36] See Beltrán,*Cartulario*, Vol. IV, pp. 120 ff. For Pérez de Mesa in relation to the chair of mathematics, see the chapter of Luciano Pereña, "Política o Educación Democrática", in Diego Pérez de Mesa, *Politica o Razón de Estado* (Madrid: CSIC, 1980), L. Pereña and C. Baciero, (eds.), with the collaboration of V. Abril, A. García y F. Maseda, pp. XIII–LXIII, in p. XV, note 6. The report of Pérez de Mesa's "oposición" exam is held in the Salamanca University Archives, "Procesos de cátedras", Ms. 970. Following Pérez de Mesa's resignation, the chair was declared vacant and competitive exams had to be held. Serrano was appointed on March 21, 1592.

[37] For Serrano, see Beltrán, *Cartulario*, Vol. IV, pp. 120 ff. For some information on Núñez Zamora, see Felipe Picatoste Rodríguez, *Apuntes para una Biblioteca Científia Española del siglo XVI* (Madrid: Tello, 1891), pp. 223–225, and E. Rodríguez San Pedro, *La Universidad Salmantina Del Barroco, 1598–1625*, 3 Vols. (Salamanca: Universidad de Salamanca, 1986), Vol. III, p. 71.

[38] *Estatutos Hechos Por La Muy Insigne Universidad de Salamanca*, Salamanca, 1595.

[39] The letter written by Serrano to Clavius, Salamanca, April 14, 1598, appears in U. Baldini, P. D. Napolitani (eds.), *Cristoph Clavius. Corrispondenza*, 6 Vols (Pisa: Quaderno del Dipartimento di Matematica, 1992), Vol. IV (1597–1601), part 1, pp. 46–54. Serrano tells Clavius in this letter that that year, he had been teaching Ptolemy's *Almagest* and had written a commentary on it. The Escorial Library houses a copy of *Tractatus de astrologia judiciaria*, MS O-III-30 by Serrano which concurs with the lessons he gave on this subject in Salamanca in 1593.

[40] Antonio Núñez Zamora, *Liber de Cometis, In Quo Demonstratur Cometan Anni 1604 Fuisse In Firmamento*, Salamanca, 1610. Although it was published in this year (1610), 1605 is the date of censure and printing given at the end of the book.

[41] According to the books of visits, Serrano and Clavius taught the same subjects as Muñoz. See Fernández, "La Universidad Salmantina".

[42] See J. M. López Piñero and T. F. Glick, "Pedro de Esquivel", in *Diccionario Histórico* (cit. note 1), Vol., pp. 310–312, and the references mentioned in this book. Also, Gonzalo Reparaz Ruiz, "The topographical Maps of Portugal and Spain in the 16th Century", *Imago Mundi*, 7: 75–82 (1980); Geoffrey Parker, "Maps and Ministers: The Spanish Habsburgs", in D. Buisseret. (ed.), *Monarchs, Ministers and Maps* (Chicago and London: The University of Chicago Press, 1992), pp. 124–153; M. E. Piñeiro, "Esquivel. Un ejemplo de la ciencia aplicada en la España del siglo de Oro", in L. J. Moreno (ed.), *La Universidad Complutense Cisneriana* (Madrid, (ed.) Complutense, 1996), pp. 261–285.

[43] According to Beltrán, *Cartulario* (cit. note 31), Vol. IV, p. 324.

[44] F. Gil Ayuso, *Historia de la Universidad de Alcalá*, quoted by A. González Palencia in the foreword to Pedro Medina, *Obras*, Vol. I (Madrid: CSIC, 1944), p. XXXVIII. In the dedicatory to his patron Gaspar de Borja and Velasco, in a Latin manuscript of *Cosmographia* (Biblioteca

Universitaria de Barcelona, Ms. 446), Pérez de Mesa states that he taught natural philosophy in his youth in Salamanca and Alcalá.
[45] In Pérez de Mesa' enlarged and revised edition of *De las grandezas y cosas notables de España* by Pedro Medina (Alcalá de Henares, 1595), fol. 226r, he highlighted amongst his recent professors at Salamanca University "the learned teacher Jerónimo Muñoz, well-versed in human and spiritual letters and in all faculties, particularly in languages, positive theology and mathematics, and in Aristotelian and Platonic philosophy, whose ingeniousness and memory, virtue and scorn for the world are wonderfully admired".
[46] See Luciano Pereña, in Pérez de Mesa, *Política o razón de Estado* (cit. note 36).
[47] Manuscripts of all these subjects written between 1595 (*Astrologia Judiciaria*) and 1603 (the *El Arte De Navegar* treatise) are housed at the Biblioteca Nacional, Madrid and in the library of Salamanca University. The manuscript entitled *Astrología Judiciaria* (Madrid, BN, Ms. 5917), states "Judiciary astrology read in Seville by Diego Pérez Mesa, Professor at Alcalá de Henares, by order of King Philip in the year 1595". The treatise on the art of navigation features the date September 6, 1603 at the end. Ms. 2294 of the Salamanca University Library concerns arithmetic, algebra, astrology and practical geometry and Pérez de Mesa, "professor of this city of Seville in the year 1598" is named as the author on the first page.
[48] A copy of *Comentarios De Sphera* by Pérez de Mesa, is housed in the Biblioteca Nacional, Ms. 8882. It is dated in Seville, September 22, 1596.
[49] Pérez de Mesa, *Comentarios De Sphera*, *op. cit.*, fols. 19v–22r. The chapter ends with Copernicus' answers to the objections about the Earth's motion.
[50] *Ibid.*, 14v ff.

ROGER ARIEW

THE *SPHERE* OF JACQUES DU CHEVREUL: ASTRONOMY AT THE UNIVERSITY OF PARIS IN THE 1620S

It has been thought that, before 1640, French scholastics did not take into account Galileo's astronomical observations of the 1610s. It may be true that the new astronomy was not normally discussed in the physics courses of French schoolmen in the 1620s, but it would not be true to assert that Galileo's observations played no role in the collegiate curriculum. The new astronomy was generally dealt with in the mathematics course, under the rubric of commentary on the *Sphere* of Sacrobosco. I discuss here the *Sphere* of Jacques du Chevreul, a professor at the University of Paris in the 1620s, 1630s, and 1640s, in order to establish the place accorded at the University of Paris in the 1620s to new astronomy and to Galileo's astronomical observations.

du Chevreul was born in Coutances in 1595 and died in Paris in 1649;[1] he was thus born a year earlier and died a year earlier than his more famous contemporary, René Descartes. Unlike Descartes, du Chevreul was always associated with a university, specifically the University of Paris and the Collège Harcourt, except for the 2 years before his death, when he was Professor of Philosophy at the Collège Royal. Like Descartes, he was the son of a magistrate; he studied humanities and philosophy, and received a Master's degree—Master of Arts from Paris (1616), not Master of Law from Poitiers. Unlike Descartes, he continued his education in the higher faculty of theology and was awarded the degree of Bachelor of Divinity in 1619. du Chevreul began teaching at Harcourt in 1620. During his lifetime, he held various administrative academic offices, including those of rector and principal. He must have taught mathematics early in his career, but he was teaching philosophy by 1622. At the time, the philosophy curriculum was divided into a year of Logic and Ethics and a year of Metaphysics and Physics. According to his manuscript lecture notes conserved at the Bibliothèque Municipale de Cherbourg, du Chevreul taught Logic and Ethics in 1623–1624, 1625–1626, and 1633–1634; he taught Metaphysics and Physics in 1628–1629 and 1634–1635. Although he did not publish his philosophy lectures, he did publish two mathematical texts, *Arithmetica* (Paris, 1622) and *Sphaera* (Paris, 1623, 1640, and 1649).

The mathematical curriculum in the early 17th century still corresponded to the traditional *quadrivium*—the four liberal arts of arithmetic, geometry, music, and astronomy—together with optics and some other mathematically based disciplines such as, trigonometry and gnomonics, geography and hydrography, or chronometry.[2] The basis of astronomical teaching was the textbook—or rather textbook

M. Feingold and V. Navarro-Brotóns (eds.), Universities and Science in the Early Modern Period, 99–109.

genre—known as the *Sphere*. Sacrobosco's original *Sphere* was itself divided into four parts. Part I consisted of an abstract mathematical discussion of the sphere, defining its center and axis, plus an exposition on the shape of the universe and the number of concentric spheres thought to constitute it. Part II contained a catalog of some of the circles, both great and small, inscribed on the surface of the various spheres. Part III incorporated an enumeration of astronomical signs, with a discussion of their risings and settings and the inequalities of the days. And Part IV contained a discussion of eclipses and the causes of motion.[3] That was the basic structure of the work as taken up by its various commentators during the Middle Ages, from the 13th through the 15th century.[4] The structure remained unchanged at the end of the 16th century and the beginning of the 17th. The most famous mathematician of the time, Christopher Clavius, professor at the Jesuit Collegio Romano, followed the same basic pattern in his first edition of the *Sphere* (Rome, 1570). He too divided the first edition into four parts. He began with the composition of the sphere, including such questions as "What is a sphere?" "What is its center?" "What is its axis?" "What are the poles of the world?" "How many spheres are there?" and "What is the shape of the world?" He continued with the circles to be identified, the matter of which the spheres are composed, and what is beyond the heavens. He then dealt with the astronomical signs, day and night, and the climates of the Earth. Finally, he discussed the motion of the planets and the causes of eclipses. It should be said that later editions of Clavius' *Sphere* were "enriched in many different places (*multis ac variis locis locupletata*)", including a *disputatio* on eccentric and epicyclic orbs.[5]

All of this changed with du Chevreul's *Sphere* in 1623. du Chevreul's commentary is divided into 10 chapters, some of them detailing traditional materials, others going well beyond them. Chapters 1 and 2, De definitionibus and De figura mundi, recapitulate, with some innovations, Part I of the traditional *Sphere*. Chapter 3, De ordine partium, consists mostly of new materials dealing with Copernicus and a Copernican topic: the arrangement or order of the spheres. Chapter 5, De circulis coelestibus, and Chapter 7, De eclipsibus, discuss the same topics as the traditional Parts II and IV. Some of the traditional Part III, about astronomical signs and such periods as days and nights, is included in Chapter 9, De accidentibus sphaerae. But du Chevreul also adds four new chapters, namely, 4—De stellis, 6—De caelorum numero, 8—De eccentricis et epicyclis, and 10—De calculo ecclesiastico.

du Chevreul's book begins, in the fashion of Euclid's geometry, with definitions of a point as that which has no parts, a line as length without breadth, and a surface as that which has length and breadth alone. He quickly proceeds with definitions of angles, circles, solid figures, and spheres.[6] He then delimits the center of the sphere, distinguishes between great and minor circles, and defines the poles as points on its surface. These geometrical beginnings cannot obscure the fact that du Chevreul's interests are not exclusively mathematical. The volume continues with a discussion of the shape of the world, in two sections. Both sections employ arguments derived from physics, or others whose origins are not strictly mathematical. The first section concludes that the world is spherical, based on four different arguments. 1) Its fitness: the sphere is a perfect shape and has a greater capacity than

Au: Is this runnung hea ok?

other shapes; it is well accommodated to circular motion. 2) Its necessity: if the world were not spherical in shape, a vacuum, or bodies without place, or interpenetration of

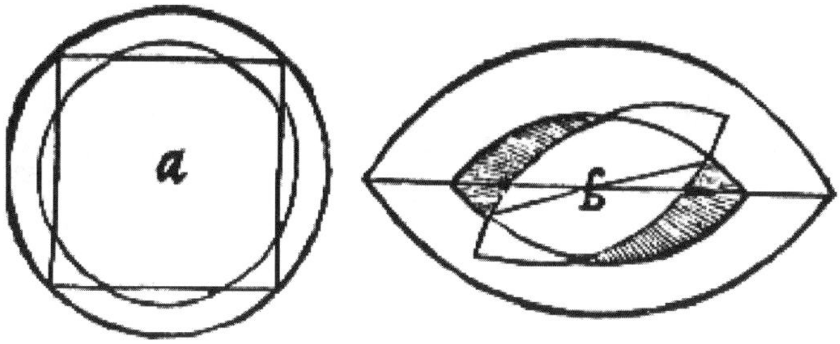

dimensions would have to be posited. 3) The nature of things: heavy bodies fall toward the center, and parts of the earth press themselves together, and thus they constitute a sphere; light bodies flee the center and also form a sphere. 4) The appearances: the fixed stars appear at the same distinct distance from us.[7] du Chevreul's second section concludes that the Earth is spherical, based on the rising and setting of stars, the fact that stars are lower on the horizon as one travels north, the circular shadow of the Earth during lunar eclipses, and other such "experiences".[8]

du Chevreul's third chapter, on Copernicus, is not unfamiliar in content or especially bold. He refers to the Pythagorean view that the Sun is the center of the world, with all the fixed and wandering stars and elements revolving around it, and says that the opinion was embraced by Copernicus, "the preeminent restorer of astronomy".[9] He then describes the Copernican system from the periphery inward, from the firmament or sphere of fixed stars, to Saturn, Jupiter, Mars, and to the terrestrial orb in which the Moon is contained "as if it were an epicycle", together with the four elements in their natural order, then to Venus, Mercury, and finally to the Sun, ruling immobile at the center of the world. du Chevreul adds that the Copernican firmament would need to be immobile and that the Earth would have a double motion around the Sun, that is, both a diurnal and an annual motion.[10]

du Chevreul then states that the Copernican system can be shown to be entirely erroneous and overly bold.[11] The reasons he adduces conclusion are multiple. First, he cites the authority of Plato in the *Timaeus* and *Phaedo*, Aristotle in the *De caelo*, and all philosophers and astrologers, who place the Earth immobile at the center of the world. Then he appeals to the authority of the Sacred Scriptures. His commentary produces such well-known Biblical passages as "The Lord laid the foundations of the Earth, that it should not be removed for ever", and "Then spake Joshua to the Lord in the day when the Lord delivered up the Amorites before the children of Israel, 'Sun,

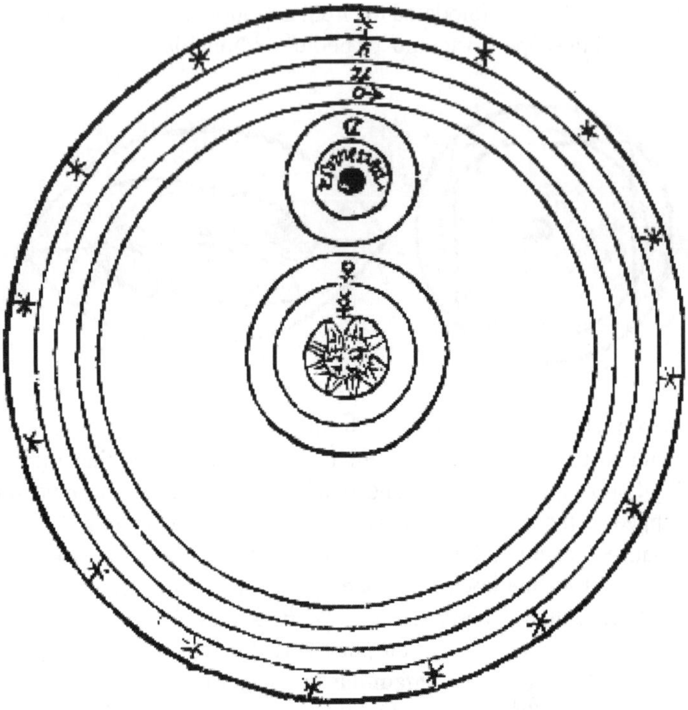

stand thou still upon Gibeon; and thou, Moon, in the valley of Ajalon.' And the Sun stood still, and the Moon stayed, until the people had avenged themselves upon their enemies".[12] This appeal to the Scriptures is consistent with the recent discussion surrounding the 1616 condemnation of Copernicanism by the Catholic Church.[13] It is also consistent with du Chevreul's own views about theorizing, even with respect to Aristotle himself: "If by chance there is anything in the latter's [that is, Aristotle's] works definitely or even potentially repugnant to our faith, it must be discredited. As Plato says in the *Gorgias*, can it be possible to let mere words stand against the truth?"[14]

du Chevreul continues his argument against Copernicus, claiming, for example, that it is inconsistent with physics for a simple body to have anything other than a single and simple motion; Copernicus' opinion, of course, would endow a simple body—the Earth—with two motions. du Chevreul also appeals to "optics", asserting that, given the Copernican opinion, we would get closer and farther to the fixed stars and would thus see changes in their brightness—something that has not been observed.[15] However, on the question of whether the Sun rules the universe at the center, which Copernicus is said to hold, he makes the startling claim that the Sun is not ruled by the Earth nor is the Earth ruled by the Sun, given that Mercury and Venus circle the Sun.[16] Apparently, he does not find the circling of the inferior planets around the Sun contrary to the appearances, to physics, or to the Sacred Scriptures.

Although he does not specifically attribute the circling of Venus and Mercury around the Sun to a reasoning based on their having phases like the Moon, he does assert that with the optical tube one can see spots on the Sun which are assumed to be Venus and Mercury.[17] Later on he also asserts that Venus and Mercury must transmit the light of the Sun (as opposed to their having their own light like the Sun), because otherwise, we would not have observed them as spots on the solar disk.[18] Clearly, these views are consistent with Venus and Mercury having phases. du Chevreul ends the chapter on Copernicus with a variety of topics, including a discussion of parallax: the parallax of near planets and the lack of parallax of the fixed stars.[19]

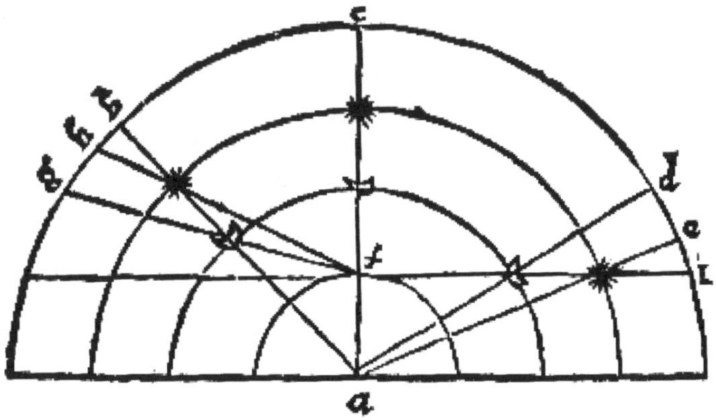

du Chevreul's numerous scientific innovations follow immediately in Chapter 4, just after his concession of the revolution of Mercury and Venus about the Sun. On questions about the matter of the heavens and its incorruptibility, he adopts a probabilistic language. He says that he does not have a certain and definite reply to the question of the material composition of the heavens and the stars, but must embrace the most plausible opinions. His probable conjecture is that the stars are denser and rarer parts of the heavens.[20] The stars are then able to propagate light in relation to their density or rarity.[21] This conjecture would safeguard the incorruptibility of the heavens, given that the contrary qualities (hot, cold, dry, and moist) would not be located in the heavens. du Chevreul also inserts a disputation with the "neoterics" who claim that the heavens are corruptible, on the basis of such astronomical phenomena as new stars (that is, novas) and comets.[22] In his replies, again couched in probabilistic language, he denies the conclusiveness of the moderns' observations and of their parallactic measurements.

du Chevreul follows tradition in dividing the stars into fixed and wandering stars. He tells us that Plato, Aristotle, and all others to the present generation observed seven wandering stars or planets: Saturn, Jupiter, Mars, the Sun, Venus, Mercury, and the Moon. But he also tells us that Galileo, that preeminent mathematician, discovered four planets circling around Jupiter, which Galileo called Medicean stars, detected two new planets concentric to Saturn, the Saturnines. Thus, du Chevreul counts 13 planets agreed by all, that is, six new ones in addition to the seven classically known ones. He then increases the count by noting that others add another 30 new planets circling about the Sun, namely those that Tardeus, that is, Jean Tarde, calls the Bourbon stars.[23]

The discoveries acknowledged by du Chevreul entail modifications in the doctrine of the number of the heavens. That is the topic discussed by him in Chapter 6—De numero coelorum. According to Plato, Aristotle, and the Aristotelians, the number of heavens, distinguished by their different motions,[24] is eight (or more). For example, the Paris-educated schoolman, Eustachius a Sancto Paulo, in his 1609 *Summa philosophica quadripartita*, counts the seven planetary heavens as those of the Moon, Mercury, Venus, the Sun, Mars, Jupiter, and Neptune, and then adds an eighth sphere for the firmament, a ninth and tenth for the crystal heaven, and an eleventh for the *primum mobile*; above that is the Empyrean heaven. Instead, du Chevreul counts only five planetary heavens: those of Saturn, Jupiter, Mars, the Sun, and the Moon.[25] Missing in this count are the heavens for the new planets and those of Venus and Mercury. In order to account for Venus and Mercury, du Chevreul rehearses an old controversy: Plato, Aristotle, and others placed Venus and Mercury above the Sun; Ptolemy, Regiomontanus, Sacrobosco, and others placed them below the Sun. But du Chevreul asserts that, *as shown by the optical tube*, Mercury and Venus circle around the Sun,

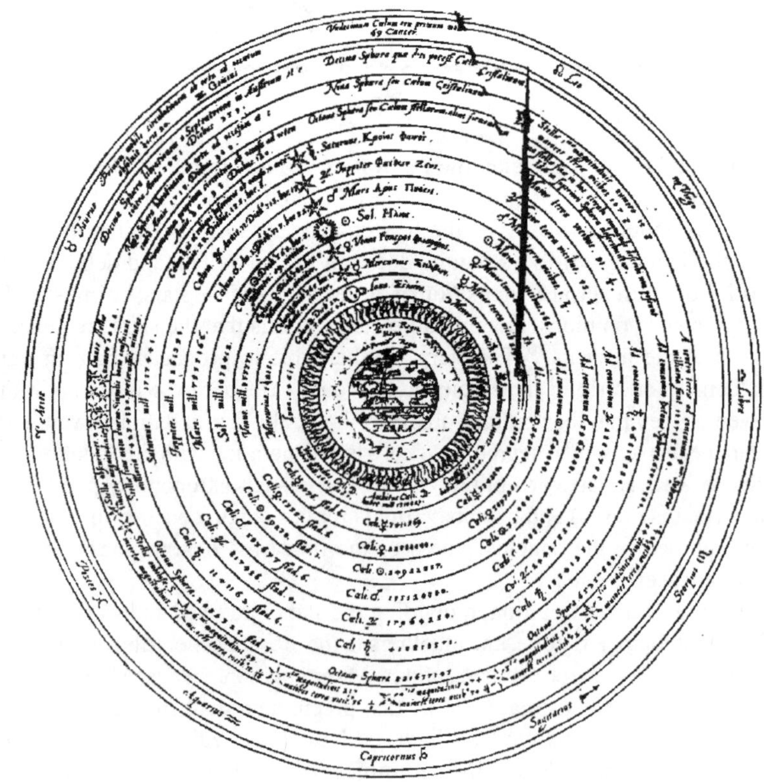

that is, they can be found above, below, and next to the Sun. Thus, the center of their orbs must be the Sun; any other arrangement would require the interpenetration of orbs, causing a vacuum—and this is impossible in nature. According to du Chevreul, only the astronomers of his generation, using an optical instrument that can detect more stars in the Milky Way and other parts of the firmament, can see that Venus and Mercury are located next to the Sun, above, and below it. Venus and Mercury thus orbit the Sun (as the Moon orbits the Earth) within the Sun's heaven.[26] The situation is similar to that of Galileo's Medicean stars around Jupiter and the two planets circling Saturn. The same is true for the thirty Bourbon planets or "shadows" around the Sun, which Jean Tarde had argued are neither sunspots nor comets in the highest region of air. If Tarde's argument turned out to be right, there would be 43 planets,[27] but only 5 planetary heavens would be needed. du Chevreul's five heavens are in order: 1) of the Moon; 2) of the Sun, consisting of the Sun itself in the middle of its heaven, surrounded by the Bourbon stars, Mercury, and Venus; 3) the heaven of Mars; 4) that of Jupiter, surrounded by the four Medicean stars; and 5) the heaven of Saturn, in the middle of which Saturn sits, with two concentric orbs or satellites.

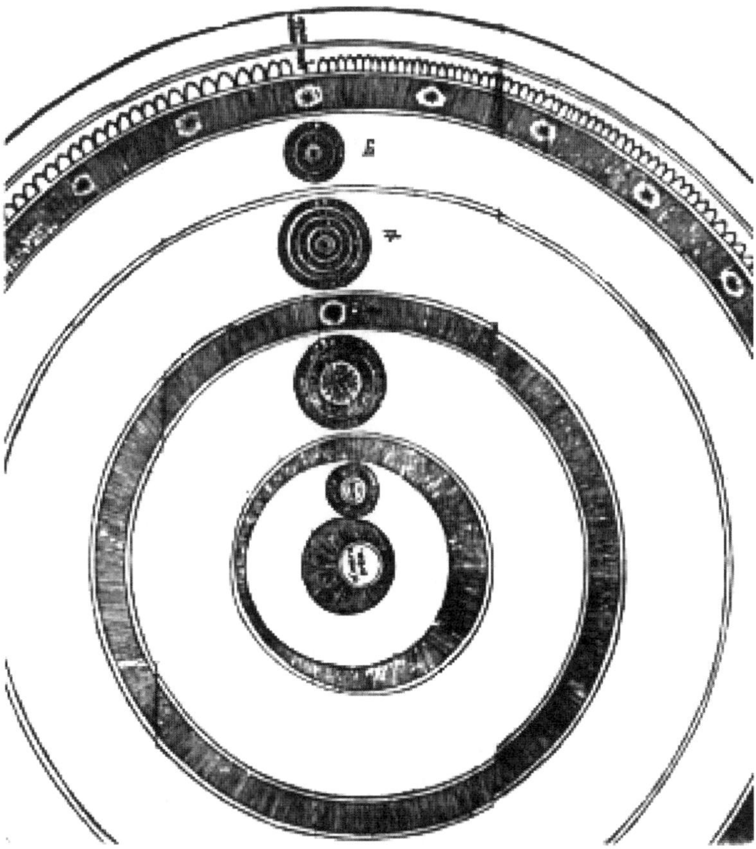

Finally, above the planetary heavens is the firmament and, as required by the Sacred Scriptures, the celestial waters above the heavens and the Empyrean Heaven, the resting place of the Blessed.[28]

It is not hard to see that du Chevreul's sphere is a post-Galilean universe. He has managed to accept all of the observations made by Galileo in 1610–1613 with the assistance of the telescope: more stars than ever seen before, four moons around Jupiter, and the "handles" of Saturn, as described in the *Starry Messenger* (1610); "sunspots" and phases of Venus, as described in the *Letters on Sunspots* (1613). Of course, du Chevreul does not specifically credit Galileo with all these discoveries; also, he interprets the "handles" of Saturn as two moons around Saturn and the sunspots as small planets circling around the Sun. He does not regard these phenomena as evidence for the Copernican opinion. Yet, it is clear that he has modified his Aristotelianism as much as it needed to be modified in order to be able to accept the observations made by the astronomers of his generation.

The one "discovery" of Galileo's that du Chevreul rejects is that there are mountains on the Moon resembling those here on Earth. Consistently with his probabilistic language about the matter of the heavens and its transmission of light, du Chevreul asserts that some parts of the Moon are rarer and denser, transmitting the light of the sun differentially, thereby making it appear as if there were gaps and concavities. If there were such mountains and valleys on the lunar disk, there would be vacuities—and nature abhors a vacuum.[29] In this way he implicitly treats Galileo's "discovery" as an inference based on a false assumption, not a direct observation using the optical tube.

It is important to emphasize that du Chevreul accepts Galileo's other observations within the framework of traditional Aristotelian astronomy, or of what Aristotelian astronomy had become in the 16th and early 17th century, before 1610. This is made quite clear in his chapter on eccentric and epicyclic orbs. There du Chevreul argues for the necessity of eccentrics and epicycles. Eccentrics are necessary because astronomical observations have shown that the parallax of the planets are changeable, that is, their distances from the center of the world vary.[30] Epicycles are necessary because we can observe the planets slowing down and speeding up.[31]

It should also be pointed out that du Chevreul does not accept the Tychonic view of the universe. In fact he formally rejects Tycho's hypothesis. He asserts that Mars cannot be below the Sun, as Tycho would have it, because that would make the heavens permeable and would go against the appearances.[32] Further, in his section on the matter of the world, he had already denied the kind of language the followers of Tycho used, that the stars wander in the heavens like fish swimming in water.[33] Adopting the Tychonic system would have entailed more adjustments to traditional astronomy than du Chevreul was willing to accept. It would have required doing away with eccentric and epicyclic orbs and thinking of the planetary heavens as liquid and permeable. And that is what in fact was to happen in the 1640s and 1650s in Paris, when the Tychonic system became popular. The universe was then described on a three heaven model[34]: the liquid and permeable planetary heaven, the solid firmament or sphere of fixed stars, and the Empyrean. [35]

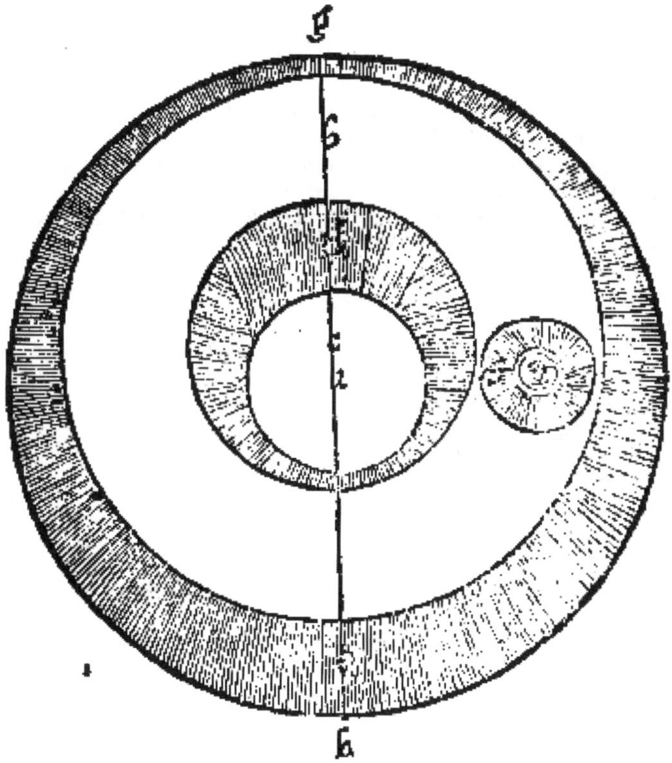

It is clear that du Chevreul produced an extremely interesting cosmology/astronomy, one not normally thought possible, that is, a non-Tychonic system that kept as much of traditional Aristotelian astronomy as possible while making adjustments for the observations of the new astronomy: moons of Jupiter and Saturn, phases of Venus and Mercury, and sunspots.[36] Obviously, eccentric and epicyclic spheres already deviated significantly from properly Aristotelian principles. However, given that thcsc had alrcady been assimilated, then du Chevreul's moves to accommodate 17th century astronomical observations could be seen as solidly in the spirit of Aristotelianism. As we may recall, du Chevreul rejected the Copernican hypothesis because it was contrary to the Scriptures and to physics, but he did not object to the arrangement of the Earth and Moon around the Sun, thinking of it as the terrestrial orb in which the Moon is contained "as if it were an epicycle", together with the four elements. Those epicycle-like systems turn out to be the key to du Chevreul's accommodation of the new astronomy. du Chevreul was a schoolman, but he was not a dogmatic Aristotelian. His attitude is well-epitomized by the following statement from his *Philosophy*: "It is in this manner that Aristotle should be supported in the schools: We should seek out his opinion [on any subject], but on the occasions when his opinions are faulty, we should only embrace what he ought to have thought".[37]

NOTES

[1] I owe many of the details of du Chevreul's biography to Lawrence Brockliss, both from personal communications and through his book, *French Higher Education in the Seventeenth and Eighteenth Centuries: A Cultural History* (Oxford: The Clarendon Press, 1987).
[2] For example, Pierre Gaultruche's *Mathematica* (Caen, 1665) was divided into Arithmetica, Elementale Geometriae, Geometria Practica, Sphaera Mundi, Chronologia, Gnomica, Sphaera Terrestris, Optica, and Musica.
[3] See Lynn Thorndike, *The Sphere of Sacrobosco and Its Commentators* (Chicago: The University of Chicago Press, 1948), pp. 76–142.
[4] See, for example, the 13th to fifteenth century commentaries of Robertus Anglicus, Michael Scot, and Cecco d'Ascoli (Thorndike, pp. 143–246, 247–342, and 343–411, respectively).
[5] *Clavius, Sphaera* (Rome, 1581), pp. 416–442. See also P. Duhem, *SOZEIN TA PHAINOMENA, essai sur la notion de théorie physique de Platon à Galilée* (Paris: Hermann, 1908), Chap. 6, and J. M. Lattis, *Between Copernicus and Galileo: Christoph Clavius and the Collapse of Ptolemaic Cosmology* (Chicago: The University of Chicago Press, 1994).
[6] du Chevreul, *Sphaera* (Paris, 1623), pp. 4–13.
[7] du Chevreul, *Sphaera*, pp. 16–21.
[8] du Chevreul, *Sphaera*, pp. 21–32.
[9] "Hanc opinionem amplexus est Copernicus egregius instaurator Astronomiae", du Chevreul, *Sphaera*, p. 32
[10] du Chevreul, *Sphaera*, pp. 33–34.
[11] "Verum hec opinio omnino coarguenda est erroris ac temeritatis", du Chevreul, *Sphaera*, p. 35.
[12] du Chevreul, *Sphaera*, p. 36. His marginal note refers to Psalms 92 and 103, Ecclesiastes 1, Joshua 10, and Isaiah 38.
[13] See, for example, Tommaso Caccini's Deposition, 20 March 1615, in Maurice A. Finocchiaro, *The Galileo Affair* (Berkeley: University of California Press, 1989), pp. 136–141.
[14] du Chevreul, B. M. Cherbourg, MS 24, fols. 334–335, as quoted in Brockliss, *French Higher Education* in the *Seventeenth and Eighteenth Centuries*, p. 374.
[15] du Chevreul, *Sphaera*, pp. 36–40.
[16] du Chevreul, *Sphaera*, p. 42.
[17] du Chevreul, *Sphaera*, p. 46. For more on this issue, see Roger Ariew, "The Phases of Venus before 1610", *Studies in History and Philosophy of and Philosophy of Science* 18:81–92 (1987).
[18] du Chevreul, *Sphaera*, p. 73.
[19] du Chevreul, *Sphaera*, pp. 47–51.
[20] du Chevreul, *Sphaera*, pp. 70–71.
[21] du Chevreul, *Sphaera*, pp. 72–74.
[22] du Chevreul, *Sphaera*, pp. 80–85.
[23] du Chevreul, *Sphaera*, pp. 80–85.
[24] du Chevreul, *Sphaera*, p. 136.
[25] du Chevreul, *Sphaera*, p. 152.
[26] du Chevreul, *Sphaera*, pp. 153–154. de Chevreul agrees that this is in keeping with Copernicus' conjecture about Mercury and Venus, p. 157.
[27] du Chevreul, *Sphaera*, pp. 154–155.
[28] du Chevreul, *Sphaera*, pp. 156–157.
[29] du Chevreul, *Sphaera*, pp. 166–168. For more on this issue, see Roger Ariew, "Galileo's Lunar Observations in the Context of Medieval Lunar Theory", *Studies in History and Philosophy of Science* 15(3): 213–226 (1984).
[30] du Chevreul, *Sphaera*, p. 176.
[31] du Chevreul, *Sphaera*, p. 181.
[32] du Chevreul, *Sphaera*, pp. 153–154.

[33] du Chevreul, *Sphaera*, p. 72.
[34] For more on this issue, see Roger Ariew, "Theory of Comets at Paris during the Seventeenth Century", *Journal of the History of Ideas* 53: 355–372 (1992) and idem, "Scholastics and the New Astronomy on the Substance of the Heavens", *Descartes and the Last Scholastics* (Ithaca: Cornell University Press, 1999), Chap. 5.
[35] See, for example, Pierre Gaultruche's *Sphaera Mundi*, in his *Mathematica* (Caen, 1665) pp. 103–107
[36] For the standard view of Galileo which allows no room for these maneuvers, see S. Drake, *Discoveries* and *Opinions of Galileo* (New York, Doubleday: 1957; see also N. Swerdlow, "Galileo's Discoveries with the Telescope and their Evidence for the Copernican Theory", in P. Machamer (ed.) *Cambridge Companion to Galileo* (Cambridge: Cambridge University Press, 1998), pp. 244–270.
[37] du Chevreul, B. M. Cherbourg, MS 24, fols. 334–335, as quoted in Brockliss, *French Higher Education in the Seventeenth and Eighteenth Centuries*, pp. 372–373.

MAIJA KALLINEN

LECTURES AND PRACTICES. THE VARIETY OF MATHEMATICAL AND MECHANICAL TEACHING AT THE UNIVERSITY OF UPPSALA IN THE 17TH CENTURY

The status of mathematical sciences in the 16th and 17th-century universities has been a matter of some discussion during recent years. The underlying issue has been the question of whether, and to what extent, early modern universities were able to participate in the mathematization of natural philosophy and in creating prerequisites for the growth of experimental philosophy. Although it is true that theology and humanistic studies dominated learning in most universities outside Italy in the 17th century, and that disciplinary structures were often hierarchic and may have been somewhat resistant to change in this respect, mathematical subjects were nevertheless an important part of their curricula.[1]

The situation varied in different parts of Europe. In England, although the official university statutes scarcely mentioned mathematics, the teaching of mathematics was in fact well established in colleges well before the Civil War. In France, on the other hand, the teaching of mathematics remained on a very modest level despite the statutes of the University of Paris in 1601, which called for lectures in arithmetic, geometry, and astronomy to be given. In continental Europe, mathematics got a much stronger foothold, especially in the Jesuit colleges, but also in Protestant Germany and in the Netherlands. Several posts for mathematicians were established in the relatively new German and Dutch universities during the 16th and 17th centuries, although the extent to which mathematical instruction was actually implemented may have varied considerably from one institute to another.[2]

In Protestant Scandinavia, mathematical studies formed an integral part of academic education. Sweden, though a major political and military power in Northern Europe during the 17th century, was still peripheral in matters of learning. When the University of Uppsala was reorganized and its quiescent life was revived in the 1620s, mathematical studies were given a strong emphasis. Despite the relatively lively interest in mathematics and mechanics in Uppsala and other Swedish universities, none of the five universities in 17th-century Sweden[3] can boast original contributions to (new) philosophy or science (with the possible exception of Olof Rudbeck's discovery of the lymphatic system in 1659). In any case, mathematical instruction during the 17th century certainly prepared the ground for the developments in the following century, when mathematical and experimental approaches to nature pervaded Swedish science.[4]

M. Feingold and V. Navarro-Brotóns (eds.), Universities and Science in the Early Modern Period, 111–125.

The significance of mathematics in 17th century Sweden lies elsewhere. The pursuit of mathematical studies was largely inspired by demands in civil life. Much of mathematical studies in Swedish universities in the 17th century was in one way or another responsive to the needs and interests of the crown and nobility. Mathematical expertise was sought after because of its utility in navigation, commerce, and architecture. Therefore, mathematical sciences, as they were understood in the 17th century, covered not only the traditional subjects of the medieval *quadrivium*: a strong emphasis was laid on practical applications of mathematical knowledge. Geodesy, navigation, architecture, and fortification were taught in universities on a more or less regular basis.

This chapter of course cannot undertake to follow the scope and extent of mathematical teaching in Sweden in detail. Rather, it will focus on certain central features of official attitudes toward mathematics, and of the practical implementation of mathematical and mechanical teaching, as well as pointing out some aspects of the availability of mathematical instruments. The main focus will be on the University of Uppsala, which was the oldest and grandest of Swedish universities, and served as a model when other universities were founded in Sweden.

Mathematical and mechanical teaching took many forms at the University of Uppsala. In the latter part of this chapter I shall discuss the mathematical teaching organized by the polymath Olof Rudbeck. Rudbeck mastered many subjects and played many roles during his lifetime. He was a great anatomist and botanist, a historian, and a musician. But one of Rudbeck's activities, rather untypically for an academic teacher, was built around his mechanical expertise. Rudbeck taught mechanics despite his position as a professor of medicine, and he organized his instruction in formats which administratively were on the fringes of the university. Recent studies in the history of universities have not adequately taken into account such semi-official teaching, which played an important role at least in some early modern universities. Bringing these semi-official practices into focus will undoubtedly add to our understanding of the educational possibilites which 17th-century universities had to offer.

MATHEMATICS IN THE STATUTES

During most of the 16th century, academic education was not available at all in Sweden, and the other levels of education were in a neglected state as well. The University of Uppsala had been founded as a Catholic institution in 1477, but only sporadic marks of any activity existed by 1500, and all its functions seem to have stopped by 1531. The University was refounded as a bastion of Lutheranism in 1593, but with inadequate funds, it was not able to function properly until the royal donations in 1622.[5]

Soon after this the University of Uppsala was given new statutes and privileges; the statutes took effect from summer 1626. They laid a strong emphasis on mathematics, including its practical applications. Three out of the ten professors in the Faculty of Philosophy[6] were to teach mathematics and its applications. First, the so-called *Euclideus* professor was appointed to teach the basic mathematical skills of arithmetic, geometry, and algebra on Ramist principles. However, mathematical instruction was not limited to arithmetic and (practical) geometry. The statutes stressed:

> Firstly, His Royal Majesty wishes the Mathematical profession not to be understood as covering only general Arithmetic and Practical Geometry, with the knowledge of the six great and four minor spheres. But to teach mathematics well and successfully requires teaching Euclid, Archimedes, Ptolemy, Copernicus, Diophantus, Nicomachus, Jordan, Apollonius, Serenus, Theodosius Menelaus, Eutochius, Pappus, Proclos, Theo, Hero, Regiomontanus, Ramus, Schonerus and other such eminent mathematicians.[7]

Thus, two other professors were appointed to take care of instruction in these further topics. The *Archimedeus* professor was expected to initiate his students into optics, music, and mechanics, based on the commentaries on Aristotle's Mechanics and other writings made by Guidobaldi del Monte, Henry de Monantheuil and a certain "Itabius". Finally, the *Ptolemaicus* professor taught not only astronomy (based on Sacrobosco and Peurbach) but also geography from Giovanni Antonia Magini's *Geographia* and architecture from Vitruvius.[8] The statutes list many more authorities for mathematics than for any other subject in the Faculty of Philosophy. In this respect the statutes illustrate the concern and enthusiasm for mathematical studies felt by the prime architect of the statutes and Chancellor of the university, Johan Skytte (1577–1645).

Although the statutes named the mathematics chairs as *Euclideus*, *Archimedeus*, and *Ptolemaicus*, these designations seem not to have come into ordinary use. Martinus Olai Nycopensis, for example, defined himself as Professor in Optics and Mechanics; more generally, the professors were referred to simply as professors of Arithmetic and Geometry, Astronomy, and so on. Sometimes the terms "superior" and "inferior mathematics" (i.e., mathematics of the supralunary and sublunary worlds, respectively) were used as well.[9]

It was not only Skytte and other advisers of the King who thought it important to promote mathematical studies; a more general agreement about the matter existed. Noteworthy is the attitude expressed by the clergy, which presented its scheme for the structure of the university in 1617. In this chapter it was suggested that a professorship in astronomy and another in mathematics should be established. An emphasis on practicality was clear in the view of the clergy as well, for the professor of mathematics was required to teach "Arithmetic and Geometry, Optics, Mechanics and to dispute publicly on optics and meteorology".[10] In Sweden, then, the clergy evidently had no principled disinclination to mathematical sciences.

Practical applications of mathematical knowledge and the ensuing public benefits were cited as justification for the instruction of mathematics at the University of Uppsala. On the one hand this emphasis was conditioned by the practical needs of the state, and on the other hand there were clearly ideological factors which also supported mathematics. The function of the universities in Sweden was mainly to educate priests (about 40–60% of students became priests of some rank), but also to train clerks for the ever expanding bureaucracy. It is notable that the reorganization of the entire educational system in Sweden in the 1610s and 1620s was accomplished simultaneously with major administrative reforms (such as the creation of Courts of Appeal, rearrangement of the fiscal chamber and the beginning of powerful centralization of the administration).[11] Mathematical skills were needed for accountancy, for surveying and the like, so that to be truly competent, civil servants had to be trained in mathematics as well. These goals did not remain unexpressed; the academic community was made clearly aware of them. The statutes make the point explicitly:

> It would be desirable to provide mathematical instruction for use in civil life, such as measuring fields, buildings, mountains, dikes and moats, fields, camps and rivers in their dimensions of altitude, depth and width; for judicial use in the division of properties; for use in architecture and military disciplines; also for merchants' use in navigation and measuring distances; and for calculating the accounts nowadays are in use in His Royal Majesty's Exchequer.[12]

Some attempts were made to maintain the standards and objectives of mathematical instruction, especially during the 1620s. Chancellor Johan Skytte made a visit to the University in 1627, during which he inquired, for example, whether the mathematicians had trained students to be "useful persons", such as bookkeepers or engineers. On this occasion he went so far as to suggest that all unavailing disciplines (except astronomy which he regarded as worthy for its own sake) should be abolished from the university.[13] It seems likely that Skytte gave up this idea, for he did not express such demands during his later years as Chancellor.

Ideological factors also played a role in the introduction of mathematics to the University of Uppsala. First, the Melanchthonian tradition still prevailing within Lutheran universities encouraged the pursuit of mathematics.[14] Secondly, mathematical studies were promoted by Ramism, which still had powerful adherents in Sweden in the 1620s, despite the growing influence of Aristotelianism. Ramus himself had stressed the practical advantages of mathematics as an antidote for what he considered the "obscurity" of scholastic speculations.[15]

Johan Skytte was one of the most prominent Ramists in Sweden, and the statutes of the University of Uppsala reveal his unequivocal support for Ramist educational ideals. For example, the tripartite structure of the mathematical profession, outlined by Ramist authors, was established in the statutes. The importance of mathematics was thought to lay in its ability to train the mind. Clarity of thought was a keyword in Ramism, because clarity brought utility. In the same way logic had to be taught without scholastic ambiguities in order to be useful for the state.[16] Skytte also had some personal contact with the mathematical and mechanical sciences. While studying in Marburg he published a eulogy on the usefulness of mechanics, which was printed together with two appendices written by Ramist mathematicians L. Schonerus and Nils Chesnecopherus (who was also Swedish), and it is notable that some formulations employed in the statutes about the right course of mathematical teaching seem to be more or less direct copies from Chesnecopherus' composition.[17]

The heyday of Ramism was over in Uppsala by the 1630s, although Skytte continued to champion it until the 1640s. In the new statutes of 1655 only vague echos of the Ramist ethos can be found; sciences were to be useful and logic clear. Only two professors for "superior and inferior mathematics" were appointed now, i.e., one for astronomy, and one for arithmetic and geometry. Despite the diminishing of posts, the range of topics to be taught remained the same. Although Ramism had such a strong influence on mathematics in Sweden, Cartesianism did not. Although in France, for example, Cartesianism helped to promote the study of mathematics during the latter part of the 17th century, in Sweden, where Cartesianism was more closely associated with medicine and natural philosophy, it did not significantly enhance the status of mathematics within the academic hierarchy.[18]

TEACHING OF MATHEMATICS IN UPPSALA

Mathematics thus prospered in 17th-century Sweden, but there were certain limitations to the professionalization of mathematicians, which affected mathematical instruction. In the first place, specialization within the field of mathematics was not encouraged; instead, a mathematician was expected to master all topics of his field. This ideal was most clearly brought forward by Chancellor Skytte, who in a sketch for the new Statutes dating from 1620 to 1621 suggested that the three professors of mathematics should regularly rotate their teaching duties.[19] Skytte expressed the same view to the professors during his inspection of the University in May 1627, when he inquired "whether the Mathematicians would not like to change their professions between themselves, so that it would not be tedious always to discuss the same things".[20] The professors, however, seem not to have been delighted with the suggestion, for according to the records they preferred not to comment.

Another kind of change of position was still usual even at the end of the 17th century. A professorship in the philosophical faculty was customarily used as a waiting post for the more respected (and better paid) chairs in the theological faculty. Some professors of mathematics were indeed advanced to the Faculty of Theology, or proceeded to high positions in the Church of Sweden.[21] Johan Bilberg, for example, who was Professor of Mathematics in Uppsala from 1679 to 1690, ended his days as the Bishop of Strängnäs.[22] Taking orders was in fact far from unusual for professors of the Faculty of Philosophy, even though Queen Christina had issued a decree which debarred theologians or members of the clergy from appointment to the philosophical faculty.[23] Assigning a parish and its tithes to a professor's use was an ordinary way of paying his salary all through the 17th century.

Waiting for a theological promotion did not, however, disqualify Bilberg and his collegues as mathematicians; most of them attended to their professional responsibilities quite decently. Nor did the prospect of becoming a clergyman discourage them from adopting radical positions. Bilberg, for example, was one of the most staunch Cartesians in Sweden. Not only were his philosophical views notorious from the clerical point of view, but his theological views were prohibited as well. Only a year before being ordained Bilberg published a thesis attacking the use of scholastic terminology in theology. This thesis was banned soon after its publication, but Bilberg nevertheless got his parish.[24]

It is well known that the actual curricula of the 16th and 17th century universities did not always correspond to the ordinances given by the statutes, and teaching practices often allowed more flexible approaches to philosophy or mathematics than the statutes would have suggest. In Sweden, however, the 1626 statutes especially put such an emphasis on mathematics that we must ask whether the instruction of mathematics could always have fulfilled the statutory requirements.

Only a very general picture of the actual teaching of mathematics in 17th century Sweden can be made. The main sources are the printed lecture announcements, which have been preserved fragmentarily from the mid-1630s on, and academic theses published under the supervision of professors of mathematics. Some notes and fragments deriving probably from mathematical lectures during the latter decades of the 17th century have survived in the Uppsala University Library.[25] It has not been possible to analyze them for this particular chapter, however, and a brief characterization of the topic will have to suffice here.

Published theses written for the master's degree are one of the best sources for our knowledge of certain subjects, for example natural philosophy, in 17th century Sweden. In mathematics, however, most of the theses do not concentrate on mathematical questions, but rather on more general themes in natural philosophy, or ethics and politics.[26] This kind of flexibility was quite usual during the 17th century, but is nevertheless an indication of the relatively nonspecialized character of the mathematician's occupation in the 17th-century academic world. Moreover, not all the theses with names which would suggest a mathematical approach, actually *use* mathematics to argue their views; physical argumentation is used instead. Thus the picture which masters' theses give of mathematical instruction at the University of Uppsala is very defective.

Lectures in mathematics were announced every year in the *Praelectiones publicae Upsaliensi*. It is not easy to determine the level of difficulty in these courses, but it seems that the students would have had a relatively good grasp of mathematic principles before they came to the university. Arithmetic, geometry, and even basic astronomy were taught to pupils in Swedish gymnasiums, some of which provided a very good level of mathematical instruction. It is probable that in some schools the elementary teaching of geography, optics, and mechanics took place as well.[27]

Topics of lectures varied from basic arithmetic to the theory of spheres or planets to optics. According to the first surviving lecture list from 1636 Martinus Olai Nycopensis "continued" to teach the "use and utility" of the octant, and much stress was put on the practical applications of geometry and mixed mathematics during later years as well.[28] The topics varied from year to year, but up to the end of the 1650s a student could participate in courses in clock-making, cartography, use of the astrolabe, civil architecture, or fortification, etc.

Instruments were thus used in the teaching of the so-called mixed mathematics to some extent all through the 17th century. Nevertheless it seems that more emphasis on basic theoretical mathematics was put on the lectures at the University of Uppsala from the 1660s on, although the occasional course on stereometry, gnomonics or—quite often—*computus ecclesiasticus* still took place. A similar pattern emerges from the University of Åbo as well, where geodesy especially was taught on a regular basis.[29] A part of this change of emphasis can undoubtedly be explained by the disappearance of Ramist support for mathematics in the 1655 statutes, where the number of mathematics professors was reduced to two.

Teaching the use of instruments as the statutes demanded, however, was not accomplished easily. There were problems in acquiring the necessary instruments especially in the early 17th century; this situation was made known to the Chancellor both in 1627 and 1638, but it is unclear whether any real improvement resulted.[30] It was not until the appointment of Bengt Hedraeus as a professor of mathematics in 1649 that we know for certain that an instrument-maker was available in the university.[31]

Hedraeus (1608–1659), was professor of "practical mathematics"[32] at the University of Uppsala in 1649–1659. His speciality was surveying, in which he had both theoretical knowledge and practical experience. He was sent by Queen Christina (or rather, by her regency) to the Netherlands to learn geodesy in 1641, and he stayed at the University of Leiden until August 1644. While in Leiden he published a book on astrolabes and quadrants and their use in surveying.[33] According to his own words, Hedraeus presented a new form of astrolabe and of an azimuth quadrant which added accuracy to their use, and the astrolabe was actually made by J. Sneewins in Utrecht.[34]

Back in Sweden, Hedraeus was granted royal privileges to manufacture instruments first in Stockholm in 1648, and soon after his nomination as Professor of Mathematics in 1649, he received the same license for Uppsala as well. The exact nature and extent of his instrument-making is not known, but he probably prepared mostly astronomical and surveying instruments. Hedraeus' instruments were undoubtedly of good quality, as they were still in use at the University of Uppsala in the early decades of the 18th century.[35] Hedraeus also had a small observatory tower built on the roof of his own house in Uppsala. He evidently intended it for instruction as well, but died without having much use of it. Hedraeus' widow was left with the remaining costs, but the University decided that it could not afford to buy the tower with its quadrant (no telescope is mentioned in this connection).[36]

One of Hedraeus' pupils, Johannes Hoffwenius, was considered a promising maker of mathematical instruments; after Hedraeus' death he received a scholarship from the University of Uppsala in 1661, followed by a royal privilege to make instruments for both Stockholm and Uppsala in 1663. Not much is known of his activity, though. Other instrument-makers of the university have remained equally obscure.[37] The availability of instruments was not always very good, despite the presence of instrument makers such as P. J. Thelott, whom Olof Rudbeck invited to Uppsala around 1670. In 1671 Jacob Fornelius announced his plan to teach about eclipses "as far as the availability of instruments allows it".[38] The university did not provide facilities for the use or storage of the instruments. A closet for the storage of instruments was planned, but nothing came of it. During the 1670s professors teaching mixed mathematics or surgery regularly applied for a permission to teach at home, because "the use of instruments was easier there".[39]

OLOF RUDBECK (1630–1702) AND MIXED MATHEMATICS[40]

Rudbeck began teaching practical mathematics in the late 1650s or early 1660s, and continued until his death in 1702. He was not an obvious person to teach mathematics at all, for he occupied one of the two chairs in Medicine at the University of Uppsala from 1660 until his death in 1702. Nevertheless, Rudbeck seems to have been as proud of his achievements as a technician as he was of the results of his anatomical and historical research. It can be said that one of the principal roles in which he presented himself was that of a technician or a practical man. For example, the only stories we know about his childhood emphasize his mechanical ability; as a child, we are told, he built a wooden clock which had a mechanism to strike on the hour. Rudbeck emphasized his practical efforts in various connections, including his work as a carpenter in the construction of the anatomical theatre and botanical garden.[41]

Rudbeck's courses covered all important technical skills: surveying, building water supplies and sewage systems, artillery, fortification, house building, mechanics, gnomonics, and geography. Copies of his notes for these courses have survived to the present. Rudbeck probably gave courses in shipbuilding, agriculture and forestry, the building of locks and mills and gardening as well, but no written material for these has survived. The manuscripts may well have been lost in the great fire of Uppsala in May 1702, when Rudbeck's house also burnt down.[42]

We have seen that Rudbeck was preceded by Hedraeus and others in the teaching of mixed mathematics at the University of Uppsala, and there was no acute lack of such instruction during the time he taught these subjects. Anders Spole, in particular (who became Professor of Astronomy on Rudbeck's recommendation in 1679), was a mathematician of wide competence who built his own astronomical instruments and also taught geography, methods for measuring time, navigation, and surveying. As a tutor of two young noblemen Spole had traveled extensively in Europe, and he mentions with pride in his autobiography his contacts with many of the famous mathematicians of the day, including John Wallis and Isaac Barrow, G. P. de Roberval and Giovanni Riccioli.[43]

Rudbeck's teaching of practical mathematics was nevertheless somewhat exceptional in Uppsala not only because of his range of topics (no one else is known to have taught shipbuilding or forestry) but also because of the extent to which he combined practical training with theory. It was indeed one of Rudbeck's main tenets that theory and practice should be intertwined in the teaching of mixed mathematics and mechanics. This principle he had learned from courses which he had attended at the University of Leiden, given by Frans van Schooten Jr. in 1653–1654. Rudbeck declared that van Schooten's classes had served as the model for his own teaching of engineering skills.[44]

It was from the Netherlands that Rudbeck claimed to have adopted the idea that the traditional skills of, say, surveying and chronology, were not the only suitable topics for academic education, but that other practical skills should be taught at universities as well. In 1685, when Rudbeck defended the necessity of teaching fencing, riding and modern languages at the University of Uppsala, he described his views as follows (in a free translation from the Swedish original): "An academy is not a trivial school where one learns how to read, neither is it a gymnasium where only such topics are studied as are useful to preachers, but an academy is built for everyone who is going to carry on some occupation or trade in the state, whether of a clerical or a secular kind, civil or military. And I really mean any kind of a trade, such as a mason, carpenter, builder, smith, or water supplier. When I was staying at the University of Leiden, the Professor of Mathematics lectured on all aspects of practical mathematics in Dutch, and men of various occupations came to learn the sound fundamentals of their craft, so that they would not so often pursue projects in vain and waste time and money".[45] Unsurprisingly, Rudbeck's ideas did not strike a chord with his colleagues.

Because only one version of the texts for Rudbeck's courses has survived, it cannot be determined how much and in what way the content of his lectures may have changed over the 40 years in which he taught them. It seems that his teaching covered the theoretical basics of these subjects, and that he did not make any major technical innovations, but rather applied existing techniques to different projects.[46] The vernacular was used in the "mechanical colleges", instead of Latin which was used in all other courses, public or private. In this Rudbeck consciously followed the example of van Schooten's courses.

It can be asked in what sense Rudbeck's courses were a part of the functions of the University of Uppsala. The courses—or "colleges", as Rudbeck preferred to call them—were given as private lectures. Private lectures as such were an integral part of the official teaching at the university, as the professors were expected to lecture not only publicly but also privately on more specialized topics. However, Rudbeck's mechanical colleges were never mentioned in the printed lecture lists, as many of the other private courses were. Rudbeck's lecture announcements were always extremely brief, seldom more than one line, whereas other professors were much more verbose.[47]

The omission of Rudbeck's courses from the lecture register does not necessarily indicate that they were not—in a sense—academic activity. Many other academic activities shared this position in Sweden: for example, none of the courses in dancing, fencing, or modern languages were included in the lecture announcements, nor were they referred to in the statutes. Yet these topics were considered an essential part of academic education (at least for the noble students) and the university paid a salary to the teachers of these exercises.[48] There was no special academy for the sons of the aristocracy in Sweden, although many of them visited such schools in Germany. The University of Uppsala was thus responsible for providing instruction for noblemen as well, and this branch of education was much intensified in the 1660s—to a great extent with the help of Olof Rudbeck.[49] No detailed analysis of the position of such exercises in the institutional structure of the university is possible here, and with regard to Rudbeck's colleges we will probably never get a detailed picture because we are simply lacking many important documents.

PRACTICAL TRAINING OF MIXED MATHEMATICS

Basic mathematical skills and theory were thus taught by Rudbeck in private lectures, but practical training took place either in the "mechanics house" or in connection with his numerous building projects. It was characteristic of Rudbeck that projects carried out for the university, the crown and the town of Uppsala meshed quite seamlessly with his private enterprises.

The mechanics house (or mill) had been built by Rudbeck in 1669–1670, and included drills, polishing wheels for copper, fulling mills for leather and cloth and turning machines—and of course the water mill itself represented the basic power supply technology of the time. The students could see the machines functioning and inspect the mechanism. At the same time the mill also functioned as a profitable workshop and manufactory employing craftsmen.[50] It is not known in what way instruction and manufacturing really functioned together, or how the daily routines of the mill were organized.

Rudbeck also planned and administered numerous building projects. He often justified these projects by referring to their utility and to the common good, and for achieving these, practical mathematics and engineering skills obviously had their uses.[51] Some tasks within these projects were assigned as exercises to the students, who at the same time provided Rudbeck with a competent workforce. For example, Rudbeck worked as master builder in charge of all the repair work done on the royal castle at Uppsala, and he designed the anatomical theatre at the University of Uppsala, as well as various private houses. He was responsible for the building of a watermill for the Chancellor of the University, the designing of a new garden for the castle of Uppsala, and the repair of the university's mill and the towers of the cathedral. It was on his initiative and under his guidance that a suspension bridge was built over the river Fyris in Uppsala, waterpipes were installed from the castle down to the town, a postal service was established between Uppsala and Stockholm, a new organ was built in Uppsala cathedral, new compasses were made for the Swedish navy, a canal-system with locks was planned connecting Stockholm with Gothenburg, and so forth.[52] Rudbeck did much business through these projects, and all these enterprises made him a wealthy man.

The idea of teaching mechanics as well as many of Rudbeck's building projects can be seen as part of his wider campaign to renew both the established practices of academic learning and the physical environment in which he lived and worked. Late in 1661 or early in 1662 he presented a list of suggestions to the Chancellor, De la Gardie, for various improvements within both the university and the town of Uppsala. These included paving the streets, building pipes for water supply, procuring fire-extinguishing equipment and building a dormitory for students. He also suggested various measures for checking on the diligence of both students and professors and for raising the standards of learning. Most importantly, he promoted the construction of the anatomical theatre and the botanical garden.[53]

Not much is known about Rudbeck's students and their background, because the participants in his private courses were not usually registered. It has been estimated that Rudbeck had some three to four technical students at a time. A few young noblemen are known to have studied under his guidance, but the majority of the students were sons of wealthier merchants and other burghers. The mechanics house on the other hand was occupied by two groups of people: university students and ordinary craftsmen.[54]

As far as the craftsmen were concerned, there was a social dimension in Rudbeck's policy. It had originally been his idea that the artisans working in the mechanics house would be recruited from the local orphanage. All the boys in the orphanage would be taught a handicraft, and as they grew big enough, they could be employed in the mechanics house or in some other form of manufacturing. It seems however, that this scheme did not function as effectively as it was supposed to. It is not altogether clear how many of the artisans were actually employed from the orphanage, if any.[55]

The system of learning in the mechanics house was very hierarchical. University students were naturally supposed to learn the theoretical background to each art, but they did not necessarily have to master all aspects of the technical performance of the work. On the other hand, the handworkers were "freed" from the duty of learning the theoretical aspects.[56] Thus, the integration of theoretical and practical teaching had its limits. Nor was the mechanics house intended to blur the borderlines between different social groups; rather it helped each social group to master the skills predetermined for them by their status.

The division of labor at the mechanics house was well in line with the educational views of Rudbeck's patron, Magnus Gabriel De la Gardie, who was Chancellor of the University of Uppsala in 1654–1686. De la Gardie expressed his views on education in a memorandum which he sent to his brother-in-law, King Karl X Gustav, in 1655. De la Gardie's aim was to renew the educational system without disrupting the hierarchic organization of the society. According to De la Gardie and most other members of the aristocracy, too many sons of the burghers came to universities and thereby gave up their productive occupations as artesans or as merchants. This was seen not only as an economic problem, but also as a threat to the balance of the estates. During the reign of Queen Christina the number of aristocratic families had risen rapidly, and many meritorious civil servants and army officers had been raised to the nobility. To stop the inflation of titles, De la Gardie strove to implement an educational policy which would limit the access of students stemming from the peasantry and the bourgeoisie to the *studium politicum*—those studies at the university which led to state offices. At the same time he argued that the prospects for life in burgher occupations must be made more attractive.[57]

The technical teaching provided in Rudbeck's mechanics house contributed to this: he trained both simple craftsmen and more educated architects and technicians for burgher careers which were considered useful to the nation's economy but from which elevation to a knighthood was improbable. On the other hand, a certain level of mathematical and mechanical skills, especially those useful in a military career, was considered fitting for a young nobleman.[58]

Rudbeck's mechanical teaching also helped to fill a gap in the Swedish educational system. There were no special schools for engineering organized by the Swedish army as was the case with the technical military academies in some other European countries. In 17th-century Sweden military technology, surveying etc., was usually learned by apprenticeship in the respective state offices. The army employed all the technicians it produced, and there was no surplus. At the same time the need for technical generalists, especially engineers and architects, was growing. The copper and iron mining industries were intensifying, and on the other hand, the Swedish aristocracy was spending more money on building manors, palaces and adjoining gardens than ever before; designing and carrying out such projects in an appropriate style, with all the necessary decorations, fountains, glasshouses, etc., was virtually impossible without proper experts and technicians. The late 17th century demand for engineers should not be exaggerated, however, and there were years when working opportunities were scarce for technicians, which slowed down the growth of the technical profession.

Rudbeck himself designed and supervised many building projects for members of the Swedish nobility. To mention just one example, Rudbeck had become quite a good gardener while constructing the Botanical Garden during the late 1650s; his skill in building greenhouses, in particular, made him a sought-after expert in garden architecture. In the 1660s and 1670s in particular, Rudbeck was practically the only person in the country who knew all the niceties of building warm greenhouses for exotic plants and of the cultivation of the plants themselves. Several orangeries were built for Swedish country houses under his direction, especially for three manors owned by his patron M. G. De la Gardie.[59] The obviously very close and fruitful patron–client relationship between De la Gardie and Rudbeck still awaits detailed study. It is clear, however, that Rudbeck's main "stock-in-trade" for his patrons, so to speak, was his technical expertise. This may also in part explain his determination to maintain the image of a practical man who fully mastered his technology.

From a more general point of view Rudbeck's "colleges" in mixed mathematics are a striking example of how such half-official activities could, within the limitations of the context, significantly complement the more official supply of instruction. With the help of these courses the university was able to adapt more flexibly to new demands. Teaching in mechanics, however, did not continue in the same way after Rudbeck's death, and the mechanics house was soon abandoned. The success of Rudbeck's teaching and his various projects within both the town of Uppsala and the University were intimately bound up with his personal abilities and his influential relations of patronage. It has been said that Rudbeck became a one-man institution, and such institutions can seldom reproduce themselves.

NOTES

[1] For an overview of 17th-century universities see J. Gascoigne, "A Reappraisal of the Role of the Universities in the Scientific Revolution", in D. C. Lindberg and R. S. Westman (eds.), *Reappraisals of the Scientific Revolution* (Cambridge: Cambridge University Press, 1990). R. Tuck, "The Institutional Setting", in D. Garber and M. Ayers (eds.), *The Cambridge History of Seventeenth-Century Philosophy*, Vol. I (Cambridge: Cambridge University Press, 1998), pp. 14–23.

[2] M. Feingold, *The Mathematicians' Apprenticeship: Science, Universities and Society in England 1560–1640*, passim. M. Feingold, "The Mathematical Sciences and New Philosophies", in N. Tyacke (ed.), *The History of the University of Oxford*, Vol. IV (Oxford: Clarendon Press, 1997), pp. 363—369; L. W. B. Brockliss, *French Higher Education in the Seventeenth and Eighteenth Centuries* (Oxford: Clarendon Press, 1987), pp. 381–382; Gascoigne, 221–224.

[3] The five Swedish universities were the Universities of Uppsala, Tartu, Åbo (=Turku), Lund and the University of Greifswald, which became under Swedish dominion in the 30 years war.
[4] S. Lindroth, *Svensk Lärdomshistoria: Frihetstiden* (Stockholm: Norstedts 1978, reprint Södertälje: Fingraf Tryckeri, 1989), pp. 48–66, 297–413.
[5] S. Lindroth, *Svensk Lärdomshistoria: Medeltiden, Reformationstiden* (Stockholm: Norstedts, 1975, reprint Södertälje: Fingraf Tryckeri, 1989), pp. 127, 137, 340–346; S. Lindroth, *Uppsala Universitet 1477–1977* (Uppsala: Almqvist & Wiksell, 1976), pp. 41–44.
[6] I'm using the term "Faculty of Philosophy" rather than the term "Faculty of Arts", because the former corresponds better to Swedish usage in the 17th century.
[7] Statutes for the University of Uppsala 1626, in C. Annerstedt, *Upsala Universitets Historia: Bihang I: Handlingar, 1477–1654* (Upsala: W. Schultz, 1877), p. 277; Principio vult S.R.M: tas Mathematicam professionem, non qvod vulgus professorum existimat esse Arithmeticam quamcumque vulgarem et Geometriam Practicam cum sphaerula sex maiorum, quatuor minorum peripheriarum sed ad mathematicam cathedram bene beateque illustrandam opus esse Euclide, Archimede, Ptolemaeo, Copernico, Diophanto, Nicomacho, Jordano, Apollonio, Sereno, Theodosio Menelao, Eutochio, Pappo, Proclo, Theone, Herone, Regiomontano, Ramo, Schonero et id genus alijs authoribus probatissimis.
[8] Statutes 1626, in Annerstedt, *Bihang I*, 277–278. I have not been able to identify the suggested "Itabius". It probably is a misprint for J. Stabius, an Austrian Cosmographer from the early 16th century. The authority for music is Freigius, and Frederic Reisner is recommended for optics. See also S. Lindroth, *Svensk lärdomshistoria: Stormaktstiden* (Stockholm: Norsteds, 1975, reprint Södertälje: Fingraf Tryckeri, 1989), pp. 468–469.
[9] *Praelectiones publicae Upsaliensi* (Uppsala, 1636–1700). On Nycopensis see the years 1636, 1639, and 1645.
[10] A report by the Clergy on how the University should be organised 10.9.1617, in Annerstedt, *Bihang I*, 141–143.
[11] Lindroth, *Uppsala Universitet*, 41; I. Mäntylä, "Suurvaltakausi", in S. Zetterberg (ed.), *Suomen historian pikkujättiläinen* (Porvoo: WSOY, 1997), pp. 185–186.
[12] Statutes 1626, in Annerstedt, *Bihang I*, 278: Praestat vero accommodare praecepta mathematica ad civilem usum in agrorum dimensione altitudinis, profunditatis, latitudinis, aedificiorum, montium, fossarum, camporum, castrorum, fluminum; ad usum judicum in divisione bonorum, architectonicae et militaris disciplinae; mercatorum quoque in navigationibus, in locorum distantiis et supputationibus rationum, qualis hodie in S.R.M:tis Camera in usu sunt.
[13] Meeting of the Senate of the University and the Chancellor on 26.–28.5.1627, in *Uppsala Universitet: Akademiska konsistoriets protokoll I, 1624–1636*, Acta Universitatis Upsaliensis, Skrifter rörande Uppsala universitet C. Organisation och historia 18:1, H. Sallander (ed.), (Uppsala, 1968), pp. 15–16.
[14] Gascoigne, The role of the universities, 221–222. On the Melanchthonian educational programme see also S. Kusukawa, *The Transformation of Natural Philosophy: The case of Philip Melanchthon* (Cambridge: Cambridge University Press, 1995), passim.
[15] P. Ramus, *Scholarum mathematicarum libri unus et triginta* (Francofurti: Andreas Wecheli heredes, 1599), Liber II.
[16] L. Schonerus, "Praefatio", in P. Ramus (ed.), *Arithmeticae libri duo: Geometriae septem et viginti* (Francofurti, Andreae Wecheli heredes, 1599), b^{r-v}. Statutes 1626, in Annerstedt, *Bihang I*, p. 280.
[17] J. Schroderus, *Dissertatio Mathematica de Mechanicae Artis praestantia, nobilitate, emolumentis ac fundamentis, adversus Aristippos & Epicureos philosophastros* (Lemgoviae, 1598) Skytte was called Schroderus before rising to nobility. See also Chesnecopherus' congratulations to J.S., *Ibidem*, E3.
[18] Brockliss, *French Higher Education*, p. 382. Gascoigne, The role of universities, p. 233.
[19] An outline for the Statutes of the University of Uppsala written by Johan Skytte 1625, in Annerstedt, *Bihang I*, pp. 166–167.

[20] Meeting of the Senate of the University 26.–28.5.1627. *Akademiska protokoll I*, p. 17.
[21] M. Kallinen, *Change and Stability: Natural philosophy at the Academy of Turku, 1640–1713* (Helsinki: Hakapaino, 1995), p. 67.
[22] A. Nilsson, "Johan Bilberg", in *Svenskt Biografiskt Lexikon*, Band 4, (Stockholm: A. Bonniers Förlag, 1924), p. 310.
[23] Queen Christina's letter to the University of Uppsala 8.11.1651, in Annerstedt, *Bihang I*, p. 398.
[24] J. Bilberg, *De theologia Christiana* (Uppsala, 1691). Nilsson, Bilberg, pp. 310–314.
[25] See, e.g., the manuscripts A501, A503, A505, and A507 in the General Collection (Allmänna Samlingen) at the Uppsala University Library.
[26] J. H. Lidén, *Catalogus disputationum in Academiis et Gymnasiis Sveciae* (Upsala: Edman, 1778). On Bilberg, 67–73, Bureus, 88, Magnus Celsius, 93–94, Fontelius, 188–190, Fornelius, 194, Gestrinius, 207–210, Nycopensis, 371–373, Stenius, 441–442, Spole, 433–436, and Vallerius, 499–505.
[27] N. V. E. Nordenmark, *Astronomins historia i Sverige intill år 1800*, Lychnos-bibliotek 17:2 (Uppsala: Almqvist & Wiksell, 1959), pp. 68–72; K. W. Herdin, "Olof Rudbeck d.ä.s födelse och tidigare ungdom", *Rudbecksstudier: Festskrift vid Uppsala universitets Minnesfest* (Uppsala: Almqvist & Wiksell, 1930), pp. 13–14.
[28] The oldest surviving lecture announcement of the University of Uppsala 20.11.1636, in Annerstedt, *Bihang I*, p. 329. *Praelectiones publicae Upsaliensi*, 1639–1659.
[29] *Praelectiones publicae Upsaliensi*, 1660–1700. *Elenchus Praelectionum publicarum* (Aboae 1659, 1665, 1667, 1673, 1688–1690, 1692). Two textbooks in chronology appeared in Uppsala at the end of the 17th century. Nils Celsius, *Computus Ecclesiasticus* (Upsala, 1688). Anders Spole, *Computus Ecclesiasticus, Multoties postquam Praelectus publice, ad Multorum desideria typis exscriptus* (Upsaliae, 1692).
[30] Annerstedt, *Bihang I*, pp. 294, 341.
[31] Lindroth, *Stormaktstiden*, pp. 472–473.
[32] This term was used in the *Praelectiones publicae Upsaliensi* 1651.
[33] B. Hedraeus, *Nova et Accurata Astrolabii Geometrici Structura . . . nec non Quadrantis Astronomici Azimuthalis, Quo non solum prima, sed & singula minuta secunda distinctè observari possunt. Una cum Utriusque usu, Claris & Perspicuis exemplis illustratio* (Lugduni Batavorum: W.C. Boxius, 1643). In addition to this publication Hedraeus has left behind a manuscript *De triangulis planis*, Uppsala University Library manuscript A501.
[34] Hedraeus, *Nova Astrolabii Structura*, "Ad Lectorem": Ne dicat quispiam cum Platone Ideas me delineasse, expertum subindicabo Artificem, qui ad divisionem hanc meam accuratum mihi confecit Astrolabium; Ultrajaceti est *J SneeWins*, qui experientiâ in aliis instrumentis Mathem. fabricandis summè laudandâ, & in hoc, cujus hîc descriptionem exhibeo, industriâ singulari. See also E. Zinner, *Deutsche und Niederländische Astronomische Instrumente des 11.–18. Jahrhunderts* (Munich: C.H.Beck'sche Verlagsbucchandlung, 1956), pp. 224, 537.
[35] A. Losman, "Hedraeus, Benedictus Christierni", *Svenskt Biografiskt Lexikon*, Band 18 (Stockholm: Kungliga Boktryckeriet P.A. Nordstedt & Söner, 1971), pp. 501–502.
[36] Meetings of the Senate of the University, 18.7.1660, 16.1.1661, in *Akademiska protokoll V, 1656–1660* (Uppsala, 1970), pp. 240–241 and *Akademiska protokoll VI, 1661–1663* (Uppsala, 1971), p. 15. Nordenmark, *Astronomins historia*, p. 47.
[37] Meetings of the Senate of the University, 17.1.1661, 6.9.1671, in *Akademiska protokoll VI*, 21–22. *Akademiska protokoll IX, 1671–1672*, (Uppsala, 1972), 141. The latter meeting suggests that one Olof Erson should get the salary which was earlier paid to Johannes Hoffwenius. See also Nordenmark, *Astronomins historia*, 47–48.
[38] Lindroth, *Stormaktstiden*, 473. *Praelectiones publicae Upsaliensi*, 1671.
[39] Meetings of the Senate of the Academy 16.2.1667, 20.4.1672, 12.3.1673, 26.10.1673, 5.11.1673, 28.3.1677, 18.11.1677, *Akademiska protokoll VIII*, 23. *Akademiska protokoll IX*, 245. *Akademiska protokoll X, 1673* (Uppsala, 1972), 62, 215, 223. *Akademiska protokoll XI, 1674–1675* (Uppsala, 1973), pp. 278, 362.

[40] This chapter is indebted to Per Dahl's and Sten Lindroth's studies on Rudbeck. Per Dahl, *Svensk ingenjörskonst under stormaktstiden: Olof Rudbecks tekniska undervisning och praktiska verksamhet* (Uppsala: TK-tryck, 1995). Lindroth, *Stormaktstiden*, pp. 284–305, 413–446.
[41] Herdin, Olof Rudbecks födelse och ungdom, 14. Rudbeck's letter to the committee of inspection on the economic state of the University 1.3.1685, in *Bref af Olof Rudbeck d.ä. rörande Upsala universitet III*, C. Annerstedt (ed.), *Upsala universitets årsskrift 1905*, 215. "I did never get the 30 ducats which had been promised for me for designing the Botanical Garden, nor the 750 thalers for being a master builder, although I had to make the capitals and plane the laths for the greenhouses with my own hands, because there was no carpenter available".
[42] Dahl, *Svensk ingenjörskonst*, 169, 178. Rudbeck, who campaigned for fire protection for the University, nevertheless kept his own library, correspondence and other collections in the wooden part of his residence, not in the more fireproof stone buildings.
[43] A. Spole, *Hans självbiografiska anteckningar*, in P. Wilstadius (ed.), Småländska hembygdsböcker XII (Tidaholm: Tidaholms tryckeri aktiebolag, 1946), pp. 13–17. *Praelectiones publicae Upsaliensi* 1661–1700. Lindroth, *Stormaktstiden*, pp. 501–506.
[44] Dahl, *Svensk ingenjörskonst*, pp. 175–176.
[45] Rudbeck's letter to the committee of inspection of the economic state of the University, 6.4.1685. *Bref III*, p. 257.
[46] Dahl, *Svensk ingenjörskonst*, pp. 178–171.
[47] *Praelectiones publicae Upsaliensi*, pp. 1659–1700.
[48] M. Klinge, "Opetus ja opiskelu" in M. Klinge, R. Knapas, A. Leikola and J. Strömberg, Helsingin yliopisto 1640–1990: Kuninkaallinen Turun Akatemia 1640–1808, (Keuruu: Otava, 1987), pp. 486–494.
[49] Lindroth, *Stormaktstiden*, pp. 41–44.
[50] Dahl, *Svensk ingenjörskonst*, pp. 95–102.
[51] *Ibid.*, pp. 169–170.
[52] A. Hahr, *Olof Rudbeck d.ä. som arkitekt*, in *Rudbecksstudier*, passim. Dahl, *Svensk ingenjörskonst*, passim.
[53] Lindroth, *Uppsala Universitet*, 58. Claes Annerstedt, *Uppsala Universitets historia*, Part II:1 (Uppsala: Uppsala Akademiska Bokförlaget, 1908), pp. 60–83.
[54] Dahl, *Svensk ingenjörskonst,* pp. 257–298.
[55] *Ibid.*, pp. 92–94.
[56] *Ibid.*, pp. 93–95, 259–266, et passim.
[57] S. Edlund, *Magnus Gabriel De la Gardies inrikespolitiska program 1655: Ett bidrag till den ståndspolitiska och pedagogiska debatten under 1600-talet*, Lunds universitets årskkrift, ny följd, Avd.1, 51:1 (Lund, 1954), pp. 52, 62, 82–86, 183, et passim.
[58] L. Canther's letter to Axel Oxenstierna concerning the education of his son Erik, 24.12.1639 in Annerstedt, *Bihang I*, 365–366. Arne Losman, *Carl Gustav Wrangel och Europa* (Uppsala: Almqvist & Wiksell, 1980), pp. 45–54.
[59] Dahl, *Svensk ingenjörskonst*, pp. 63, 71–84.

MARIA TERESA BORGATO

MATHEMATICAL RESEARCH IN ITALIAN UNIVERSITIES IN THE MODERN ERA

Italian mathematical research was essentially linked to the university throughout the modern age, even in periods of crisis such as the second half of the 17th century, and despite outdated teaching programs which were not seriously reformed until the 18th century. We shall focus our attention particularly on the Universities of Bologna and Padua, where there was a greater flow of students and teachers from other European centers, and on two other universities of the Po Valley, Pavia, and Ferrara, which were founded or re-founded at the end of the 14th century.

THE UNIVERSITY OF ARTS AND MEDICINE: TEACHING OF MATHEMATICS, ASTRONOMY, AND ASTROLOGY

The University of Bologna, the oldest in Europe, originated at the end of the 11th century as a center of legal studies: Irnerius and Gratianus, author of the famous *Decretum*, the basic text over the centuries for the teaching of canon law, were among its most celebrated teachers. Later on, besides the University of Jurists there began to develop teaching of liberal arts (arithmetic, geometry, and astronomy) and of medicine: at the beginning of the 14th century the most famous Bolognese doctor and anatomist, Mondino de' Liuzzi, taught in Bologna, and during this century the students of arts and medicine organized into an independent university. Less smooth was the organization of the third one, the Theological University, for which Paris remained the main European reference (Thomas Aquinas was a teacher in Paris).[1]

The Bolognese model was taken as a reference by the first universities of the Iberian Peninsula: Salamanca, Alcalá, Lérida, and Huesca. The Bolognese statutes of Medicine and Arts of 1405 provided for a 4 year course of astrology which also included the teaching of Euclidean geometry, the use of mathematical and astronomical instruments, the study of the Sacrobosco's *Sphaera Mundi*, and the use of the Alphonsine *Tabulae Astronomicae*.[2]

The language of the university teaching was Latin, in many cases until the end of the 18th century. The sources for the teaching of mathematics and astronomy were almost exclusively translations from Arabic into Latin carried out in Spain in the 12th and 13th centuries. Scholars from everywhere in Europe came to Toledo and Seville to learn, from the Muslims, the ancient sciences which the European monastic collections had not been able to preserve. Adelard of Bath translated Euclid's *Elements* into Latin, a

M. Feingold and V. Navarro-Brotóns (eds.), Universities and Science in the Early Modern Period, 127–139.

version which was revised and integrated with comments many times and which also formed the basis of the first printed edition of the *Elements*. Robert of Chester and Gerard of Cremona translated the astronomical and astrological works by Ptolemy into Latin. Only the *Tetrabiblos*, however, became part of the university teaching, whereas the *Almagest* was abridged by Sacrobosco, a professor of the University of Paris, in the middle of the 13th century, and its compendium, the *Sphaera*, was the standard textbook of astronomy for many centuries.

Owing to the precession of the equinoxes, the astronomical tables, which served for the calendar, the horoscopes, and the prediction of the eclipses contained in the *Almagest*, had to be brought up to date, and this, too, took place in Spain, first in Toledo and then under the supervision of Alphonso El Sabio King of Castile. The *Tabulae Alphonsinae*, which would be one of the first textbooks to be set in print,[3] were popularized by several Parisian professors around 1320: Jean de Lineriis, Jean de Muris, and John of Saxony. There is evidence of direct derivations from Spain, one has only to mention Gerard of Cremona and Plato of Tivoli, but the main line of dissemination of Greek and Arabic mathematics and astronomy, through the university network, follows the direction from Spain to Padua by way of Paris. In some cases the name of Oxford must be added to that of Paris, from which the former derived.

The University of Padua was founded after a secession of students from Bologna in 1222; in the 14th and 15th centuries its students had lecturers of mathematics and astrology like Pietro d'Abano, Biagio Pelacani da Parma, and Prosdocimo de' Beldomaldi, all of whom carried out original research work.[4]

At Bologna astrology was taught by Guido Bonatti, Cecco d'Ascoli, Giovanni Aurispa, and Domenico Maria Novara who was the teacher of Copernicus; at Ferrara by Battista Piasi and Pietro Bono Avogaro.

In the 14th and 15th centuries practical mathematics, especially commercial arithmetic, was taught in the Italian abacus schools which made great progress in Tuscany, Siena, and Florence in particular. These schools were linked to commerce and the banking system that sustained it (the Medici were bankers).

Bologna was among the first Italian universities to address the demand for adequate training in arithmetic, particularly for those who were involved in administration of public money, tenders, and salaries and payments. At the end of the 14th century a second teaching position in Mathematics was established; the first holders of the appointment were two Florentine teachers of abacus, Antonio da Firenze and Checco Fiorentino (from 1384 to 1440). Without wishing to belittle the work of the abacus teachers, which is described in the ongoing publication of many unedited documents,[5] the traditional view that little progress was made in arithmetic and algebra between Leonardo Pisano (*Liber abaci*, 1202) and Luca Pacioli (*Summa de arithmetica*, 1494) would appear to be justified. It was only in the university environment that the limit of the general solution of second degree equations was overcome. At the end of the 15th century the Chair of Arithmetic at the University of Bologna was entrusted to Scipione del Ferro who, around 1516, discovered the general formula for the solution of third-degree equations, which had been sought, in vain, by Arabian mathematicians for centuries, and by teachers of abacus for generations.

Girolamo Cardano (1501–1566) and Ludovico Ferrari (1522–1565) were also university academics. Cardano inserted Scipione del Ferro's formula into a corpus of doctrine making it public for the first time in *Ars Magna* (Nuremberg, 1545); Ferrari discovered the formula to solve the general equation of the fourth degree, a critical advance in the study of algebraic equations.[6] Egnazio Danti (1537–1586), one of the creators of modern Italian cartography, taught Mathematics at the University of Bologna. A similar position was held by Giambattista Benedetti (1530–1590) at the University of Turin; Luca Valerio (1553–1618) was lecturer in Mathematics and Greek at the University of Rome from 1591, and Galileo Galilei, who was lecturer at Padua, gave rise to a school which became deeply rooted within the university. The University was important not only for its lecturers: Nicolaus Copernicus is perhaps the most illustrious example of a foreign student who completed his university studies in Italy, graduating in Canon Law at the University of Ferrara in 1503. Federigo Commandino, to whom, above all others, modern mathematics is indebted for his editions of the most important Greek works on mathematics—Euclid, Apollonius, Archimedes, Ptolemy, and Pappus—studied at the University of Padua in 1530s and graduated in medicine at the University of Ferrara, together with the celebrated physician Antonio Musa Brasavola. Torquato Tasso, too, was a lecturer in mathematics at Ferrara for some years.[7]

With the Council of Trent and the Counter-Reformation the privileged bond which linked mathematics to its practical application in astrology was officially severed. The practice of astrology was eventually withdrawn from official university teaching.[8] Mathematics remained as a discipline of the University of Arts and Medicine, but its relation with the Aristotelian corpus (that is Natural Philosophy, Logic and Metaphysics which constituted the main parts of the philosophical disciplines) had to be clarified. There was a serious crisis in the second half of the 16th century as witnessed by the famous *Quaestio de certitudine mathematicarum* which was the theme of various debates.[9]

Astronomy, however, continued to be a "necessary discipline" for the calendar, the prediction of eclipses, the new discoveries which the telescope had revealed at the beginning of the 17th century. The detachment from astrology did not turn out to be the academic disaster that it was first expected.

THE GALILEAN SCHOOL IN THE ITALIAN UNIVERSITY

Galileo began his university teaching in Pisa; in 1592 he was called to Padua. He left Padua in the month of September 1610. Antonio Favaro, the great scholar and editor of Galileo, wrote three richly detailed volumes devoted to Galileo's time at the University of Padua.[10]

Galileo delivered his inaugural lecture on 7 December 1592: the text has been lost, but we know that it was highly considered. On 13 December he began his course, but owing to the great number of students he was obliged to move the lectures to the great hall of the University of Jurists, the largest in Padua. Some university notices (*rotuli*) containing the subjects of Galileo's lessons are known:

1593–1594 *Sphaera* and Euclid
1594–1595 *Almagest* by Ptolemy
1597–1598 Euclid's *Elements* and *Quaestiones Mechanicae* of Aristotle
1599–1600 *Sphaera* and Euclid
1603–1604 *Sphaera* and Euclid
1604–1605 *Theorica planetarum*

Galileo has left us an unequivocal statement about his teaching years at Padua: "the best eighteen years of my life". These were the years of the geometric and military compass, physical equations, and astronomical observation, the telescope and the celestial discoveries announced in the *Sydereus Nuncius* (1610).

The unpretentiousness of Padua's official teaching programs in mathematics had long been compensated for by the private teaching at university level, beginning in the time of Biagio Pelacani. The system was explicitly approved by the Veneto authorities which continued to support it with ordinances and decrees up to the 18th century. Among Galileo's private teachings there is mention of Euclid and the *Sphaera*, as well as the use of the geometric and military compass, fortifications, perspective, mechanics, and cosmography. As university students were often joined by other pupils in the private lessons, it is no easy matter to establish the teacher–student relations within the university, particularly because the biographical references that testify to their presence are not confirmed by the official list of students. However, it is possible to reconstruct a teacher–pupil genealogy stemming from Galileo's teaching in Padua and Florence:

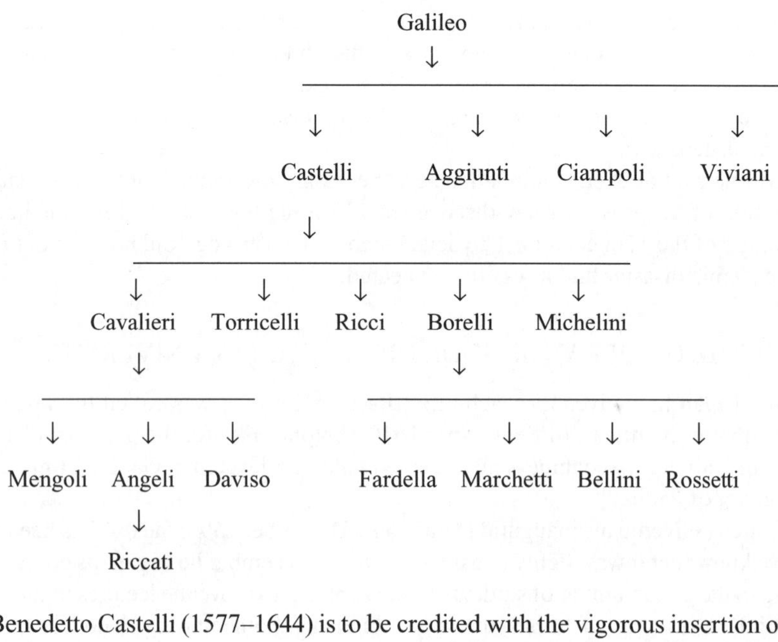

Benedetto Castelli (1577–1644) is to be credited with the vigorous insertion of the Galilean school into the 17th century universities. He studied under Galileo at Padua, and then in Florence. He then went on to become professor at the Universities of Pisa and Rome. At Pisa he had, as a pupil, Bonaventura Cavalieri (1598–1647), and at Rome Giovanni Alfonso Borelli (1608–1679), Evangelista Torricelli (1608–1647), Michelangelo Ricci (1619–1682), and Fabiano Michelini (1593–1666). Under the protection of the Barberini family of Pope Urban VIII, Castelli was removed from Rome during the period of Galileo's condemnation, which Castelli refrained from criticizing publicly. This allowed him to continue teaching mathematics; some of his pupils, particularly in the Piarist order, became highly influential.

Bonaventura Cavalieri became professor at the University of Bologna and among his pupils were Pietro Mengoli (1626–1686) and Stefano degli Angeli (1623–1697). The latter finished his career at the University of Padua where Jacopo Riccati (1676–1754) was his pupil. Giovanni Alfonso Borelli taught for many years at the Universities of Messina and Pisa where he, too, had important pupils who were to go on to university teaching: a few examples are Michelangelo Fardella (1650–1718), Alessandro Marchetti, Lorenzo Bellini, and Donato Rossetti. Galileo himself, when he returned to Florence, continued to do some teaching at the Medici court; among his pupils were Niccolò Aggiunti (1600–1635), Giovanni Ciampoli (1589–1643), and Vincenzo Viviani (1622–1703).[11]

The Galilean school, therefore, had a widespread and lasting influence on the Italian University. In comparison, the experiences of the Lincei Academy, the Cimento Academy, and the Delia Academy of Padua, although important, appear more restricted in time.

A JESUIT SCHOOL

In the 17th century the Jesuits were excluded from teaching in the two oldest Italian universities, Bologna and Padua. In Ferrara they filled the Chair of Mathematics only in 1675 when Francesco Lana Terzi was appointed; but then they maintained it for almost a century. The alliance between local powers and other religious orders restricted the influence of the Jesuits in the Italian universities in the period of their greatest expansion. Notwithstanding this, a few Jesuit colleges, like the Collegio Romano, were to become true universities able to confer professional qualifications (with the exception of jurisprudence). In other places, such as Sassari and Cagliari, there was a sort of osmosis between the universities and the Jesuit colleges. This was so also in Parma, where the Society of Jesus had found protectors in the Farnese family (Paul III Farnese had given official recognition to the Society in 1540). In Parma, indeed, the university became an appendage of the Jesuit College. For almost half a century, after the expulsion of the Jesuits from the Venetian Republic, throughout the Veneto province of the Society of Jesus, the Jesuit College continued to offer university level teaching. A Jesuit scientific school extended from Parma throughout Emilia and then to Rome, and it may be compared in influence to that of the Galilean school. It is true that it mostly operated outside the university, but there was a structured network of higher education colleges which, in the middle of the 17th century, was comparable to the universities of the same region.[12]

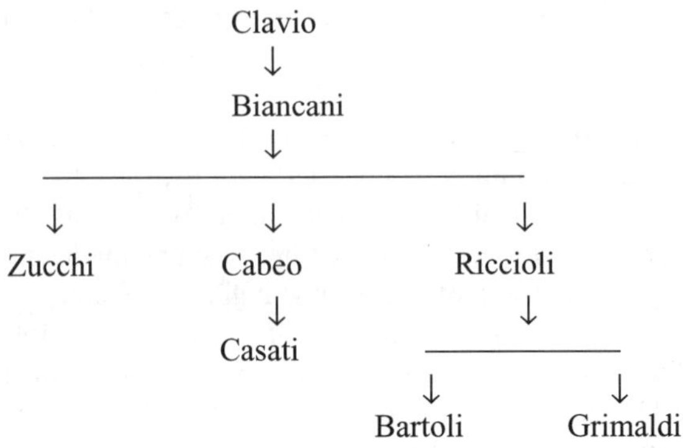

The creator of what may be called the Emilian school of the Society of Jesus was Giuseppe Biancani (1566–1624), a pupil of Cristoforo Clavio (1537–1612) at the Collegio Romano, whose initial relations with Galileo had been good. Although Biancani was not an eminent scientific personality, during his 20 years of teaching at Parma he established a method of study in which experimentation in astronomy and physics was supported by a serious critical study of sources. Biancani's pupils included Niccolò Zucchi (1586–1670), Niccolò Cabeo (1586–1650), and Giambattista Riccioli (1598–1671); Riccioli, in his turn, taught Francesco Maria Grimaldi (1618–1663) and Daniello Bartoli (1608–1685). In Rome and Parma, Paolo Casati (1617–1707) taught for many years and his scientific activity continued almost until the following century.

In the Jesuit College of Parma an early approach to Descartes' *Géométrie* developed, facilitated by the fact that Descartes had been a pupil of the Jesuits in La Flêche, and was taken up by Guglielmo Weilhamer with his pupil Antonio Maria Costantini (1640). The subsequent condemnation of the *Meditationes metaphysicae* and their annotation in the *Index* interrupted this important process of assimilation.

LEIBNIZ, THE INFINITESIMAL CALCULUS AND THE UNIVERSITY

Even if the Jesuits were less hostile to the new analytic methods than was the Galilean school, which was tied to geometric methods, they did not take a great part, at the beginning of the 18th century, in the revival of mathematical research in Italy which took place essentially in the university environment. The trigger of this revival was Leibniz and his differential calculus which radically changed mathematical research in less than 15 years, starting from the publication of his memoir *Nova methodus pro maximis et minimis* in *Acta Eruditorum* of 1684.[13] Having come to Italy to carry out historical research on the common origin of the Dukes of Este and Brunswick, Leibniz stayed in that country from 1689 to 1690 and established profitable relations with members of the intelligentsia in various Italian cities.[14] Of particular interest were those with the University of Padua, where two of the best Bolognese professors were invited to teach: Domenico Guglielmini (1655–1710) and Giambattista

Morgagni (1682–1771). In Padua Leibniz could rely also on the friendship of Michelangelo Fardella (1650–1718), whom we have already mentioned as one of Borelli's pupils.

Leibniz succeeded in obtaining the Chair in Mathematics in Padua University for Jacob Hermann, a pupil of Jacob Bernoulli. In his public and private teaching from 1707 to 1713, Hermann was able to arouse interest in differential calculus in a group of scholars of the Veneto area who were destined to play an important role in Venetian culture and the University of Padua: Giovanni Poleni (1683–1761), Jacopo Riccati (1676–1754), Bernardino Zendrini (1679–1747), Antonio Conti, and Apostolo Zeno. Although Riccati did not have an official position in the university, he was teacher to two important professors of Padua, Giuseppe Suzzi and Lodovico Riva, as well as of Father Ramiro Rampinelli. The latter was the teacher of Maria Gaetana Agnesi (1718–1799), the celebrated authoress of the *Instituzioni analitche ad uso della gioventù italiana* (Milan, 1748), a work which was translated into French and English. Suzzi was the teacher of Giuseppe Toaldo (1719–1797) who, with his nephew Vincenzo Chiminello, controlled the teaching and research of astronomy in the University of Padua for many years, and of Simone Stratico (1733–1824) then professor of mathematics and navigation at the University of Padua and an important scholar of river hydraulics.

Jacopo Riccati also taught three of his own sons, Vincenzo, Giordano, and Francesco, and encouraged them in mathematical research.[15] Vincenzo Riccati (1707–1775), in his turn, had several pupils who were to become distinguished mathematicians: Gianfrancesco Malfatti (1731–1807), a professor at the University of Ferrara for several decades, Girolamo Saladini (1731–1813), active in Bologna also during the Napoleonic period, and finally Virgilio Cavina.

At the beginning of the 18th century the University of Bologna, too, was stimulated by demands for instruction in the new mathematical disciplines. Under the protection of Domenico Guglielmini, a group of young scholars, supporters of this renewal, gathered to learn the techniques of Cartesian geometry and differential calculus and to carry out experiments and astronomical observation: they included the brothers Eustachio (1674–1739) and Gabriele Manfredi (1681–1761), Vittorio Francesco Stancari (1678–1709), and Giuseppe Verzaglia (1669?–1730). Eustachio Manfredi and Stancari each obtained a university chair and began to teach the new analytic methods. The most advanced scholar in this research, however, Gabriele Manfredi, the author of the first European work on differential equations (*De constructione aequationum differentialium primi gradus*), remained on the fringe of the university. Verzaglia spent a period of study in Basel with Johann Bernoulli and then returned to teach in the small University of Cesena, his hometown.

A private institution in Bologna, named simply Instituto, was founded in 1714 thanks to the generosity and participation of Luigi Ferdinando Marsigli; in the first half of the 18th century it was more receptive to scientific progress than the university. The Instituto included the Accademia delle Scienze of Bologna (founded 1690 by a private initiative of the Manfredi family) and the Accademia Clementina (an academy

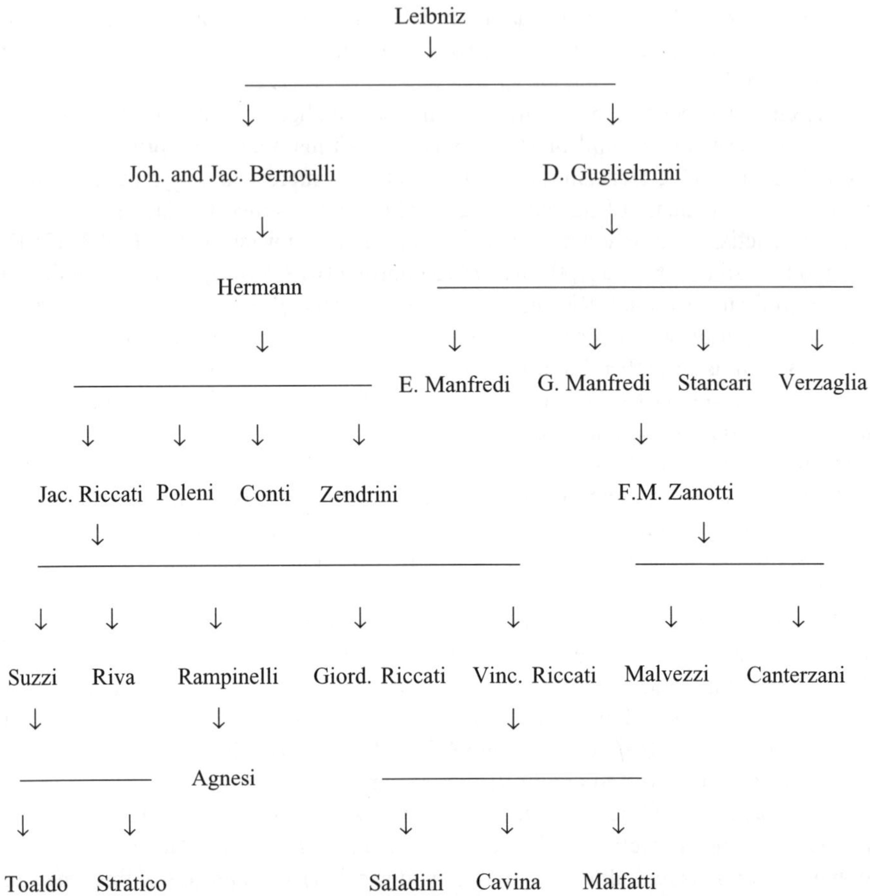

of fine arts) and provided rooms to exhibit objects of natural history and scientific instruments from the Marsigli collection. An observatory was one of its main achievements: its tower was built in 1726 and in 1741 new instruments were acquired from the English maker Jonathan Sisson, and further purchases of telescopes were made from the Dollonds in London in 1788. The observatory was directed by prestigious astronomers like Eustachio Manfredi, Eustachio Zanotti (1709–1782), and Petronio Matteucci (1717–1800). In 1744 the physics laboratories were improved with new equipment, produced by Dutch makers for the teaching of Newtonian physics, and in 1790 the entire famous collection of George Cooper was acquired.[16] Once it was clear that the statutes of the University of Bologna could not be reformed, Benedict XIV (formerly Prospero Lambertini, cardinal of Bologna) donated his important library to the Institute, which became the center of the present University Library. Throughout the 18th century the Institute continued to prepare the best university students for research and had as professors many qualified teachers from the University of

Bologna: besides the Manfredis mention should be made of Francesco Maria Zanotti (1692–1777) and Sebastiano Canterzani (1734–1819).

MATHEMATICS, UNIVERSITY AND REFORM IN THE 18TH CENTURY

When speaking of university and reform in the 18th century a distinction must be made between the general institutional setting and the lectures and their contents.

As far as the institutional background is concerned, almost all universities, like the corporations of arts in the late Middle Ages, resisted every serious reform. The core of this resistance was represented by the privileges of the professional colleges which conferred degrees and licensed practice in medicine and law. Attempts to reform the Universities of Bologna, Padua, Pisa, and Rome failed. In Rome, "La Sapienza" was essentially governed by the Corporation of the Consistorial Advocates of which Benedict XIV was a member.

In Turin, on the other hand, and, above all, in Pavia, with the reform carried out by Maria Theresa of Austria and the subsequent overambitious plans of reform by Joseph II, there was a serious attempt to transform the universities into a real state structure. In Turin this reformism resulted in a bureaucratization of the university and its subjection to the state.

The analysis of the contents and at times the titles of the university lectures, in contrast, are more encouraging: they show a constructive attempt at keeping up with current advances in modern science. We shall examine a few of these changes by going through the notices (*rotuli*) of the University of Bologna, which were publicly displayed at the beginning of every academic year. In this century Bologna was not in the avant-garde of reform, but it was still a reference point of highly qualified teachers for Italy as a whole.

In 1700, beside the teaching of natural philosophy (*Parva naturalia*), metaphysics (books I, VII of Aristotle's *Metaphysics*), moral philosophy, and Scholastic theology (book I of the *Sentences* by Peter Lombard) there were lectures in mathematics (Ptolemaic astronomy, the theory of planets), lectures in arithmetic (which were held at the professor's house), and a course in hydrometry devoted to the study of the movement of waters, which was added in the second half of the 17th century to prepare the jurists and technicians for the regulation of rivers (in particular the Reno): these subjects offered no opening for analytic methods. These were introduced with the Leibnizian calculus in the years 1700–1725.

In 1725–1726 differential calculus and Cartesian geometry were taught in courses on algebra and infinitesimal analysis, whereas Aristotelianism remained in the teaching of natural philosophy (*De physico auditu, De coelo et mundo, De generatione et corruptione*), and logic (*Analytica posteriora*); other courses of mathematics continued to be devoted to the theory of planets. The transformation of natural philosophy into particular physics and general physics occurred in the years 1750–1751: this meant an opening, if not to the Newtonian physics, at least to Cartesian physics; metaphysics survived (pneumatology and ontology) as did the teaching of mathematics and astronomy "juxta antiquiores", which was perhaps one way of introducing a Copernican approach in the form of Pythagorean astronomy. The teaching of analytic geometry and calculus was confirmed and a new teaching of Chemistry was entrusted to Iacopo Bartolomeo Beccari (1682–1766). A teaching of physics was held by a woman, the famous Laura Bassi (1711–1778). The teaching of the Newtonian physics, even if it developed late and through the Cartesian physics which was not quantitative but, however, showed the pre-eminence of mechanics and the role of mathematics, took place in Italian universities earlier than in Jesuit colleges.[17]

There were fewer new developments in the second half of the 18th century until 1798–1799, when theological teaching, which traditionally belonged to the University of Arts and Medicine, was suppressed. In the last decades of the century preceding the arrival of the French troops and the great institutional changes, since the university reform of Benedict XIV had failed, the University of Bologna withdrew into itself, as it had done a century before when it was abandoned by its best teachers.

Also the University of Rome, La Sapienza, had only been partially touched by the reform, whereas slightly more had been carried out in the minor universities of Perugia and, above all, Ferrara, which, in those years, the foreign travelers Lalande and Goethe did not fail to praise.

With the Reform of 1771 three Chairs of Mathematics had been created: one in Algebra and Geometry for Gianfrancesco Malfatti, one in Mechanics and Hydrolics for Teodoro Bonati (1724–1820), and the third in Practical Geometry for Ambrogio Baruffaldi (?–1776).

Simone Stratico's attempts at reform failed in the University of Padua and the decline in studies was partially made up for by the institutionalization of an old academy of Padua, re-founded as the Academy of Sciences which vaguely echoed the Academy of Berlin. Moreover, some changes were carried out in the University of Padua: a second chair of Mathematics was established in 1741, Experimental Physics in 1739 and Experimental Chemistry in 1759.[18]

In Pavia, where the Jesuits had not been excluded from university teaching, (Girolamo Saccheri, Ruggero Giuseppe Boscovich), Maria Theresa of Austria's reforms had a profound effect on the university with positive results on university studies. Following this Gregorio Fontana (1735–1803), Lorenzo Mascheroni (1750–1800), Mariano Fontana (1746–1808), Pietro Paoli (1759–1839), and Angelo Lotteri (1760–1840) were called to teaching positions. Gregorio Fontana was also entrusted with the position of curator of the university library. The most significant fact for mathematical research after the reforms in Pavia was the systematic introduction in Italy of the study of Leonhard Euler's treatises on mathematical analysis. In Pavia, Fontana had the *Institutiones calculi differentialis* published in 1787; Mascheroni wrote some celebrated and original *Adnotationes ad calculum integralem Euleri* (Pavia, 1790–1792). Joseph II also took an interest in the university and created a new College (German-Hungarian) where he tried to establish a permanent teaching for engineers despite tenacious resistance on the part of the professional classes of Pavia and Milan.

The limitations of 18th century reformism must have been many also in Pavia, if one considers that practically all the professors there, starting with Gregorio Fontana and

Lorenzo Mascheroni, enthusiastically supported the new order imposed by Bonaparte in 1796. Among the professors of Pavia we must not forget Alessandro Volta, who terminated his experiments on the battery during the years of the French dominion (1799).

APPENDIX

We have so far described the main lines of mathematical research in Italy from the end of the 15th to the end of the 18th century, mostly carried out in the University of Bologna, Padua, and Pavia. Focusing on the most advanced results obtained in scientific research in the fields of algebra, geometry, and mechanics in the 16th and 17th centuries, and on the analytical methods of Cartesian geometry and Leibnizian calculus in 17th and the 18th centuries, we have neglected several age-old universities which in those three centuries also had teachers of mathematics of great esteem: Naples, founded in 1225, Rome in 1303, Pisa in 1343, and Turin in 1405.

In the 17th century the University of Pisa had many teachers who were exponents of the Galilean school, which was considered in this paper: Benedetto Castelli, Famiano Michelini, Giovanni Alfonso Borelli, Alessandro Marchetti, and in the 18th century, scholars like Guido Grandi and Paolo Frisi. However the University of Pisa, as well as that of Turin, which was the seat of an important School of Artillery and Fortifications in the 18th century, became central to mathematical research above all after the Unification of Italy.[19]

The University of Rome was, at times, a cultural center of mathematics (for example in the 18th century with Paolino Chelucci, François Jacquier, and Gioacchino Pessuti), but in the city of Rome "La Sapienza" had an atypical role of "primum inter pares" among numerous prestigious colleges (the Collegio Romano of the Jesuits, the Clementino of the Somaschi, the Nazareno of the Piarists, the Propaganda fide, etc.). The only prerogative which differentiated the Sapienza was that it could give qualifying degrees to the legal professions.[20]

The University of Naples enjoyed several moments of glory in the field of mathematics (Agostino Ariani, Niccolò and Pietro di Martino, Giuseppe Orlandi, etc.) but, as in Rome, it competed with rival institutions which limited its influence: in the 17th century a powerful Jesuit college, in the 18th century the military schools and the Liceo Convitto del Salvatore. Moreover, in Naples the tradition of private studies was also of great importance: a private *studium* of mathematics was opened in 1753 by Vito Caravelli who had as pupils Vincenzo Porto, Niccolò Fergola, and Filippo Guidi; Fergola, in his turn, opened a famous private *studium*, and another one in 1792 had Carlo Lauberg and Annibale Giordano as teachers.[21]

NOTES

[1] A. Sorbelli, *Storia dell'Università di Bologna. Volume I: Il Medioevo* (Bologna: Zanichelli, 1940).

[2] E. Bortolotti, *La storia della matematica nella Università di Bologna* (Bologna: Zanichelli, 1947).

[3] Venice, 1483.
[4] L. Pepe, "Astronomia e matematica nelle università italiane del Quattrocento" *Luca Pacioli e la matematica del Rinascimento* (Città di Castello, Petruzzi, 1998), pp. 29–40; L. Pepe (ed.), *Copernico e la questione copernicana. Opere della Pubblica Biblioteca di Ferrara* (Ferrara: Corbo, 1993). A description of the astonomical teachings in Bologna can be found in E. Baiada, F. Bonoli, and A. Braccesi, *Museo della Specola* (Bologna: Università degli Studi, 1995).
[5] Thanks to the studies of Gino Arrighi, Angelo Procissi, Ettore Picutti, Laura Toti Rigatelli, and Raffaella Franci of the University of Siena. For these works one can refer to F. Barbieri and L. Pepe (eds.), "Bibliografia italiana di storia delle matematiche", 1969–1990, *Bollettino di storia delle scienze matematiche*, 12: 3–181. See also W. Van Egmond, *Practical Mathematics in the Italian Renaissance: A Catalog of Italian Abbacus Manuscripts and Printed Books to 1600* (Firenze: Istituto e Museo di Storia della Scienza, 1980).
[6] With algebraic operations no general formula for equations above fourth degree can be obtained.
[7] W. Moretti and L. Pepe (eds.), *Torquato Tasso e l'Università* (Firenze: Olschki, 1997).
[8] Although it maintained a large following as shown by the publications of the 16th century.
[9] G. Cesare Giacobbe, "Il 'Commentarium de certitudine mathematicarum disciplinarum' di Alessandro Piccolomini", *Physis* 14: 357–374 (1972) n. 4; P. Freguglia, *Ars analytica, Matematica e methodus nella seconda metà del Cinquecento* (Busto Arsizio: Bramante, 1988).
[10] Published in 1883 by the successors Le Monnier, they have been reprinted: A. Favaro, *Galileo Galilei e lo Studio di Padova* (Padova: Antenore, 1966), 2 Vols followed by A. Favaro, *Galileo Galilei a Padova: ricerche e scoperte, insegnamento, scolari* (Padova: Antenore, 1968).
[11] P. Galluzzi and M. Torrini (eds.), *Le opere dei discepoli di Galileo Galilei, Carteggio*, 2 Vols (Firenze: Giunti-Barbera, 1975–1984).
[12] U. Baldini, *Legem Impone subactis. Studi su filosofia e scienza dei Gesuiti in Italia, 1540–1632* (Roma: Bulzoni, 1992). In the last few years important volumes of syntheses on Jesuit scientific activity have been published: A. Romano, *La contre-réforme mathématique: constitution et diffusion d'une culture mathématique jésuite à la Renaissance* (École française de Rome, 1999); M. T. Borgato (ed.), *Giambattista Riccioli e il merito scientifico dei gesuiti nell'età barocca* (Firenze: Olschki, 2002); G. Paolo Brizzi and R. Greci (eds.), *Gesuiti e università in Europa (secoli XVI–XVIII)* (Bologna: Clueb, 2002); M. Feingold (ed.), *The New Science and Jesuit Science: Seventeenth Century Perspectives* (Kluwer, 2003); M. Feingold (ed.), *Jesuit Science and the Republic of Letters* (Cambridge Mass: MIT Press, 2003).
[13] L. Pepe, "Il calcolo infinitesimale in Italia agli inizi del secolo XVIII", *Boll. Storia Sci. Mat.* 1:43–101(1981), n. 2.
[14] A. Robinet, *G. W. Leibniz Iter Italicum* (Firenze: Olschki, 1988); S. Mazzone and S. Clara Roero, *Jacob Hermann and the Diffusion of the Leibnizian Calculus in Italy* (Firenze: Olschki, 1997).
[15] G. Piaia and M. Laura Soppelsa (eds.), *I Riccati e la cultura della Marca nel Settecento europeo* (Firenze: Olschki, 1992).
[16] Clueb (ed.), *I materiali dell'Istituto delle Scienze* (Università degli Studi di Bologna, 1979); See in particular E. Baiada—E. Braccesi, "Astronomia e Istituto delle Scienze", pp. 248–258; G. Dragonì—V. Pallotti, "Strumenti, didattica e ricerca: la fisica sperimentale nell'Istituto delle Scienze", pp. 217–225.
[17] L. Pepe, "Matematica e fisica nei collegi del Settecento", *Studi Settecenteschi* 18: 407–420 (1998).
[18] P. Del Negro, "I 'Pensieri' di Simone Stratico nell'Università di Padova (1760)", *Quaderni per la storia dell'Università di Padova* 17:191–229 (1984).
[19] C. Maccagni, "La matematica", *Storia dell'Università di Pisa (1343–1737)*, 2 Vols (Pisa: Pacini, 1993), 1st Vol. pp. 331–362; L. Giacardi and S. C. Roero (eds.), *Bibliotheca mathematica. Documenti per la storia della matematica nelle biblioteche torinesi* (Torino: Allemandi, 1987); M. Teresa Borgato and L. Pepe, *Lagrange. Appunti per una biografia scientifica* (Torino: La Rosa, 1990).

[20] F. M. Renazzi, *Storia dell'Università degli Studi di Roma, detta comunemente La Sapienza*, 4 Vols (Roma: Pagliarini, 1803–1806); L. Pepe, "Matematica e fisica nei collegi del Settecento", *op. cit.*
[21] F. Amodeo, *Vita matematica napoletana. Parte prima e Parte seconda* (Napoli: Tipografia dell'Accademia Pontaniana, 1924); R. Gatto, *Tra scienza e immaginazione. Le matematiche presso il collegio gesuitico napoletano* (Firenze: Olschki, 1994). Mathematical teachings in others seats, such as Messina, Modena, Catania, Cagliari, etc., must not be neglected. We refer you to the following volumes: R. Moscheo, *I gesuiti e le matematiche nel secolo XVI* (Messina: Società Messinese di Storia Patria, 1998); O. Montaldo and L. Grugnetti (eds.), *La storia delle matematiche in Italia* (Cagliari: Università di Cagliari, 1984); F. Barbieri and F. C. Degani (eds.), *Pietro Riccardi (1828–1898) e la storiografia della matematica in Italia* (Modena: Università degli Studi, 1989).

LUIGI PEPE

UNIVERSITIES, ACADEMIES, AND SCIENCES IN ITALY IN THE MODERN AGE

In order to explain the centrality of the university system in Italy it may be useful to examine the relations between the universities and the many academies, which were in a somewhat precarious position since they had been unable to obtain continuous public financing. From this point of view the crucial period in Italy was to be in the two decades of Napoleonic government (1796–1814), when the whole education system was financed by the State; not only the universities, but also the academies, particularly the Istituto Nazionale and the Atenei civici. Because of these public funds an increasing number of scientists were able to live on their research and their professions as engineers, doctors, university professors, or public functionaries. In this way the contemporary Italian scientific community was created. During the Restoration of the old governments, however, many scholars were obliged to leave Italy; some, like Ottaviano Fabrizio Mossotti in Argentina and Agostino Codazzi in Venezuela, took part in furthering science in South America.[1]

MATHEMATICAL RESEARCH AND PUBLIC FUNDS

In book VII of the *Republic* (*Rep.* VII, 528) Plato discussed the problem involved in the development of research in solid geometry. The two interlocutors of the dialog (Socrates and Glaucon, Plato's younger brother) agreed that State intervention was necessary in order to advance this research:

> "After plane surfaces," said I; "we went on to solids in revolutions before studying them in themselves. The right way is next in order after the second dimension to take the third. This, I suppose, is the dimension of cubes and of everything that has depth." "Why, yes, it is," he said; "but this subject, Socrates, does not appear to have been investigated yet." "There are two causes of that," said I: "first, inasmuch as no city holds them in honour, these inquiries are languidly pursued owing to their difficulty. And secondly, the investigators need a director, who is indispensable for success and who, to be with, is not easy to find, and then if he could be found, as things are known, seekers in this field would be too arrogant to submit to his guidance. But if a State as a whole should join in superintending these studies and honour them, these specialists would accept advice, and continuous and strenuous investigation would bring out the truth.

M. Feingold and V. Navarro-Brotóns (eds.), Universities and Science in the Early Modern Period, 141–151.

> Since even now, lightly esteemed as they are by the multitude and hampered by the ignorance of their students as to the true reasons for pursuing them, they nevertheless in the face of all these obstacles force their way by their inherent charm and it would not surprise us if the truth about them were made apparent." "It is true," he said, "that they do possess an extraordinary attractiveness and charm. But explain more clearly what you were just speaking of. The investigation of plane surfaces, I presume, you took to be geometry?" "Yes," said I. "And then," he said, "at first you took astronomy next and then you drew back." "Yes," I said, "for in my haste to be done I was making less speed. For, while the next thing in order is the study of the third dimension or solids, I passed it over because of our absurd neglect to investigate it, and mentioned next after geometry astronomy, which deals with the movements of solids." "That is right," he said. "Then, as our fourth study," said I, "let us set down astronomy, assuming that this science, the discussion of which has been passed over, is available, provided, that is, that the State pursues it".[2]

This link between state structures and mathematical research was not new even in Plato's times.[3] Pythagoras had founded a school in Crotone which also took over the government of the city; Archimedes worked closely with the Tyrant of Syracuse; the School of Alexandria, which was the center of mathematical studies for seven centuries (Euclid, Archimedes, Ptolemy, Theon, Pappus), was linked to state structures: the Museum and the Library. Similar considerations apply to arithmetic in the Middle Ages: its transmission to Arabic civilization took place in Baghdad, in a school created under government protection. It was brought to Western Europe through Spanish institutions and the European university network, thanks to the works of Leonardo Pisano. His father was not a merchant, as has often been thought, but an official of the Republic of Pisa, representing its interests in a large commercial city of the Arabic world.[4] Merchants and bankers, then, popularized Arabic arithmetic, but they did not contribute directly to the development of new scientific fields.

In Italy mathematical research was linked to the only educational institution with public financing that is the universities. It is known that the academies in Italy had difficulty in obtaining continuous financing from the State; the Accademia dei Lincei in the 17th century and then the Accademia del Cimento are examples which were limited in time. Throughout the 18th century, in Turin, Naples, Padua, and Bologna the academies attempted—in general with little success—to obtain public financing; not necessarily from the State but also from local administrations. The creation of the Istituto Nazionale and the public financing of the Società Italiana founded by Lorgna in the Napoleonic period were exceptions, which were confirmed by the "refounding" of the Accademia Nazionale dei Lincei after the Unification of Italy. The Italian situation thus differed from that of other European countries, like France and Russia, where the academies were maintained by public funds.

UNIVERSITIES AND ACADEMIES IN ITALY

Italy is home to many of the oldest universities in Europe, beginning with 11th-century foundation of the University of Bologna. Initially the professors were paid by the students, but as time went on public contributions were provided by town communes and Renaissance rulers. Italian universities attracted students and famous teachers from all over Europe. Erasmus graduated from Turin, Copernicus from Ferrara, after having studied at Bologna and Padua. Ludovico Ariosto was a student at Ferrara University and Torquato Tasso a lecturer of mathematics there. Many privately financed academies also sprang up in the 16th century, starting with Florence, Naples, and Rome; they were private gatherings in which learned men met regularly for reading and literary discussion. The academies were almost exclusively privately financed, often by patrons who were also of political importance, such as the Medici in Florence; in Venice the Accademia della Fama was supported by Senator Federigo Badoer. In the 16th century, however, Florence and Venice did not have a university. In university towns also many academies sprang up, some of which became famous: in Bologna the Gelati, in Siena the Accademia ole' Fisiocritici, and in Padua the academies degli Infiammati and de' Ricovrati. Not all of these academies were of a literary nature. In Ferrara, for example, in Tasso's times there were nine academies which were mainly juridical, scientific, and medical.[5] They were often organized by professors of the Studium. Thus, throughout the 16th century in Italy it would be mistaken to contrast the universities and the academies. They were, in some ways, complementary, and the university towns were animated by the same people. University teaching was conducted in Latin, which served for the professional training of jurists, doctors, and theologians. Lectures in the academies, almost always in Italian, were less formal and, on principle, oriented toward the increase of knowledge and the pleasure of communication among scholars. This freedom within the academies also had its negative side. Although they were established with great enthusiasm, the regularity of the meetings, the burden of work that fell to the organizers, at times hostility from without and, with the counter-reformation and foreign dominance in Italy, the ever vigilant and adversarial prying of the Inquisition and the police, and last but not least, the conflicts among the academicians themselves, all contributed to the short lives of most of the academies.[6]

The 17th century in Italy was the age of Galileo, Torricelli, Cavalieri, Borelli, Malpighi, Guglielmini; of eminent Jesuit scientists like Cabeo, Riccioli, Grimaldi, Lana; and of the academies Lincei, Cimento and Arcadia. Yet in the 17th century, Italy lost its earlier pre-eminence in literary and scientific culture, falling behind by at least 20–30 years compared to other European countries. The 17th century universities in Italy ceased to attract illustrious teachers for lack of adequate salaries, while political and religious divisions considerably reduced the flow of foreign students. In Italy itself the rivalry between the colleges of the various religious orders, in particular the Jesuits, which had already begun in the previous century, became even greater. The Italian nobility, who did not require university accreditation for the juridical and medical professions, continued to desert the universities, whose attendance became restricted to those wishing to enter the professional classes of the towns, in large part themselves the sons of lawyers, doctors, apothecaries, and a few men of letters. Very often the sons or relatives of the teachers themselves fostered a closed academic class by accepting selection only on the basis of seniority. The institutional paralysis into which Bologna University gradually fell in the 17th century and the unsuccessful attempts at reform reflect that situation, which was not limited to Italy. The biography of Leibniz provides a striking example. Leibniz' father was a professor at Leipzig University who had left him an orphan; as a young man, Leibniz showed great talent in all subjects. However, his law degree and inclusion in the doctoral colleges were not accepted by the University of Leipzig since this degree could have allowed him to pass over others who had claims of seniority. Leibniz therefore went on to graduate in the small University of Altdorf.[7]

In the 17th century the academies did not suffer as much as the universities from the overall reduction in available public resources, which were almost exclusively used for the construction of churches and palaces and for entertainment purposes. In some cases the academies were able to secure public financing, as in Ferrara where the Intrepidi undertook to organize theatrical shows, and in Florence where the Accademia del Cimento was protected and funded by the Medici family from 1657 to 1667. Most academies, however, survived on private contributions. In either case, their survival was usually precarious; an example is the celebrated Accademia dei Lincei, founded in Rome in 1603 by the 18-year-old Federico Cesi, Marquis of Monticelli, his relative Count Anastasio de Filiis of Terni, Francesco Stelluti, a nobleman of Fabriano, and the Dutchman Joannes van Heeck.[8] Van Heeck, a Catholic, had left his country to avoid religious persecution; after graduating in Medicine in Perugia, he practiced his profession under the protection of two noble families, the Caetani and the Orsini. After a conflict with an officer, he had been incarcerated in the Savelli prisons, from which he was released by Federico Cesi. The Cesi family's reaction to their son's initiatives was swift and violent, since he threatened to squander his patrimony and incur the displeasure of the Inquisition. Federico's father, the Duke of Acquasparta, sought the help of Cardinal Borghese (later Pope Paul V) who gave orders to imprison Heeck once more, while Stelluti and De Filiis were persuaded to return to their country properties, and Federico was exiled to Naples where he met Della Porta.

In his 5 years' absence from academic life, Federico began to write a voluminous, still unpublished, series of regulations for his academy, the *Lynceographum*[9] and kept up a correspondence with his colleagues. In 1610 the academy resumed its public activity with the addition of Della Porta, and the following year the Lincei were joined by Galileo, Johann Schreck,[10] and Johann Faber, a professor of the University of Rome (the 'Sapienza') and director of the Papal Botanical Gardens. Several Lincei fellows were also university professors: Galileo in Padua, Faber and Valerio in Rome, Stelliola in Naples, Neri in Perugia, and Achillini in Ferrara. In 1616 Heeck, who was the first to add the qualification of "Linceo" to the titles of his publications, was expelled from the Academy on the grounds of insanity. In the same year also Luca Valerio, a professor in the Sapienza, left the Lincei when they refused to take disciplinary measures against Galileo, after the latter had been warned not to profess the movement of the Earth. The Academy continued, however, to flourish and was financed privately by Cesi (there were 32 Lincei members in 1625). The Academy maintained its support for Galileo, on the basis that he defended the heliocentric system as a mathematical hypothesis. It also supported Galileo's experiments and observations and financed the printing of the *Saggiatore* (Rome, Mascardi, 1623). Two other printed works, at least, must be credited to the Academy: the translation of the *Persius* by Francesco Stelluti (Rome, Mascardi, 1630), and Cesi's *Rerum Medicarum Novae Hispaniae Thesaurus* (Rome, Mascardi, 1651), completed by Stelluti. After Cesi's death in 1630, and the definitive condemnation of Galileo in 1633, the books, the natural history collections, the scientific instruments, the iconographic collection, and the archives of the Lincei were dispersed even though Cardinal Francesco Barberini, nephew of Pope Urban VIII, had himself once been a member of the Lincei. Some of these materials were saved thanks to the private initiative of Francesco Stelluti and Cassiano del Pozzo, who made them part of his own collection.

The first academy of the Lincei had, therefore, more affinity with the Accademia degli Umoristi, founded in Rome by Paolo Mancini, whose members included Alessandro Tassoni, Battista Guarini, Cardinal Sforza Pallavicino, Cassiano del Pozzo, and two popes (Clement VIII and Alexander VII) than with the Accademia Nazionale dei Lincei, founded in Rome by Quintino Sella after the Unification of Italy. The name of the Accademia dei Lincei was forgotten in Rome but was taken up again in the middle of the 18th century by a medical doctor, Giovanni Bianchi, who refounded it in his own city, Rimini. Thus, the history of the Lincei exemplifies the lack of continuity in the Italian academies, connected to the lack of continuous public funds.

An important episode in the relations between the universities and academies in the 17th century may be seen in Bologna, when Anton Felice Marsili set out to establish an ecclesiastical academy in 1686 and another for experimental philosophy. Already at the beginning of the century Pietro Antonio Cataldi had failed to create a scientific academy due to the hostility of the Bolognese Senate, which guaranteed the cultural monopoly of the city's university.[11] Marsili's attempt was initially well founded; he was the Archdeacon of the Cathedral and, *ex officio*, the most important Chancellor of the Bolognese University. Marsili believed that the creation of the academies would serve to reform the university, in which the number of lecturers had come to exceed that of the enrolled students. He recognized three main defects in the University of Bologna:

1. Failure to enforce attendance requirements as a condition of receiving a degree.
2. Automatic confirmation of temporary teachers (in practice an annual nomination became permanent).
3. Seniority of enrolment in the doctoral colleges as the only criterion of priority in appointments to teaching posts.

The long and bitter conflict between the Archdeacon and the doctoral colleges in Bologna, which, in support of their members, opposed every change, lasted from 1689 to 1694 (a period in which no academic qualifications were conferred) and ended in defeat for Marsili; at the end of the century, the University of Bologna lost some of its best professors.

In 1690, shortly after Marsili's attempt, Eustachio Manfredi founded, in Bologna, a new private academy, the Inquieti, which aimed to bring Italian mathematical and astronomical research up to the level of other European countries, by introducing the methods of Cartesian geometry and Leibniz's calculus. Marsili's project was taken up by his brother Luigi Ferdinando, who, during his military career in the service of the Emperor, had acquired substantial natural history collections. He succeeded in establishing a new, stable structure of teaching and research which made use of his collections and which he funded for many years. In this way the Istituto delle Scienze, which also housed the Accademia degli Inquieti, was established in Bologna in 1714.[12] Pope Benedict XIV (formerly Prospero Lambertini, Archbishop of Bologna) realizing that, in spite of all his powers as Head of the Church, he could not bend the will of the colleges in their resistance to university reform, extended his protection to the Istituto by donating his prestigious library and ensuring its continuous funding by the town's government.

Events in Turin around the middle of the 18th century also affected the relation between universities and academies. In 1757 three young scholars, Count Giuseppe Angelo Saluzzo, the mathematician Giuseppe Luigi Lagrange, and the doctor Gianfrancesco Cigna, began to meet in Saluzzo's residence in order to discuss mathematics and natural philosophy and carry out scientific experiments; these meetings developed into a small Società privata, which was highly regarded by several of the city's leading cultural figures. The University of Turin, which jealously guarded its monopoly in the management of public instruction, remained hostile to this new initiative, which was hindered by the university's most illustrious teacher of sciences, the physicist Giambattista Beccaria, who was adverse to the opening up of new ways other than his own innovative teaching. It was not until 1783 that the Reale Accademia delle Scienze of Turin emerged from this Società privata, but unlike other 18th century academies (in Berlin, Naples, and Padua) it was not permitted to include humanistic and juridical subjects, which remained a monopoly of the university. Moral sciences did not become part of the Academy of Turin until the annexation of Piemonte to France.[13]

THE ACADEMIES AND THE FRATERNITY OF SCIENCES

Among the frequently cited merits of the academies (freedom of research, rejection of the authority of the Aristotelian system, care in experimental methods, practice of astronomical observation with precision instruments), there is one of great importance that is often underestimated; that is, the destabilization of the hierarchy among the various sciences, and, in particular, between mathematics and physics.

Hierarchy, "Hiérarchie" as the *Encyclopédie* by Diderot and d'Alembert informs us, is a concept that derives from theology: "Il se dit de la subordination qui est entre les divers choeurs d'anges qui servent le Très-haut dans les cieux." This "celestial" hierarchy has its counterpart in a terrestrial hierarchy in the organization of the Church itself: liturgical ceremonies are distributed among three orders: the diaconate, the priesthood, and the episcopate.[14]

When the Liberal Arts became part of university teaching (in the 12th and 13th centuries) or of teaching in the colleges of religious orders (Jesuit, Benedictine, Piarist) they were obliged to follow the hierarchic spirit, which characterized these institutes. The Muses became the "deacons" of natural philosophy, which was, in its turn, subordinated to theology.

In the Jesuit philosophy, physics and metaphysics were *Scientiae reales* required to reflect nature in an objective way. "Natural" was every body whose existence was asserted regardless of the fact that it was conceived: nature consisted not only of "substances" but also of "accidents": whiteness, acidity, hardness.[15] On the other hand, the cosmos of Jesuit *physica* was doubly contingent: the creation was a voluntary act both in that it was a materialization of one of the possible worlds, and because God could have created no world at all. Mathematics, on the other hand, was not a "natural" discipline, it did not explain phenomena but restricted itself to describing them. When *mathesis* was used to investigate the "quantitas continua" or the "quantitas discreta" as *mathesis pura*, or to describe phenomena, as *mathesis mixta*, what mattered was not to investigate reality but to "save the appearances". It is in this light that we should also interpret Roberto Bellarmino's famous request that Galileo should treat the Copernican system as only a "mathematical hypothesis".

This different status between *mathesis* and *physica* was also reflected in the remuneration given to university professors. In general the salary of a professor of natural philosophy (*physica*) was 3–10 times higher than that of a lecturer in mathematics.[16] It was by no accident that the first breach in this rigid hierarchy of subjects was achieved by the academies, ever attentive to disciplines on the fringe of the university, such as poetry, observation of natural phenomena, and mathematics.

In the academies, unlike the colleges and universities, there grew a sort of "fraternity" among the disciplines. A singular example may be seen in the Collège Royal founded by Francis I in Paris. In this case the fraternity between arts and sciences was affirmed but at the cost of considering attendance at their courses as completely voluntary and in no way a condition for academic qualifications. Not surprisingly, in the 18th century the universities, like the one destined to fame in Göttingen, looked to the Collège Royal and the academies as a model for their reorganization.

The Académie des Sciences, founded in 1666, marked no hierarchy among their astronomers, physicists, mechanics, and anatomists. On 22nd December 1666, it was established that mathematicians and physicists were to meet twice a week (on Wednesday and Saturday) in order to discuss mathematics on Wednesday and physics on Saturday, "comme il y a une grande liaison entre ces deux sciences". On the occasion of the foundation of the Académie des Sciences a medal was struck to commemorate the event depicting Louis XIV on the back, and Minerva on the front sitting among instruments for anatomy, astronomy, and chemistry and showing no preference among them. The regulation of the Académie des Sciences of 26th January 1699 distinguished "les honoraires, les pensionnaires, les associés et les élèves" but no precedence was made among the disciplines: "Les pensionnaires seront tous établis à Paris: trois géomètres, trois astronome, trois chimistes, trois mécaniciens, trois botanistes, trois anatomistes, un secrétaire et un trésorier."[17] Anxious to establish perfect equality and open debate among academics, the regulations of 1699 prohibited nomination, except as "académicien honoraire", of any "Régulier, attaché à quelque ordre de religion".

MATHEMATICAL RESEARCH AND THE UNIVERSITY IN ITALY IN THE NAPOLEONIC PERIOD

The Napoleonic government in Italy brought about considerable change in university regulations and scientific research, through the creation of a state education system, including schools of every level, through the university reform, and through the Istituto Nazionale. The driving force of renewal which pervaded an army of "sons of the Revolution", was combined in Italy with the great scientific and organizational qualities of Gaspard Monge, who was the first Commissioner for the requisition of works of art from 1796 to 1797, and then envoy to Rome by the French Directoire for the constitution of the Roman Republic in 1798.[18] In this capacity, in the few months preceding his departure for Egypt (February–March 1798), Monge created the Istituto Nazionale of the Roman Republic, whose duty it was, under penalty of abolition, to make plans for the primary and secondary schools. The *Progetto per le scuole superiori* was worked out by a commission formed by a mathematician (Gioacchino Pessuti, 1743–1814), a jurist (Daniele Francesconi, 1761–1835), a Greek scholar (Luigi Lamberti, 1759–1813), a naturalist (Carlo Giuseppe Gismondi, 1762–1824), and a physician (Domenico Morichini, 1777–1836). The *Progetto* provided for the creation of a polytechnic school for every chief town of a "Dipartimento", and a central polytechnic school in Rome, to replace the universities of the Papal State, which had sprung up over the centuries.[19]

While the *Progetto* was being worked out in Rome, in Milan the Gran Consiglio of the Cisalpine Republic initiated discussion of the *Plan* for state education drawn up by a commission composed of Gregorio Fontana (1735–1803), Lorenzo Mascheroni (1750–1800), Francesco Antonio Alpruni (1732–1804), Luigi Valeriani Molinari (1758–1828), Giuseppe Compagnoni (1754–1833), Francesco Gianni (1750–1822), and Ottavio Morali (1763–1826).[20]

The *Plan*—known as the *Piano Mascheroni*, after its spokesman in the Gran Consiglio—dealt with the entire educational system, from primary schools to universities. The universities were to be reduced to two, Bologna and Pavia, each divided into faculties of Medicine, Law, Letters and Arts, Mathematics and Physics. Theology, as well as other teachings related to religion such as canon law, were suppressed, engineering was reinforced, while doctoral colleges were suppressed to give the faculties the right to confer academic degrees. The *Piano Mascheroni* also provided for the creation of an Istituto Nazionale in Bologna for didactic tasks and scientific research.[21]

The institutional fragility of the Cisalpine Republic prevented the *Piano Mascheroni* from becoming law. In 1799 the Austrian and Russian armies put an end to the 3-year republican experiment. Few changes had direct effect but the legislative projects, which were worked out and approved played an essential role in the later constitution of a state education system (1802), which could no longer be neglected in a modern state.

THE BIRTH OF A SCIENTIFIC COMMUNITY

The rise of the first Republics in the years 1796–1799, and the active role played by university professors of scientific subjects such as Pessuti, Fontana, Mascheroni, Franchini etc.; the creation, after the victory of Napoleon Bonaparte at Marengo, of a State of considerable size (the Republic and then the Kingdom of Italy); the annexation of Piedmont, Liguria, Umbria, Lazio, and Tuscany to the French Republic, created the conditions for the foundation of a modern State, a state system of education and a state body for scientific research, namely the Istituto Nazionale.

The law for state education, one of the first laws of the Italian Republic, established a new faculty of mathematics, which conferred a diploma in land surveying (2 years), in architecture (4 years), and in engineering (5 years). Moreover, a Corps of Engineers was created for the roads and waterways, opening up careers for new engineers in public administration.

The same law created the *licei* (grammar schools), one in each Dipartimento, whose teaching programs initially coincided with those of the first years of university, and were gradually separated from the latter. In 1810 algebra and geometry became an established part of teaching in the *licei*, which were expected to educate the leaders for the administration and the army. Many teachers in the Napoleonic *licei* were highly qualified in scientific subjects.

The Italian Republic with its capital in Milan, which became the Kingdom of Italy extending into the Veneto in 1806, had three universities: Pavia, Bologna, and Padua. Pavia was to play a key role by providing the two general directors of Education, namely, Pietro Moscati and Giovanni Scopoli.

The premature death of Mascheroni (1800) and Fontana (1803) deprived Pavia of its most original mathematicians; in Vincenzo Brunacci, however, it found a teacher who was capable of carrying out original research and fostering students. Born in Leghorn, Brunacci had been a supporter of the Republic in 1799 and in consequence exiled to Paris where he met Lagrange and was able to complete his mathematical training under leading figures of the Institut. On his return to Italy he became a professor at Pavia and promoted the reform of the teaching of infinitesimal calculus through a modified version of Lagrange's theory of analytical functions. Brunacci wrote a treatise in four volumes (later abridged to two) for the university teaching of the calculus, as well as a manual for the teaching of algebra and geometry in the *licei*.

In Pavia, Brunacci had several students who went on to teaching and research work, not only in the field of mathematics. They included Ottaviano Fabrizio Mossotti, Gabrio Piola, and Antonio Bordoni: the latter succeeded his master in Pavia, and remained faithful to the Lagrange model in his teaching of calculus even after Cauchy's criticisms; he also wrote an important work on mathematical methods of hydraulic engineering. Francesco Brioschi was one of his students at Pavia. Mossotti's university career came to an end on the fall of the Kingdom of Italy in 1814, with the cutting down of the science faculties that had been set up in the Restoration. For some years he found a post in the Brera Observatory in Milan, but due to his connections with liberal circles and the journal *Il Conciliatore*, he was obliged to leave the country to avoid arrest. Mossotti at first stayed in London and then went to Argentina where he played a dynamic role in the studies of physics and mathematics, while maintaining contact with the advanced milieu of European research. When he returned to Italy he was refused the Chair of Astronomy in Bologna because of his liberal views but finally found a post in the University of Corfu. When teaching in the University of Pisa was reorganized, he was invited by Vittorio Fossombroni and Gaetano Giorgini to hold the Chair of Mathematics in that university. Among some of Mossotti's students here were Enrico Betti, who fought under his command in the Tuscan battalion against the Austrians at Curtatone and Montanara (1848). Betti carried out fundamental research in the fields of algebra and mathematical physics. Betti and Brioschi played an important role in the creation of the State education system and the academies after the Unification of Italy, as well as in the formation of a group of university professors of mathematics, physics, and engineering, who occupied the chairs in almost all Italian universities.

NOTES

[1] M. Longhena (ed.), *Memorie inedite di Agostino Codazzi nei suoi viaggi per l'Europa e nelle Americhe (1816–1822)*, (Milano: Alpes, 1930).

[2] Plato, *The Republic with an English Translation* by P. Shorey, II, Books VI–X, (Cambridge Mass.: Harvard University Press, 1987), pp. 175–179.

[3] On reflection it may be said that also Egyptian geometry developed not thanks to commerce but to the needs of the State when changes in taxation were required for the changing boundaries of cultivated land as a result of the flooding of the Nile. The Egyptian mathematicians in charge of these measurements were public officials. Not even Babylonian mathematics developed through private initiative: as is well known, it was closely linked to astronomy, a field of investigation reserved to a closed sacerdotal caste, which could scarcely be distinguished from the governors.

[4] Bugia, now Bougie in Algeria.

[5] W. Moretti and L. Pepe (eds.), *Torquato Tasso e l'Università*, (Firenze, Olschki, 1997), pp. 92–94.

[6] G. Tiraboschi, *Storia della letteratura italiana*, Vol. VII p. I (Venezia, Zatta, 1796), pp. 98–185. L. Boehm and E. Raimondi, (eds.) *Università, Accademie e Società scientifiche in Italia e in Germania dal Cinquecento al Settecento*, (Bologna: il Mulino, 1981). "Accademie scientifiche nel Seicento", *Quaderni storici*, 16:48 (1981).

[7] E. J. Aiton, *Leibniz. A Biography* (Bristol: Hilger, 1985).

[8] D. Carutti, *Breve storia dell'Accademia dei Lincei* (Roma: Salviucci, 1883). G. Gabrieli, "Il carteggio linceo della vecchia accademia di Federico Cesi (1603–1630)", *Memorie della R. Accademia dei Lincei, Classe di scienze morali, storiche e filologiche* 7(6):1–993 (1939). *L'Accademia dei Lincei e la cultura europea nel XVII secolo* (Roma: Accademia Nazionale dei Lincei, 1992).

[9] Rome, Accademia dei Lincei, *Arch. Linc.* 4.

[10] Who later resigned on becoming a Jesuit and subsequently died as a missionary in China, where he had worked on the reform of the Chinese calendar.
[11] G. Libri, *Histoire des sciences mathématiques en Italie*, 4 Vols; Vol. IV (Paris: Rénouard, 1838–1841), pp. 90–91.
[12] E. Bortolotti, *La storia della matematica nella Università di Bologna* (Bologna: Zanichelli, 1947). *I materiali dell'Istituto delle scienze* (Bologna: Accademia delle scienze, 1979).
[13] *I primi due secoli dell'Accademia delle scienze di Torino*, 2 Vol (Torino: Accademia delle scienze, 1985–1987).
[14] The priests can consecrate the bread and wine, a duty that is denied the diaconate, but they are allowed to do everything that is conceded to the diaconate; on the other hand, the priests are not permitted to ordain new priests, a function carried out by the bishops who may do everything that priests have a right to perform.
[15] U. Baldini, *Legem impone subactis: studi su filosofia e scienza dei Gesuiti in Italia 1540–1632* (Roma: Bulzoni, 1992).
[16] There was not, however, a corresponding ratio between the salaries of a natural philosopher and of a theologian because the former was generally a layman who lived on his salary, whereas a professor of theology was normally a friar who lived in a monastery, often a Dominican or a Franciscan.
[17] E. Maindron, *l'Académie des Sciences* (Paris: Alcan, 1888), p. 19.
[18] L. Pepe, "L'Istituto Nazionale della Repubblica Romana", *Mélanges Ecole Française de Rome Italie et Méditerranée* 108:703–730 (1996).
[19] L. Pepe, "Gaspard Monge in Italia: la formazione e i primi lavori dell'Istituto Nazionale della Repubblica Romana", *Bollettino di Storia delle Scienze Matematiche* 16(1):45–100 (1996).
[20] L. Pepe, "La questione delle università minori in Italia nel periodo napoleonico", in G. P. Brizzi and J. Verger (eds.), *Le università minori in Europa (secoli XV–XIX)* (Soveria Mannelli: Rubbettino, 1998), pp. 425–442.
[21] L. Pepe, "Università e Grandes Ecoles: il Piano Mascheroni e il dibattito al Gran Consiglio della Repubblica Cisalpina", in A. Romano (*ed.*), *Università in Europa* (Soveria Mannelli: Rubbettino, 1995), pp. 511–523. R. Simili (ed.), *Ricerca e istituzioni scientifiche in Italia* (Bari: Laterza, 1998).

LUÍS MIGUEL CAROLIN[1] AND HENRIQUE LEITÃO[2]

NATURAL PHILOSOPHY AND MATHEMATICS IN PORTUGUESE UNIVERSITIES, 1550–1650*

INTRODUCTION

Recent historiography of the so-called "Scientific Revolution" of the 16th and 17th century makes clear that no simple characterization of this momentous event can offer a satisfactory description of it. Nevertheless, despite numerous re-appraisals, certain aspects remain central to our comprehension of this complex cultural phenomenon, namely the modifications to the body of knowledge, the role of practitioners, and the epistemological role of natural philosophy and mathematics. This being the case, some distinguished historians of science have tended to see universities as traditional centers of opposition to new forms of knowledge, explaining the conflicts between universities and the new cultural and scientific institutions as the scientific societies as overt evidence of such university conservatism.

It is not our intention to analyze the complex relationships which developed between universities and scientific societies in the early modern period; nor do we wish to appraise the role of universities in the "Scientific Revolution" of the 17th century.[3] However, we believe it would be useful to analyze the specific case study of the Portuguese situation, for during that period, specific historical circumstances were in evidence in Portugal which, unlike other European countries, would profoundly determine the Portuguese social and cultural setting. On one hand, Portugal was a small country with a serious demographic problem. The constant demands of maritime expansion had to be met which, at their peak, required the control of military and trading matters in an area stretching from the Western coast of Africa to the Southern coast of China. On the other hand, being a catholic country participating in the Counter Reformation movement and experiencing some structural problems with regard to institutions of higher education, it was a country in which the Society of Jesus seems to have played a fundamental and even unique role in education, at all levels throughout Europe.

THE JESUITS AND UNIVERSITY TEACHING OF NATURAL PHILOSOPHY

One of the main manifestations of Portuguese engagement in the Counter Reformation movement, perceived more particularly from the reign of João III (1502–1557), in parallel with the establishment of the Inquisition in lands governed by the Portuguese

M. Feingold and V. Navarro-Brotóns (eds.), Universities and Science in the Early Modern Period, 153–168.

(1536), was the arrival of the Society of Jesus in Portugal. The Jesuits arrived in 1540 and immediately launched apostolic and educational activities both in Portugal and in Portuguese possessions outside Europe. Cherished and supported by the King, the Society of Jesus would develop rapidly in Portugal, its members engaging in many religious, apostolic, and educational initiatives. Starting with only two members in 1540, by the end of the 16th century, Jesuits in Portugal numbered more than 600 and this figure grew steadily throughout the 17th century. In Portugal, the Jesuits established a few dozen colleges and one University (in Évora), but the whole Portuguese Assistance also included 17 learning centers (colleges, seminaries, etc.) in Brazil, 30 in India (in the Provinces of Goa and Malabar), and around 10 in East Asia (Japan, Macao, China).[4]

The teaching of natural philosophy took place in some of these colleges. By the mid-17th century, although this subject was taught irregularly, it always formed part of the academic program in the Colégio das Artes (Coimbra) and the University of Évora as well as in colleges in Santo Antão and Braga in Metropolitan Portugal, São Salvador da Baía and Rio de Janeiro in Brazil, and Goa, Cochim, and Macao in Asia. However, the only two institutions that lectured the subject at university level were the former Colégio das Artes and the University of Évora.[5]

The Colégio das Artes was inaugurated in 1548, at the initiative of King João III, in order to train pupils who aimed to study at the University of Coimbra, the only university that existed in Portugal at the time. There, in accordance with the medieval model, pupils would obtain the preparation in natural philosophy which would enable them to attend the Higher Faculties of the university, those of Theology, Law, Canon-Law, and Medicine. As they encountered difficulties in carrying out this educational goal, the Colégio das Artes was transferred to the Society of Jesus in 1555 and integrated into the University of Coimbra[6] in 1561. In this manner, its members, rector, teachers, and students were all granted the privileges of the University of Coimbra.[7] In addition to this sort of medieval Faculty of Arts, in the second half of 16th century (1559–1779) there was another institution that provided philosophical training, the University of Évora. This university, which would be the only Jesuit University in the Portuguese Province, was started at the wishes of cardinal Henrique, brother of King João III, who wanted to establish public classes at a Jesuit college in the city of Évora. This college was founded in 1553 and the teaching of Arts started in 1556. In 1559, with royal agreement and papal blessing, this college was raised to University status. As it was a Jesuit university where the final goal was the training of theologians, it consisted of four faculties—Humanities, Arts, Theology, and Cases of Conscience (Moral Theology)—where Medicine, Civil Law, and the contentious part of Canon-Law were excluded. Thus, it is clear that by the mid-16th century onward, the teaching of philosophy in Portugal was not only mostly influenced by the Society of Jesus, but one could even go to the point of asserting that, in Portugal during this period, university natural philosophy was essentially a *Jesuit* natural philosophy.

This meant that during studies, Loyolan's standing rule would be followed, which decreed that natural philosophy would be taught in the doctrine of Aristotle and in theology, that of Saint Thomas Aquinas would be followed. This is in fact, what happened in Portuguese universities with regard to natural philosophy and in particular to cosmology.[8] Therefore, lectures were on matters such as the perfection and finitude of the world, the centrality and immobility of the Earth, the distinction between the celestial and terrestrial regions, the changelessness of the former and constant corruptibility of the latter, an Aristotelian concentric orbs model together with Ptolemaic eccentric and epicycle circles, among many other cosmological issues.

However, the question is what kind of knowledge was promoted by these university institutions? Since it would be impossible given the limited space of this paper, to provide a general characterization of the university teaching of natural philosophy, we will concentrate briefly on what was a significant question for Portuguese 16th century teachers of philosophy with the aim of highlighting some general trends in the philosophy taught in this country. We are referring to the theory of the influence of the celestial region on that of the terrestrial. This subject played a major role in Portuguese philosophical tradition. In fact, this question was the object of much more extensive analysis than was the case with other philosophical teaching contexts. This subject was developed by the Coimbra Jesuits (Conimbricenses), whose philosophical commentaries on the works of Aristotle were widespread over all 17th century Europe, such as *Agantne Corpora Coelestia in Sublunarem Mundum, an non*. In addition, this applied to other European Jesuit lectures and courses which were part of the Coimbra philosophical course and was also the case with regard to philosophical teaching practices. In their *Commentarii Collegii Conimbricensis Societatis Iesu in quatuor libros De Coelo Aristotelis Stagiritae* (Lisbon, 1593), which were widespread in 17th century Europe, not only was the supposed evidence of celestial influence and its associated agents discussed, the limits of such celestial influence were also examined in detail. A number of arguments on the anti-astrological debate were thus brought to the plan of philosophy.[9] This was a much more detailed approach than what was carried out, for example, by Roderigo de Arriaga (1592–1667) in his *Cursus Philosophicus*[10] or even by Thomas Compton-Carleton's (1591–1666) *Philosophia Universa*,[11] courses which, after briefly referring to the influence of the stars on the terrestrial region, finished synthetically by condemning the position of judicial astrology. Probably for that reason some other European Jesuits, such as Niccolo Cabeo (1586–1650), took the Coimbran philosophical course as a reference-work on the anti-astrological controversy.[12]

In effect, this is not the moment to study scholastic theory on the influence of celestial bodies on the terrestrial region,[13] nor is it the time to discuss its specificity among 17th century Portuguese philosophers. One should just note that this very theory on celestial influence over the Earth was an essential pillar of Aristotelian cosmology, where the heavens, which were made of an incorruptible substance, were thought to be more perfect and superior to the terrestrial region. According to the Coimbran philosophers, some natural phenomena were proof of supremacy over the terrestrial region. This included the apparent incorruptibility of the heavens, superior location, the circular motion of celestial bodies and above all, the governing power which emanated from celestial bodies over the Earth.[14]

With regard to the action of celestial bodies, this was thought to take place through the four primary qualities—hot, cold, dry, and moist. As the natural philosophers put it, according to Aristotelian theory *On Generation and Corruption*, those qualities were self-organizing, were the origin of the four main elements that constitute terrestrial matter and therefore, were responsible for the configuration of terrestrial life. Since this causality was processed through the physical qualities of the planets observed naturally as motion and light, the influence of celestial bodies was conceived as being the efficient cause of the terrestrial phenomena and the four elements as their material cause. In addition to this kind of action, the effect of celestial bodies was also felt through hidden qualities, which were named *influentia* or *influxus*, which in spite of not being obvious or directly cognoscible, were verifiable in practice. Thus, for example, if Diogo Lopes S.J. attributed the quadripartite division of the year to the motion of Sun and sunshine, with the two equinoxes of the spring and autumn and with the consequent alternation between cold and heat that produced terrestrial life.[15] According to Bento Rodrigues S.J., the generation of gold and other metals inside the Earth, the phenomenon of magnetism, the tides, and the origin of cold[16] were attributed to the *influentia.*

Nevertheless, at the end of the 16th and beginning of the 17th century some astronomical observations could potentially and profoundly affect this cosmological system and thus, the theory on celestial influence over the terrestrial realm. In fact, after 1572, a number of new stars and comets were observed and placed by Thyco Brahe and other astronomers above the heavens of the Moon, where supposedly no change would occur, facts which together with the Galilean observations circa 1610 seemed to reveal the corruptibility of the heavens and their similarity to Earth. Therefore, if the theory under scrutiny was based on the hierarchy between the celestial region with its celestial bodies composed of a fifth perfect and incorruptible essence and the corruptible terrestrial region, how did Portuguese teachers of philosophy react to these new developments? Let us analyze very synthetically, the philosophical debate on the comets observed after 1577.[17]

It is clear that, until the 1640s, the position that would dominate philosophical teaching was the Aristotelian conception of comets as durable masses of incandescent vapors made of terrestrial and maritime exhalations and moving in the upper area of the terrestrial region. Teachers such as Luís de Cerqueira S.J. (1552–1614), philosophy professor at the Colégio das Artes between 1581 and 1585, and future Bishop of Japan,[18] Baltazar do Amaral S.J.,[19] teacher of the philosophy course at the College of Arts between 1618 and 1619, and Francisco Rodrigues S.J. (1594–1654), teacher at the College of Santo Antão at the end of 20s,[20] among many others, followed this very Aristotelian position during their teaching activities.

This position on comets certainly dominated philosophical teaching during the first decades of the 17th century. However, by 1642, Baltazar Teles S.J. (1596–1675), philosophy teacher at the College of Arts between 1630 and 1634, and later, Francisco Soares S.J. (1605–1659) called *Soares Lusitanus* to distinguish him from his Spanish homonymous confrere Francisco Suarez, defended a different position, formally recognizing the celestial nature of these phenomena. According to Baltazar Teles', comets were not exhalations nor did they circulate in the region of the air, because, in addition to other arguments, as the distance from the supreme area of the air to the Earth was thought to be approximately 20 leagues, the comet would not be able to follow its trajectory over our hemisphere in 12 h, as mathematicians had clearly calculated.[21]

Although we need not go into the specifics of the cosmological consequences of the celestial location of the comets, it is important for the present to stress that for Teles and Soares as well for others following these professors of philosophy, it was an important argument to recognize the corruptibility of celestial matter and its essential identity in relation to terrestrial matter.[22] Given this philosophical position, one would have expected from the 40s critique about celestial matter and, consequently, about the border between the heavens and the Earth, different positions on the theory of the influence of celestial bodies on the terrestrial region. However, this did not happen. In the same way as had been discussed by the Conimbricenses on their Commentary on *De Coelo*, the theory of the influence of celestial bodies on the terrestrial region continued to be discussed in great detail and to the same extent by authors such as Francisco da Cruz S.J. (1629–1706),[23] philosophy teacher in Coimbra (circa 1660) and of theology in Lisbon before becoming censor of books in Rome, and by António Cordeiro S.J. (1640–1722),[24] philosophy teacher at the College of Arts between 1676 and 1680 and author of a *Cursus Philosophicus Conimbricensis* printed in the 18th century.

Thus it seems clear that Portuguese philosophical teaching was grounded in the context of the Aristotelianism typical in the Late Scholastic period, both strongly speculative as well as eclectic. As has been stressed in recent historiography, this philosophical teaching was far from barren and was still demonstrating a significant dynamism and capacity to integrate different, recent, and non-Aristotelian aspects[25] in the first half of the 17th century. In fact, abandoning a strictly Thomist basis that characterized 16th century philosophical teaching—singularly materialized in the commentaries to Aristotle's books produced in the College of Arts at the end of 16th century—it will turn into a rich synthesis that assimilated, under an Aristotelian influence, a number of important and decisive cosmological theses such as those of Tycho Brahe on the celestial location of comets.

Therefore, a pertinent question to ask would be: How can we explain this speculative nature that characterized philosophical teaching in Portugal during the 16th and 17th centuries? If the insistence on the more speculative aspects can certainly be attributed to different factors, as for instance to the propaedeutical function that natural philosophy had in relation to theology, the profile of Portuguese philosophy teachers should not be forgotten. What kind of teacher was charged with philosophical training in Portuguese universities? Let us briefly look at those referred to during the controversy on comets: Luís de Cerqueira, Baltazar do Amaral, Francisco Rodrigues, Baltazar Teles, and Francisco Soares. Judging from the existing information, all these Jesuit professors had very similar curricula at the beginning of their teaching activities: they had concluded their training, which included the study of theology, had been teachers of humanities or Latin for some time and at the time of reading the material on comets they were all in their 30s: Luís de Cerqueira, 32 years old; Baltazar do Amaral, around 30 years old; Francisco Rodrigues, approximately 35 or 36 years old;

Baltazar Teles, 38 years old; and Francisco Soares, 33 years old.[26] With regard to the future, one can conclude that they would indeed continue in different ways. Luís de Cerqueira, after obtaining a doctorate in theology, would become bishop of Japan; Francisco Soares would become rector of the University of Évora, while Baltazar Teles would dedicate himself to writing a history of Portuguese Jesuit Assistance. However, the main fact now is that none of them returned to the teaching of philosophy. Therefore, it seems that these professors had only briefly been in contact with philosophical teaching during a transitory phase of their lives. Moreover, this was not a specific case, but rather the pattern among those professors who were involved in such philosophical teaching activities.

Analyzing the training of the teachers of philosophy at the University of Évora and Colégio das Artes, between 1555 and 1650, conclusions can indeed be drawn on the secondary nature of philosophy within the hierarchy of university knowledge. In fact, as a general rule in the 16th century, one finds professors teaching philosophy when they are still completing their training. Generally, once teachers had concluded their humanistic and philosophical studies, and before concluding their theological studies and ordination as a priest, a Portuguese teacher started "reading" the course of philosophy for the three and half years[27] such a course lasted in Portugal. From early on, it is probable that some of the best students would go on to read philosophy at the most prestigious colleges, especially the College of Arts or the University of Évora, while those that had not been so successful would teach it in other colleges. This being the case, it becomes clear why the average age of these teachers is so low, generally around 24 or 25 years old in the mid-16th century. Although this was the normal process, which was certainly followed by a great number of professors, various types of circumstances forced different paths of evolution.

From the beginning of the 17th century onward, with a more stable and developed situation within the Society of Jesus, philosophy teachers gradually began to start teaching after concluding their theological studies and ordination. Therefore, they were approximately 34 and 35 years old.[28] This was already the situation of those Jesuits mentioned above on the theory of comets.

Therefore, these professors found themselves reading philosophy in a transitory phase, a phase in which they were preparing for a much more significant challenge to their opinions. More than philosophers, their objective was certainly to become theologians and/or missionaries. In fact, once their theological studies were concluded and having received sacred orders, the destiny of many of these Jesuits was missionary work in Asia, Africa, or Brazil. A considerable number of other ex-teachers of philosophy would devote themselves to a more complete study of theology, finally obtaining a Doctorate in Theology. Natural philosophy was therefore a secondary consideration, truly a medieval *ancilla theologiae*, in which the task was above all, to support the study of the Holy Scripture.

There are other clues which support the position of lesser importance occupied by philosophy within the hierarchy of the University teachings, namely the almost exclusivity of the Portuguese in the teaching of this discipline. During the century under analysis, the only exceptions are seven Spanish professors: three at the University of

Évora and four in Coimbra.[29] If one bears in mind the fact that during a great part of this period, more precisely between 1580 and 1640, the Portuguese and Spanish crowns were closely linked as part of the same dynasty, this is a remarkable fact. On one hand, this certainly reflects an abundance of teachers considered fit to lecture on natural philosophy, but on the other hand, can also demonstrate a lesser solicitude to philosophy than to theology and mathematics, the importance of which seems to have demanded the recourse to better professors, from outside Portugal. For example, with regard to mathematics, as we shall see, important numbers of non-Portuguese Jesuits taught in Portugal. Therefore, for a markedly speculative teaching of natural philosophy which did not require specific skills—for example, such as the techniques involved in astronomical observations—a sound preparation in Aristotelian tradition would be sufficient.

NAUTICAL DEMANDS AND THE TEACHING OF MATHEMATICS

In addition to philosophy, Portuguese universities, as with other European university institutions, provided mathematical training. In fact, the Chair of Mathematics had been established in 1537 and was occupied by the leading Portuguese mathematician Pedro Nunes (1502–1578), in the years 1544–1562. In the University Statutes of 1559, the areas to be taught in mathematics and the duties of the Professor of mathematics are very briefly described. The Chair of Mathematics is included in the Arts Course and it is stated that the professor should "read" the traditional tracts on Arithmetic, Geometry, Perspective, and Sphere. Candidates to the position of teacher of mathematics should also be competent in the more advanced topics of Euclid and the Theory of Planets. Furthermore, the professor of mathematics was obliged to test students on the Theory of Music.[30] The Statutes determine the salary of the professor of mathematics to be fifty thousand *reis* per year.[31] This number tells us something about the real importance attributed to mathematical teaching duties. Professors of any Chair of the Higher Faculties earned substantially more. The most important Chairs in Theology, Law or Medicine had salaries that were roughly double that stipulated for mathematics: one hundred or one hundred and twenty thousand *reis* per year. However, when compared to the Chairs of the Arts Course, the position of the professor of mathematics is generally on par with other reputed Chairs: the salary of the professor of mathematics is less than the salaries of most professors of Latin but is identical to the salary for the professor of Greek and the professor of Hebrew and greater than the salary of the professor of Music (forty thousand *reis* per year).

The structure of the teaching of mathematics, the privileges and duties of the Professor of mathematics remained essentially unchanged in the text of the following University Statutes. It appears, though, that the Chair of mathematics was relatively autonomous from an early date, despite the fact that it is often presented in parallel with the other Chairs of the Arts Course.[32]

From a strictly formal point of view, the study of mathematics at the University of Coimbra enjoyed conditions which were not much different from those at other European Universities. However, when reality is inspected in more detail, it becomes obvious that appearances are misleading. In fact, one of the most striking observations that awaits the historian interested in this period of Portuguese science is the manner in which the University almost complete neglected mathematical studies during the late 16th and early 17th centuries. Despite the fact that a Chair of mathematics existed and its importance mentioned in all versions of the University Statutes, the proper functioning of mathematical classes at the University of Coimbra was always very rare. A good example is provided by Pedro Nunes himself. By the mid-16th century, Pedro Nunes dominated the mathematical scene in Portugal and maybe throughout the Iberian Peninsula. Although Nunes was appointed professor of mathematics at Coimbra in 1544, his action as a teacher of mathematics never rose to the level of his reputation as a mathematician. In fact, Nunes' classes in Coimbra seem to have been constantly plagued by lack of attendance, both by students and by Nunes himself. His teaching at the University failed to create a circle of disciples, and nothing resembling a "school" of mathematics. His concerns in Lisbon as Chief-Cosmographer (Cosmógrafo-Mor) seem to have interested him more than his obligations at the University of Coimbra—a fact that must be understood in view of Portugal's intense commitment to nautical activities at the time. As a consequence of these duties in Lisbon, Nunes was frequently away from Coimbra and others had to replace him in the teaching of mathematics, but none of the men that substituted Pedro Nunes in Coimbra became noted for their knowledge of mathematics.[33]

After Nunes' retirement the situation worsened considerably. The mathematics Chair was occupied only irregularly and a general climate of indifference to mathematical studies seems to have become the rule.[34] With the possible exception of the immediate successor of Nunes, André do Avelar (b. 1546) who, although an intellectual of much lesser caliber than Nunes, was a competent teacher of mathematics, no professors of mathematics in Coimbra are worth noting until the end of the 17th century. Furthermore, there were long periods when no professor of mathematics was appointed at all. According to Francisco de Lemos, the rector of the University of Coimbra in the period 1772–1777, who was in charge of implementing Pombal's rulings, the mathematics Chair had been vacant for 41 years from 1612 to 1653. After that only three professors were appointed. Writing in 1777, Lemos stated that for the past 60 years there had been no classes of mathematics at Coimbra.[35] The statements of Francisco de Lemos cannot be taken at their face value for Lemos was himself involved in the process of University Reform and the situation before 1772 is always described by him in the worst possible tones. It is known today, for example, that the periods of complete vacancy of the Chair of mathematics were not extended as Lemos claimed. Nevertheless, the general picture of decadence and lack of interest in the pursuit of mathematical studies in Coimbra during this period is correct.

In practical terms (but not in the sense of the legal apparatus or programmatic determinations) the situation was not much different at the other University in Portugal, the Jesuit University in Évora. For more than one century after its foundation in 1559, there was no teaching of mathematics at the University of Évora. Some discussion of Sacrobosco's *Sphere* was included in the Course of Arts, but even this elementary subject was often neglected.

This means that in the period from 1550 to 1650 there was hardly any University-based teaching or practice of mathematics in Portugal. These are puzzling historical facts, not only because the situation in other European universities was so markedly different, but also because it is clear that Portuguese historical circumstances themselves, namely the effort associated with overseas expansion, certainly required personnel with some mathematical expertise. Although the Portuguese were able to provide new and adequate solutions for many of the nautical problems associated to navigation, one should note the peculiar context of this intellectual and technological activity, and how the University was to a great extend alien from it. In fact, the relationship between the University and maritime enterprise—and the new knowledge it generated—has been analyzed by several authors. Although some authors have defended the existence of an interaction, it is acknowledged today that, on the contrary, no influence is discernable.[36] That is, the scientific and technical legacy of the "Discoveries" was fundamentally of an empirical nature, and was developed in the restricted circles of practical sailors and seamen,[37] with little or no debt to University discussions. This fact was especially perturbing some decades ago, when the accepted historiography described the Portuguese nautical contributions as scientific progress promoted by "research centers" such as the School of Sagres.[38] However, even after it became generally accepted that the technological progress associated with the maritime enterprise had a highly empirical and practical nature, it remained obvious that nautical pilots and seamen had to have some training somewhere, and this clearly had not been a task performed by the University.

The mathematical career of Pedro Nunes is almost like a parable of the scientific interests pursued in Portugal at the time. Not only did he seem to have influenced his countrymen much more as a teacher of nautical themes than as a mathematician; he also seems to have been well aware of the audience he could expect in Portugal. One cannot perhaps stretch this kind of analysis too far, but it would be insensitive not to observe that the very choice of language used by Nunes in his work betrays the audience he was intending to address. Thus, he wrote his famous commentary on the Sphere in Portuguese (*Tratado da Esfera*, 1537), but he used Spanish to present the printed version of his work on Algebra (*Libro de Algebra*, 1567), and he resorted to Latin when writing what is perhaps his most sophisticated work (*De Crepusculis*, 1542). If some conclusion is warranted from these facts, one is led to observe that Nunes was well aware that mathematical studies in Portugal at that period were hardly more than what was required for the training of seamen.[39]

By the second half of the 16th century Royal support for the study of mathematics was mostly concentrated in the so-called "Aula do Cosmógrafo-Mor", an informal class of mathematics attached to the duties of the Chief-Cosmographer.[40] However, although one can count men of true mathematical talent among the Chief-Cosmographers, such as Nunes himself or at a later date João Rodrigues Lavanha (ca. 1550–1624), all evidence shows that the classes they taught were irregular and that the scientific level was low. More than a teacher, the Chief-Cosmographer was an administrative official with the responsibility of certifying the quality of nautical charts and navigational instruments. He had also the duty of examining candidates for the position of nautical pilots.

Therefore, if what one seeks is an institution that maintained a regular teaching of mathematics, at a level above the merely elementary, it is again to the Society of Jesus that one's attention is directed. All through the late 16th and the 17th century in Portugal, the Jesuits were the only institution that pursued mathematical teaching with a certain degree of consistency. Mathematical classes are known to have taken place most importantly in the "Aula da Esfera" (Class on the Sphere), in the *Colégio de Santo Antão*, in Lisbon.[41] Classes of mathematics in other colleges did take place, for example in Coimbra, but were somewhat irregular, and mostly private, that is, only to Jesuits. In Évora, public courses on mathematics started only in the beginning of the 18th century. Starting by mid-17th century, the College in Elvas held a public class of mathematics for about 20 years, with particular attention being given to problems of fortification and military architecture.[42] Therefore, until around 1650 the only Jesuit institution that could have been the source of any impact was the "Aula da Esfera" in the College of Santo Antão.

The College of Santo Antão was the leading educational center of the Jesuits in Lisbon. From its inception, it was an ambitious project, both in dimension and in the expected quality of teaching. Of particular relevance in this college was the "Aula da Esfera", established in response to an appeal by the King, for the instruction of sea-pilots. It is believed that the "Aula da Esfera" existed as of 1574,[43] first only to Jesuit students, and as a public course from 1590 onward. This class held public mathematical courses continuously from 1590 to 1759. Many of the men (both lay and Jesuits) that would make important contributions to mathematics, fortification, or architecture in Portugal, were students at Santo Antão.[44] The Lisbon earthquake of 1755 severely damaged this college, and completely destroyed its astronomical observatory that had been built in the 1720s.

Until the mid-17th century, many of the teachers of mathematics of the "Aula da Esfera" were non-Portuguese. Many of them had been especially called from other Provinces of the Society to teach in Portugal. Although different reasons explain this recruitment of non-Portuguese teachers, the main reason is merely the fact that Portuguese Jesuits that were competent in mathematics could not be easily found in sufficient numbers. The first teachers (from 1590 until roughly 1610) were almost all Portuguese, but afterwards they are nearly all foreigners.[45] Of the Portuguese, both João Delgado (ca. 1553–1612) and Francisco da Costa (1567–1604) were reputed teachers. João Delgado was trained in mathematics in Rome, under Christopher Clavius, and is perhaps the only Portuguese Jesuit (until the 18th century) of whom we can assume an advanced mathematical training.

Inspection of the lecture notes of the teachers of the Aula da Esfera shows the great importance attached to nautical questions: elements of cosmography, rules of nautical astronomy and navigation, uses of nautical instruments (astrolabe, quadrant, etc.), the design and construction of nautical charts and globes, etc. In the latter decades, however, the (mostly foreign) professors sometimes introduced more advanced and theoretical material: Geometry, theoretical astronomy (including the Theory of Planets), some algebra, etc.[46]

To understand the presence of so many non-Portuguese teachers—a pattern completely distinct from that of the teachers of philosophy—as well as the evolution of the contents of the courses taught, one needs to recall the scientific activities that the Jesuit missionaries would carry out in China. Although the need to send "mathematicians" to China is present at a very early stage in the China mission, its true urgency would take some decades to unfold. By the first decades of the 17th century it was already obvious to the missionaries in China that involvement in astronomical activities in the Ming Court, and specifically the reformation of the Chinese calendar, was a very promising strategy. However, work of this kind required the presence in China of a considerable number of Jesuits with advanced training in mathematics. The China missions were directly subordinated to the Portuguese Assistance, but the difficulties of the Portuguese Province in responding to these demands soon became obvious. The teaching of mathematics in Portugal had always been subordinated to nautical needs and had generally been developed within the spirit of the *De Sphaera* literature or in relation to cartographic problems. In these themes it can be assumed that the Portuguese Jesuits had a sound training. Indeed, one can observe that some influential commentaries on the *Sphere* published in East Asia, were written by Portuguese scholars [for example, the *Tianwen lüe*, (1615) by Manuel Dias, in China[47]]. However, a tradition of more sophisticated astronomical or mathematical problems did not exist in Portugal. Faced with the lack of Portuguese teachers that could provide more advanced mathematical training, the Society gradually recruited them in other European countries.

If the presence of these foreign teachers reveals the deficiencies in the mathematical culture in Portugal, it was also the occasion for the introduction in Portugal of some of the latest scientific findings. In this sense, it represented a unique moment of contact with European scientific culture. As a direct consequence of the presence of teachers coming from some of the best Jesuit colleges, namely from the Roman College, from 1615 onward, students at the Colégio de Santo Antão were taught the latest developments in astronomy and the cosmological consequences they entailed.

The "Aula da Esfera", fulfilled the basic purpose for which it had been established—the training of nautical pilots—but from the mid-17th century onward, a growing uneasiness with the quality of the teaching of mathematics in the Portuguese colleges is recorded.[48] This would culminate with the decisive action of the Jesuit General Thirsus Gonzalez to reform mathematical studies in the Portuguese Province. Starting in 1692, Gonzalez issued a series of Ordinations to the Portuguese Province, with detailed instructions directed at promoting and increasing the studies of mathematics there.[49] The context and the consequences of these Ordinations go beyond the historical period we are considering here and a discussion of these events will not be considered in this paper. It is important at this point to relate what has been said before about the studies of natural philosophy, with the characterization made above of the practice of mathematics. What appears to a modern observer as the defining trait of the mathematics practiced at this period in Portugal is its essentially practical goal. University attention to mathematical subjects was very limited; the King promoted the creation of a very elementary class for nautical applications and even when mathematical studies were pursued in Jesuit institutions it was not with the objective of complying with the indications of the *Ratio Studiorum*, but with the desire to respond to Royal demands for trained personnel in applied aspects of mathematics. Exceptions to this general pattern occurred, but the overwhelming majority of practitioners in this period are connected to very practical applications of mathematics.

This state of affairs could not fail to create a profound divorce between mathematical studies and natural philosophy. Practiced as an applied and essentially technical body of knowledge, mathematics could not claim epistemological value and intellectual status comparable to that attributed to philosophy. All testimonies of the period refer to the low position of mathematical studies in the hierarchy of the university social organization. Therefore, in addition to the fact that mathematical production in Portugal was scarce, it suffered from a subordinate position that made it very difficult for mathematical arguments to be incorporated into the crucial scientific discussions of the period. It is no wonder that philosophers in Portugal were slow to incorporate "scientific" or "mathematical" developments into their teaching. During the critical decades of the cosmological controversies, throughout Europe, tension between mathematicians and their philosopher colleagues was constant. In Portugal, this divorce was certainly much more pronounced. Mathematical teaching was, as we have described, pursued with practical applications in sight and never rose to the level of an intellectual discipline that could match the excellence of philosophy.

CONCLUSION

Even bearing in mind the limitations of our approach, a number of points seem sufficiently evident allowing us to use them as conclusions of this work. First of all, one should note the prominent cultural and academic role played by the Society of Jesus in Portugal. In other European Catholic countries, the Jesuits also had a leading position, but the distinguishing feature in Portugal is that in several instances they were largely unaffected by intellectual challenge from other institutions and thus shaped Portuguese cultural life in a profound way.

Although pursued in an Aristotelian mould that, in the long run, would create some problems toward new conceptions of nature, natural philosophy in Portugal was taught at a University level, in a speculative and flexible way. As was the case with other European Jesuit colleges, Portuguese natural philosophers discussed and gradually incorporated the latest celestial developments and their cosmological consequences into their eclectic philosophical corpus. Mathematical studies, surprisingly, were nearly disregarded at the University and when taught and practiced, it was with very practical applications in sight, or to serve the needs of a specific professional group and context, namely, the navigational problems to be solved by nautical pilots. Although one Portuguese mathematician early in this period, Pedro Nunes, rose above these limitations, this was an exception and a mathematical tradition cannot be found during this period. Perhaps the most important observation lies in the fact that by the mid-17th century in Portugal, mathematics and natural philosophy were treading very different paths. It is difficult to escape the suggestion that this wide separation of the context, aims and institutional setting of natural philosophy and mathematics would have important repercussions in the way the developments of the *Nova Scientia* were received in this country.

NOTES

* Abbreviations used in this chapter:

A.R.S.I.—Archivum Romanum Societatisè Iesu
B.A.—Biblioteca da Ajuda, Lisbon
B.G.U.C.—Biblioteca Geral da Universidade de Coimbra
B.N.L.—Biblioteca Nacional, Lisbon

[1] This study was partially carried out as part of the Fundação para a Ciência e a Tecnologia/ Centro de Estudos de Història e Filosofia da Ciència—University of Éuora research project: POCTI/HCT/37709/2001.
[2] Financial support form the Fundação Oriente, Lisbon, is acknowledged.
[3] See, especially, M. Feingold, "Tradition versus Novelty: Universities and Scientific Societies in the Early Modern Period", in P. Barker and R. Ariew (ed.), *Revolution and Continuity. Essays in the History and Philosophy of Early Modern Science* (Washington, 1991), pp. 45–59 and J. Gascoigne, "A Reappraisal of the Role of the Universities in the Scientific Revolution", in D. Linberg and R. Westman (eds.), *Reappraisals of the Scientific Revolution* (Cambridge, 1990), pp. 207–260.
[4] The best overview of the Jesuit enterprise in Portugal is still the massive: F. Rodrigues, *História da Companhia de Jesus na Assistência de Portugal*, 4 Tomes in 7 Vols. (Porto: Livraria Apostolado da Imprensa, 1931–1950). See also D. Alden, *The Making of an Enterprise. The Society of Jesus in Portugal, its Empire, and Beyond, 1540–1750* (Stanford, California: Stanford University Press, 1996).
[5] Philosophical courses were also held regularly in Lisbon, at the Colégio de Santo Antão, and in Braga, at the Colégio de São Paulo. However, philosophical courses at these two colleges would never reach the level and reputation of those taught in Coimbra and Évora. In fact, only in Coimbra and Évora, were philosophical studies were part of a university structure.
[6] See M. Brandão, *O Colégio das Artes* (Coimbra, 1924).
[7] See R. de Carvalho, *História do Ensino em Portugal. Desde a Fundação da Nacionalidade até o fim do Regime de Salazar-Caetano* (Lisbon, 1986), pp. 269–299 and F. Rodrigues, *História da Companhia de Jesus* : t. I, Vol. II, pp. 336–400.
[8] *Estatutos ordenados pelo muy alto Princepe e sereníssimo senhor Dom Anrique... pera a universidade que ordenou e fundou na cidade d' Evora*, B.N.L., Cod. 8014, pp. 181–182.
[9] *Commentarii Collegii Conimbricensis Societatis Iesu In Quatuor Libros De Coelo* (Lisbon, 1593), Lib. 2, cap. 3, pp. 155–197.
[10] See R. Arriaga, *Cursus Philosophicus*, Antuérpia, ex Officina Plantiniana Balthasaris Moreti, 1632, *Disputatio unica Caelestis*, section 6, p. 507.
[11] See Thomas Compton-Carleton, *Philosophia Universa*, Antuérpia, Apud Jacobum Meursium, 1649, *In libros de Coelo*, disp. 2, section 4, p. 404.
[12] Indeed, according to him, "In hac conclusione est controversia inter Auctores, et huic aduersatur Auer. et alij apud Conimb.2 Coeli, cap. 3 quaest.3 et Picus Mirand. contra Astrol. qui putant esse asylum ignorantiae recurrere ad Coeli influentias ubi causa alicuius effectus sit obscura. Est tamen communior inter Philosophos, ut ostendunt Conimbr. art.2 et probatur primo..."—Niccolo Cabeo S.J., *In quatuor libros Meteorologicorum Aristotelis Commentaria et Quaestiones...*, Roma, Typis Haeredum Francisci Corbelletti, 1646, Lib.1, textus 5, quaest.2, p. 34.
[13] E. Grant, "Medieval and Renaissance Scholastic Conceptions of the Influence of the Celestial Region on the Terrestrial", *Journal of Medieval and Renaissance Studies* 17:1–23 (1987) and

J. North, "Celestial Influence—The Major Premise of Astrology", in P. Zambelli (ed.), *Astrologi hallucinati'. Stars and the end of the World in Luther's time* (Berlin, 1986), pp. 45–100.
[14] *Commentarii Collegii Conimbricensis Societatis Iesu In Quatuor Libros De Coelo*, Lib. 1, cap. 2, quast. 5, art. 2, pp. 39–40.
[15] "(...) Motus solis affert quadripartitam anni distinctionem, quae duo aequinoctia Vere et Autumno, ac totidem solistitia aestate et Hyeme complectitur."—Diogo Lopes S.J., *Compendium Totius Philosophiae ...*, 1623, B.G.U.C.—Ms.2314, *Liber secundus Coelorum*, cap. 3, quaest. 1, fl. 76.
[16] Bento Rodrigues S.J., *Caelorum liber, Generaois, et aliae simul traditur, et explicatur a sapientissimo Patre Benedicto Rois Societatis Jesu in Collegio diui Antonij eiusdem Societatis, anno domini 1664,* 1664, B.A., Cód. 50—III—17, *Tractatus De Caelo et Mundo*, quaest. 3, art. 3, p. 24 and the following ones.
[17] For a more detailed examination on the controversy on comets in Portugal see: L. M. Carolin, "Philosophical teaching and mathematical arguments: Jesuit philosophers *versus* Jesuit mathematicians on the controversy of comets in Portugal (1577–1650)", *History of Universities* 16 (2001), pp. 65–95.
[18] L. de Cerqueira S.J., *Commentarii in Libros Meteorum Aristotelis*, B.G.U.C., Ms. 2414, cap. 6, fls. 223–223v.
[19] Baltazar do Amaral S.J., *Doctrina Philosophica Auctore Ludovico Dias Franco*, Lisboa, ex Officina Petri Craesbeek, 1618, *De Meteoris*, quaest. 2, cap. 3, 198. (Baltazar do Amaral published this book under the pseudonym of Luís Dias Franco).
[20] Francisco Rodrigues S.J., *Compedium Philosophycum De Metheoris, Parvis Naturalibus, De Coelo, item de Generatione, et Corruptione, De Anima coniuncta materiae et ab illa separata Denique De Ethicis traditum a ..., Ullyssipone Ano Domini 1629 ...*, B.G.U.C.—Ms. 2316, *Metheoris*, cap. 10, fls. 4–4v.
[21] Baltazar Teles S.J., *Summa Philosophiae ... Pars secunda in Libros Physicorum et in Libros de Coelo ac Meteorum*, Lisboa, ex Officina Laurentij de Anveres, 1642, *De Meteoris*, disp. 1, section 2, p. 380.
[22] Baltazar Teles S.J, *Summa Universae Philosophiae* ... cit., *Physica*, disp. 22, sectio 4, pp. 199–201; Francico Soares S.J., *Cursus Philosophicus* ... cit., *Physica*, disp. 2, sectio 6, pp. 23–25.
[23] Francisco da Cruz S.J., *Institutiones Phisicorum. 3ª Pars: De Corpore animato, s.e et anima Tradita a Sapientissimo P. M. Francisco da Crux Societatis Jesu, Accepit Simon dos Sanctos eiusdem Societatis. Anno a reparata salute 1665,* 1665, B.G.U.C.—Ms. 2367, *Sinopsis 1ª: Pro Mundo et Coelo*, section 4, art. 1, fls. 232v–235v.
[24] António Cordeiro S.J., *Cursus Philosophicus Conimbricensis auctore ... in tres partes distributus: Prima Logicam amplectitur; Secunda Physicam, cum corpoream, tum spiritualem; Tertia enucleabit Metaphysicam*, Lisboa, ex Officina Regia Deslandesiana, 1714, tract. III, disp.2, quaest.2, art. 2, pp. 633–635.
[25] See, among others, R. Ariew, *Descartes and the Last Scholastics* (Ithaca, 1999).
[26] These ages were established using the *Catalogi breves et triennales*, A.R.S.I.: Lus. 39, fl.7; Lus. 44, II, fls. 310v, 446v, 491v, 578.
[27] The statutes of the University of Évora state that: "Avera na universidade como fica dito quatro cursos de Artes dos quais hum comecara cada anno o segundo dia de Outubro, e durará quatro annos cada hum dos tres primeiros sera de des meses de leitura, e o quarto de seis mezes, que acabara no derradeiro de Março." *Estatutos ordenados pelo muy alto Princepe e sereníssimo senhor Dom Anrique(...)*, B.N.L., Cod. 8014, p. 181.
[28] Cfr. J. P. Gomes, *Os Professores de Filosofia da Universidade de Évora* (Évora, 1960), p. 41.
[29] Pedro Luís, Domingos de Araoz and Pedro Freire in Évora and Pedro Gomez, Nicolau Gracida, Inácio Tolosa and Luís de Molina in Coimbra.
[30] "(...) o oppositor de Mathematicas lerá duas lisõis de ponto, huma em Euclides e outra na Theorica dos Planetas; e na opposição da cadeira de Musiqua não averá lição de ponto, porem

o tal oppositor será examinado na Theoriqua de Musiqua pollo catedratico de Mathematicas", in S. Leite, (ed.), *Estatutos da Universidade de Coimbra (1559)* (Coimbra: Por ordem da Universidade, 1963), pp. 114–115.

[31] S. Leite (ed.), *Estatutos da Universidade de Coimbra (1559)*, p. 94.

[32] After 1559, new University Statutes for the University of Coimbra were presented in the years of 1591, 1597, and 1654 and the corresponding texts have survived to our days. Judging by the number of new University Statutes in the period 1537–1772, the desire of the Reform movement seems to have been constant in this period. See J. F. Gomes, "Os vários Estatutos por que se regeu a Universidade Portuguesa ao longo da sua história", *Revista Portuguesa de Pedagogia* 20:3–62 (1986).

[33] In fact, of the various men that are known to have substituted Nunes in his absences—António de Sousa, Francisco Calado, Manuel de Pina, Pedro de Sousa, Pedro da Cunha, Nicolau Coelho do Amaral—only N. Coelho do Amaral published work on mathematics—a rather uninteresting *Chronologia*, 1554.

[34] T. Braga, *História da Universidade de Coimbra*, 4 Vols. (Lisboa, 1892–1902). See esp. Vol. 2, pp. 812–835.

[35] "Por que consta que desde o anno de 1612 até o de 1653, em que se passou o longo intervalo de 41 annos, esteve vaga a Cadeira de Mathematica sem Professor que a regesse: E que desde este anno até a prezente Reforma so fora regida por tres Professores, Gaspar de Mere, O Padre João Coning Jezuita, e o Padre Fr. Ignacio de Atayde Beneditino, havendo longas vacancias entre huns e outros; e sendo esta ultima de mais de 60 annos". Francisco de Lemos, *Relação Geral do Estado da Universidade, 1777* (Coimbra, por ordem da Universidade, 1980), p. 80.

[36] For an authoritative explanation of the divorce between the University and the Maritime Expansion, see: Luís de Albuquerque, "A Náutica e a Cartografia em Portugal nos séculos XV e XVI", in *A Universidade e os Descobrimentos* (Lisboa: Imprensa Nacional-Casa da Moeda, 1993), pp. 91–101.

[37] There is a considerable body of literature to back this assertion. A particularly penetrating analysis can be found in Francisco Contente Domingues, "Horizontes mentais dos homens do mar no século XVI. A arte náutica portuguesa e a ciência moderna", in *Viagens e Viajantes no Atlântico Quinhentista* (Lisboa: Colibri, 1996), pp. 203–218.

[38] The most irrefutable and clear destruction of this myth is W. G. L. Randles, "The alleged nautical school founded in the 15th century at Sagres by Prince Henry of Portugal, called the 'Navigator' ", *Imago Mundi*, 45:20–28 (1993).

[39] The difficulties in establishing a mathematical community in Portugal can be detected at earlier dates. By the beginning of the 16th century, a number of Portuguese students received some mathematical training in Paris, among the many foreigners there. Of these, men such as Álvaro Tomás, Francisco de Melo, Pedro Margalho, or João Ribeiro were certainly exposed to some mathematical training. However, this was of a different nature to that of many of their Spanish colleagues in Paris and upon their return to Portugal, very few of them pursued any type of mathematical career. Of all these scholars, only Francisco de Melo (1490–1536) is known to have acquired any degree of fame as a mathematician after returning to Portugal. Whether in the first decades of the 16th century, or by mid-16th century, the existence of isolated men competent in mathematics was never sufficient to create circles of practitioners in mathematics.

[40] The Cosmógrafo-Mor and, in general, the problem of nautical teaching are analised in A. Teixeira da Mota, "Os Regimentos do Cosmógrafo-Mor de 1559 e 1592 e as origens do ensino náutico em Portugal", *Memórias da Academia das Ciências de Lisboa (Classe de Ciências)* 13:227–291 (1969).

[41] F. Rodrigues, *História da Companhia de Jesus*; Luís de Albuquerque, "A Aula da Esfera do Colégio de Santo Antão no século XVII", in *Estudos de História* (Coimbra: Universidade de Coimbra, 1974), Vol. 2, pp. 127–200 [Originally: *Anais* (Academia Portuguesa de História), 21 (1972) pp. 337–391]; U. Baldini, "As Assistências ibéricas da Companhia de Jesus e a actividade científica nas Missões Asiáticas (1578–1640). Alguns aspectos culturais e institucionais",

Revista Portuguesa de Filosofia, 54 (1998) 195–245, and U. Baldini, "L'insegnamento della matematica nel collegio di S. Antão a Lisbona, 1590–1640", *Saggi sulla cultura della Compagnia di Gesù (secoli XVI–XVIII)* (Padova: CLEUP, 2000) pp. 129–167.

[42] This does not mean that the quality of the mathematical classes was necessarily inferior to those given in Lisbon. For example, some competent mathematical teachers taught in Coimbra: J. Raston, J. König, A. Buckowski, A. Thomas, M. Soares, and A. Castelo-Branco.

[43] Precise documentation on the foundation of this *Aula* is lacking, but all sources attribute its foundation to King D. Sebastião. Such as, for example, the opinion of a later Jesuit-mathematician, Manuel de Campos, in the preface of his *Elementos de Geometria Plana e Sólida* (Lisboa, 1735).

[44] Such as Manuel de Menezes, Luís Serrão Pimentel, the Infant D. Theodosio, António Pimenta.

[45] Lists of teachers of mathematics at the Colégio de Santo Antão have been established by Luís de Albuquerque "A Aula da Esfera do Colégio de Santo Antão no século XVII", and Ugo Baldini, "As Assistências ibéricas da Companhia de Jesus e a actividade científica nas Missões Asiáticas (1578–1640)" and also U. Baldini, "L'insegnamento della matematica nel collegio di S. Antão a Lisbona, 1590–1640".

[46] See Luís de Albuquerque, "A Aula da Esfera do Colégio de Santo Antão no século XVII".

[47] On the *Tienwen lüe*, see P. M. D'Elia, *Galileo in China. Relations through the Roman College between Galileo and the Jesuit-Scientist Missionaries* (1610–1640), (Cambridge, 1960).

[48] On the other hand, the need to send mathematically trained missionaries to China was as demanding as ever, or maybe more so, because as of 1668 Louis XIV had deliberately "short-circuited" the accepted sovereignty of the Portuguese Padroado, and sent "mathematicians" directly to China. That problems in the mathematical training of missionaries were becoming more and more acute, can be found in some of the contemporary correspondence, but measures to put an end to this situation were wanting all through the 17th century. One cannot fail to get the impression that the status quo was becoming an accepted reality, until the General Thyrsus Gonzalez tried to put an end to it.

[49] **1.** *Ordinatio Rev. P. N. Thyrsi Gonsales Praepositi Generalis ad suscitandum, fovendumque in Prov. Lusitana studium mathematicae directa ad P. Emmanuelem da Sylva Provincialem.* (12 April 1692). **2.** *Ordinatio R. P. N. Thyrsi Gonzales Praepositi Generalis de forma et legibus examinis Mathematici in Provincia Lusitanorum* (17 January 1693). **3.** *Confirmatio et extensio R. P. N. Thyrsi Gonzales cc.*^{a} *ordinationem de forma et legibus examinis Mathematicae* (1 August 1693). **4.** *Declarationes Praepositi Generalis circa studia Mathematicae* (4 February 1702).

PIERO DEL NEGRO

VENETIAN POLICY TOWARD THE UNIVERSITY OF PADUA AND SCIENTIFIC PROGRESS DURING THE 18TH CENTURY

In this chapter, I first consider the institutions through which the Venetian Republic in the 18th century exercised its governance over the University of Padua, especially in relation to the teaching of the sciences, and through which the Republic formed its policy in that area. The most important of these institutions was a body, elected since 1517 from among the Senators of the Republica, named the *Riformatori dello Studio di Padova*. The office of the *Riformatori* had been set up in; but only in part did their activities first reflect a consistent project of the reform. In the second part of the chapter, this project, its preconditions, and its effects on the scientific disciplines in the University, are considered, from an institutional and political standpoint. For the purposes of this chapter, I include under "scientific disciplines" all the subjects, except theology and classical studies that were taught in the Faculty of Arts.

The aim is to draw attention to the peculiar conception of scientific progress which took shape in the research structures of the University of Padua in the second half of the 18th century. In brief then, this chapter looks beyond the history of science in the strict sense of the word and instead outlines the cultural policy of an ancient State—exceptionally ancient, considering the one-thousand-year history of the Serenissima—while also seeking to assess the influence of this state on specific academic structures.

If it appears anachronistic or exaggerated to speak of "scientific policy" anywhere during the 18th century, in the Venetian case this is particularly true, because the archaic and woolly constitutional system of the Repubblica—a City–State on the eve of its fall which still kept much of its medieval imprinting—favored the expansion and complication of political and administrative roles, which in turn produced floods of interactions between the government of the Serenissima and the scientific community. The *Riformatori dello Studio di Padova*, who will concern us here, constituted one of the about 400 administrative authorities (*magistrati*) appointed to supervise diverse aspects of the public life of the Venetian city and territory. A part of these *magistrati* was concerned with the scientists and inevitably they overlapped in their responsibilities and in the scientific expertise they required.

For example, after the foundation of the Academy of Science, Letters, and Arts in Padua, which served as the national Academy of the Venetian State (to be discussed below), its members were required to serve—up to 1790—on 12 public commissions received from the same number of Venetian magistrati, sometimes assembled in "conferences": *Savi alla mercanzia* [councillors for trade]; *Deputato alle strade* [highways authority]; *Deputati alla regolazione delle tariffe mercantili* [trade

M. Feingold and V. Navarro-Brotóns (eds.), Universities and Science in the Early Modern Period, 169–181.

tariff regulation authority]; *Provveditori all'Adige* [Adige river authority]; *Deputato* of Verona valleys reclamation; *Savi esecutori alle acque* [executive water authority councillors]; *Provveditori* of artillery; *Riformatori dello Studio di Padova*; *Capitanio e vice-podestà di Padova* [one of the two Venetian governors of the town]; *Savio alla scrittura* [Secretary of War]; *Provveditori ai beni inculti* [common land authority]; *Deputati* to agriculture; *Deputati alla provvision del denaro* [tax authorities]; *Inquisitore all'Arsenale* [Inquisitor of the dockyard].[1]

The complex interactions between government and academic science are illustrated, for example, in the list of the commissions received by the scientist Giovanni Arduino. He was the surveyor for agriculture from 1769 till his death in 1795 and as such a technical, scientific, and bureaucratic pillar of the Magistrato of the Deputati all'agricoltura, but due to his vast knowledge he contributed to a great number of other offices of the Venetian government such as the *Provveditori* of artillery, the *Inquisitore* of arts, the *Provveditori* of salt, the *Provveditori* of common land, the State Inquisitors, the *Deputati* for mines, the *Censori* of the art of glass, the *Provveditori* and Patrons of *Arsenale*, the *Provveditori* in mint, the *Deputati* of Montona Woods, the *Savi* of trade, the *Provveditori* of the crops, the Inquisitors of gold and coins, and the *Deputato* of Verona valleys reclamation.[2]

Clearly, it is impossible to distil any Venetian scientific policy from a study of the initiatives of these 22 *magistrati*, all the more since they are the only one example, although a meaningful one, of a network of relations between the politico-administrative machine of the republic and the scientists which was certainly wider and more articulated. It is true that many of the recommendations of these *magistrati*, particularly the most important ones which involved expenses other than those foreseen for ordinary administration, were to be sifted by the Senate and thus brought, more or less, into line with more general policies. We also know that the items on the agenda of the Senate were determined by the *Consulta dei savi*, a sort of council of ministers of the Venetian republic at the head of which were appointed six *savi grandi*, an office which guaranteed a relatively high degree of institutional stability and therefore made possible, at least in the mid-term, a consistent and reasonably wide-ranging policy.

In practice, although the Senate usually followed the recommendations of the *Consulta*, a coherent policy was by no means guaranteed. The *Consulta* was chaired in rotation, from week to week, by one of the six *savi grandi*, who often asserted his power without caring very much for what had been decided previously by his colleagues. This *despotisme hebdomadaire*, as it was named by the patrician Leopoldo Curti[3] on the eve of the fall of the republic, would often bring about contradictory and short-lived decisions (hence the proverb: "Venetian law lasts a week"). This zigzagging political line is difficult to reconstruct and even more difficult to interpret. In addition, the *Consulta*, and consequently the Senate, intervened only rarely in the measures submitted to them; if we compare the papers submitted by the *magistrati* with the Senate's relevant legislative decisions, we find that the official serving as secretary of the Senate typically confined himself to summarizing the paper and to drafting a law. The few exceptions are rarely attributable to interventions made by the *savi grandi*, more often to Senate's resistance to the measures proposed by the *magistrati*.

Moreover, it must be remembered that usually the Venetian government—that is to say the *Consulta*, the Senate, as well as, earlier in the process, the various *magistrati*—took its decision in answer to the single "issue", without regard to the development of general principles. This loose empiricism however was backed by a well-established cult of tradition based on minute research and on respect for precedent. The result was a strange hybrid, a sort of traditionalist pragmatism, in which any form of planning was very unlikely. In brief, the Senate, to a large extent, merely backed the decisions taken by the single *magistrato* without embodying them in a more far-reaching policy, while for their part, the *magistrati* very rarely left narrow, parochial perspectives.

However, having underlined the institutional limits of Venetian policy, it is worth considering also that the rapid turnover of officers and, above all the dense network of informal relations among aristocrats, who held the most important offices, favored the flourishing of a common political sense, which, in turn, could offer a more or less solid base for the development of strategies which had some relative consistency and continuity.

It is clear that scientific policy also was drawn within these co-ordinates. The *magistrato*, whose authority in this field was by far the greatest, above all because of its relation with the University, belonged to the three *Riformatori dello Studio di Padova*. As described around 1777, some of its responsibilities lay in the city of Venice: the supervision of "the art of print, the control of books, [the academy of] painting, sculpture [and architecture], the nautical school, the schools of the sestieri [city primary schools reformed in 1774], the state schools [secondary schools, which had replaced the Jesuit college in 1773], the Professor [Francesco] Paiola surgical school, the doctoral degrees [in medicine] in Venice, and more". In Padua: "the University, professors and chairs, the chemical laboratory, the astronomical observatory, the thermal baths [in Abano], the herb garden [that is the botanical garden], the Professor [Pietro] Arduino School of agriculture, the Delia Academy [a riding school for gentlemen], and more".

"All these things should require a *magistrato* working every day" was the sorrowful conclusion of the officer, who drew up this list[4]—very probably Davidde Marchesini, the secretary of the *Riformatori*. Instead, the office of the *Riformatori* was an additional one: the majority of the *Riformatori* belonged to the "order" of the *Savi grandi* and so that a wide range of cumulative responsibilities for "all these things" could be held at the same time as even more daunting tasks. As a consequence, the patrician *magistrati* very rarely had the chance to do any serious planning. At the same time, precisely because they were holding such important offices in Venetian politics, the *Riformatori* were in a position to impose their recommendations on the *Consulta* and the Senate, or to tailor their recommendations according to the balance of power within government.

Three *Riformatori* served actively in the *magistrato* at any one time, for a term of 2 years (longer than the average in other offices) followed by 2 years out of office, "*in quarantena* ", after which they were normally reappointed, so that a *Riformatore* held in fact a lifetime appointment. The presidency rotated among the three; from 1737 on, each served as president for 2 months (again, a relatively long term). These features favored continuity; but they were often compromised by the fact that the three serving *Riformatori* took office at different times, and that any of them might have to leave before serving his full term—if, for example, he received another appointment that required him to leave Venice, or was elected Doge, or died or became incapable from old age. Very rarely, in fact, did a set of three remain stable even for one whole year; and since, as a rule, only senior senators with distinguished political careers became *Riformatori* (the average at first election was just under 60), a patrician rarely served more than three 2-year terms.

Also, although the Republic was by no means free of political conflicts, and one party or clan might gain the upper hand in the Senate at any given time, the rights appertaining to a *nomenklatura* of individuals who had reached high office were really jeopardized; so that it often happened that senators with contrasting ideologies served in the same team of three *Riformatori*, and this in turn might paralyze the *magistrato*. This occurred, for example, in the 1760s and 1770s, when the struggle between statists (*giurisdizionalisti*) and supporters of the Church (*curialisti*) was at its height.[5]

Bertolt Brecht, in his play *The Life of Galileo*, took several liberties with the *Riformatori*, in particular, misled by the name, he transplanted them in Padua. In fact the *Riformatori* resided permanently in Venice: in the whole 18th century they officially visited Padua only once, in 1771. Their secretary also avoided visits to Padua which would have given him the opportunity to judge personally the situation of the University. Information came to the *Riformatori* from Padua, sometimes from formal sources, including the officials of the University itself (till 1738 two students and since 1738 two professors represented and also "ran", apparently, the *Studio*, but in fact the whole of power was in the hands of the *Riformatori* and the professors did not enjoy any degree of self-governing), the Venetian governors of the city of Padua and its bureaucracy, but more often from informal sources, and was processed in Venice. Examination of the initiatives of the *Riformatori* in the 18th century shows that their policy as a rule was shaped not according to official institutions, but depended on the initiative of the individual working for that month and followed strategies which did not know well enough what was really going on in Padua.

In 1797, when a democratic municipality had been installed in Padua by France and the University could safely distance itself from the old régime, Melchiorre Cesarotti, (from the 18th to the 19th century the most internationally renowned professor of Padua University), argued that "Venetian aristocracy had ruled that institution without really understanding its object or importance".[6] Actually, the policy of the Venetian government with regard to the University—and particularly to the professors—generally followed rules which, according to Giacomo Nani, a former representative of Venice in Padua, were designed to exhibit "love and esteem" to the professors, "by honoring them as individuals, by attending their table, by not entering into their rivalries and arguments, by having provided them with some money [. . .], by having discussions with them upon each science, by attending their sessions and academies". The fact that this attitude respectful only of good manners was sufficient to guarantee professors' respect is proved by the fame of Nani himself, who was universally celebrated as "a lover of letters and of those people professing them".[7]

It is worth considering also that the "natural" inertia of the *magistrato* was sometimes counteracted by the initiative of those individual *Riformatori*, who were able to take advantage of their brief institutional pre-eminence; having taken some expert advice, they took care to present their recommendations in a form respectful of the powers and political prejudices of the moment, and managed to gain the necessary backing of the colleagues and the Senate. This occurred quite often in the two decades following 1760, which was the only period in which the *Riformatori* acted in the spirit of their name. This was made possible by an unrepeatable situation, marked by three deeply correlated factors.

In the late 1750s, when a number of aged men left their office as *Riformatori*, at almost the same time some senators in their fifties were able to gain a seat in the *magistrato*. Several of them had an education which was open to cultural influences from across the Alps and to the lessons of the early Italian Enlightenment. Because of the its relatively young age and the its strong influence in the Senate, this group would condition the decisions of the *Riformatori* for a very long period, till early in 1780s.[8] Secondly, in the long term and mid-term among this new generation of politicians, despite internal conflicts, the "progressive" and statist side led by Andrea Tron and by Francesco Morosini prevailed. They were the two major members holding power who, above all in the 1770s, encouraged a vigorous policy of reforms. Finally, these patricians' ideologies supported a cultural and scientific policy, which combined some aspects which, as we shall see, were very advanced even when compared, using a European standard, to others more respectful of Venetian idiosyncrasies.

The recent renewal of interest in the Veneto of the 18th century (including the works of Marino Berengo, Gaetano Cozzi, Gian Franco Torcellan, and Franco Venturi)[9] has also made it possible to set aside once and for all those stereotyped views of Venetian policy which dominated the historical horizon for a long time and still condition popular images of the last times of the Republic. These stereotypes also affected the traditional view of Venice's policy toward Padua University. In 1888, for example, Antonio Favaro claimed that in the 18th century "studies had greatly worsened", but added also that "such decadence was noticed first by the *Riformatori* " who "put more effort into the care and measures necessary to face the decline".[10] Fifty years later, the opinion of Attilio Simioni was far more severe: "the decline of the University simply mirrored the rapid decline of the Republic".[11]

Both of these scholars took for granted the inexorable and pervasive decadence of the Republic during its last decades. Decadence in the political and military sphere is undeniable, but it cannot arbitrarily be extended, for example, to the economic, much less to the artistic sphere, or to culture in general. In the Veneto, after all, this was the century of Giambattista Tiepolo, of Carlo Goldoni, and of Antonio Canova, to mention no more. According to both Favaro and Simioni, the crisis of the University depended on the political crisis of the Venetian Republic, although Favaro paid respect to the good will of the *Riformatori*.

These preconceived opinions appear largely unjustified in the light of new data, which—although they do indicate inconsistencies within the Venetian policy guided by the *Riformatori*—demand that we dismiss the image of general and inexorable decline, in favor of a more complex curve, nearly horizontal in the early half of the century (and, in certain respects, even later); but which rises in the 1760s and 1770s and reaches stability in the last decades of the century at much higher levels than earlier. As a matter of fact, many indicators—from the University budget to the quantity and quality of scientific "departments" and the renewal of the range of courses—show that the 18th century was on the whole (despite the limits emphasized by Cesarotti and by others), a century of progress for the University of Padua; and this is particularly true for the two decades of active reform which brought it back, at least in Italy, to its former prestige.

For example, in the second half of the century University expenditure increased—substantially, considering the low rate of inflation: 25,000 ducati on average between 1736 and 1750, but more than 30,000 in 1755, around 37,000 in the 1760s, more than 46,000 in 1770, and at around 40,000 for the following 13 years. It is worth noticing that the composition of the expenditure greatly changed, as the portion for the salaries of professors expressed as a percentage of the total dropped from 89% in 1736 to 80% in 1750, to 68% in 1755, to 79% in 1760, and to 60%—the lowest—in 1770; later the percentage rose again to 63% in 1775 and 74% in 1783.[12]

The total for the salaries of professors depended more on their longevity than on the number of chairs held; there were no age limits, and as a rule salaries increased greatly on reappointment, that is to say every 6 years. It might happen, as with to Giambattista Morgagni and to Giovanni Poleni, that professors with a long academic career received an annual amount of money equal to that paid to nearly 10 younger colleagues in their first posts put together. It is clear then that the retirement of one such figure might greatly affect the so-called "sum for lecturers", so that these variations in the expenses of the University are not always necessarily to be ascribed to changes in cultural policy.

It is clear from the case in point, that the great increase in the portion of the University budget assigned to other expenses, as compared with expenditure on salaries for professors, recorded in the two decades of reforms, reflected the considerable resources invested in the "departments" of scientific research. On February 27th 1779, the *Riformatori*, including Lorenzo Morosini, the leader of Venetian cultural policy for 20 years, presented to the Senate a "proposal" which suggested the foundation of "national" Academy of Science, Letters, and Art previously mentioned, they recommended Padua as the seat of the new institution on the grounds of the fact that the town "provides a great number of men of doctrine and letters inside and outside the University, and above all in view of all the other institutions which have lately, thanks to magnificent liberality, been created for the use of that University and which might also be of use for the Academy".

In their proposal, the *Riformatori* recorded with great precision a dozen of these existing "institutions" but without noting the dates of foundation: "Public Library, Botanic Garden, Museum of Natural History, Anatomy Theater, Experimental Theater, Astronomical Observatory, Chemical Laboratory, Agrarian School, Obstetrics School, Medical, and Surgical Schools in the Hospital *degli infermi*, School of Veterinary Medicine".[13] Two of them, the Botanic Garden and the Anatomic Theater, dated back to the 16th century; the Public Library, a University Library in all respects, the first in Italy, was a 17th century foundation. The Museum of Natural History and the Experimental Theater [a laboratory of physics], had been inaugurated in 1734 and in 1740. The remaining seven "institutions" dated to "these times"; specifically, to the 15 years before 1779.

The progressive accumulation of the wealth of scientific university institutions shows that, although the 17th century had been ushered in Padua by Galileo Galilei, it had not left a great institutional heritage for the next; and also that the early 18th century had made only a poor contribution compared with the latter part. The increase in fact was even steeper, since the list given in the proposal omitted two further "institutions" of the second half of the 18th century: the very small laboratory of Nautical Science and Architecture, founded by Poleni in the late 1750s,[14] and the School of Practical Architecture, founded in 1771.[15]

This double exclusion is understandable. Since 1756, the professor of physics (Poleni was to be succeeded by Giannalberto Colombo and Simone Stratico) had been charged with both the chair of his subject and the Nautical Science Laboratory; as a result this laboratory was as yet poorly equipped and scarcely a distinct "institution". The School of Practical Architecture had been founded at the request of the Paduan corporations of the carpenters, bricklayers, and stonecutters and, although the teacher of architecture Domenico Cerato held the title of University professor, as a technical school for artisans, the School of Practical Architecture was not counted as an institution "created for the use of the University", or as offering a scholarly resource to the Academy.

The Museum of Natural History had a different origin from all the other institutions. It had been acquired, exceptionally, by barter: Antonio Vallisneri junior had offered his father's—Antonio Vallisneri senior—valuable scientific and archaeological collection to the *Riformatori*, in exchange for the creation of a new chair of *altri semplici* to be occupied by himself. The title of "other simples" was to cover natural elements of use in medicine, other than those whose nature and uses were taught by the professor of botany.[16] The *Riformatori* could not slip the opportunity of endowing the University, at no expense, with its first museum and ensuring also the service of a teacher who should take care of it.

The foundation of the laboratory of physics on the other hand had been quite normal. It was approved in 1739 and opened in the following year.[17] This had not been Padua's first recognition of modern science: Brendan Dooley shows, for example, that although the presence of Aristotle and Avicenna continued well into the 18th century to frequent the "rotuli" of the University (that is to say the list of disciplines published every academic year), yet the courses of the greatest men of science of the time in Padua, from Vallisneri senior to Poleni, took account of the discoveries and the research methods of Galilei and Descartes as well as of Newton and Leibniz.[18] Indeed, the chair of mathematics was held from 1707 to 1719 by two followers of Leibniz who came from abroad, Jakob Hermann and Nicolas Bernoulli.[19]

As early as 1715, the *Ricordo per la riforma dello Studio* drawn up by the marchese Scipione Maffei of Verona had focused on the need to modernize the scientific disciplines taught in Padua[20]—implying that the chairs of chemistry, physics, and astronomy should be endowed with laboratories or observatories—and this choice had been shared with the other advisers of the *Riformatori*, from the physician and naturalist Vallisneri senior to the two Venetian patricians Antonio Conti and Francesco Grimani Calergi. But at that time, the *magistrato* had turned a deaf ear to plans intended to give a new form to the University and to replace "disciplines which were studied in the darkness of 300 and 500 years ago" with "new sciences".[21]

In 1733, a conversation partner of the young patrician Bernardo Nani pointed out succinctly that "the University of Padua was one of the most important in Italy, but since the method of studying was changing (everything was to be done according to experience), the principles of the method of teaching in their University were also changed by states. Venetians kept the old method, few are of the opinion to change it, holding the opinion that what they changed of the old was for the worst and those few do not mean to move as they could".[22] But, it was not only because of a stainless conservatism that powerful patricians were induced to maintain an "old method" and consequently to hinder "new sciences".

In the 1730s Nicolò Donà made it clear that in Venice the University was valued above all as a "center of commerce"—a source which, if opportunely run, could contribute to "enriching" the Republic. Donà was convinced that the University was endowed with "perfect laws within" and that the laws in force should not be altered; the aim of rendering the University "more and more flourishing with a rising number of students" and thus a productive "center of commerce" could be achieved, in his opinion, by accommodating to the students' requests and views without caring too much for the validity of studies.[23]

From the early decades of the century, the University of Padua continued to register a diminishing number of attending students and consequently a deterioration of its "profitability" (in about 1720 there was a teacher for every five students). This commercial view of the University required that its balance sheet be improved by reducing the number of chairs and by avoiding the creation of new ones. In this regard, however, it is to be noticed that the University roll was "pyramidal", implying that the chairs were differentiated in *primo, secondo*, and *terzo luogo* [first, second, and third place] and in "ordinary" and "extraordinary", thus many chairs were dedicated to the same subject. For example, early in the 18th century the Faculty of Arts included seven chairs of theoretical medicine, six chairs of practical medicine, five chairs of philosophy, three chairs of logic, two chairs each of botany, of anatomy, of surgery, of sacred theology, and metaphysic. The single chairs were only four (or five including the appointment *de pulsibus et urinis*, which was "temporary" owing to the low esteem in which the medical clinic held): holy scripture, moral philosophy, mathematics, Greek, and Latin humanistic disciplines. This situation made it possible to abolish many chairs without endangering the variety of disciplines or impoverishing the *curriculum* of studies.

Until 1738, however, loyalty to "old method" discouraged the *Riformatori* and the Senate from eliminating more than a few chairs, half a dozen in all; in that period they made appointments only to two new chairs, one of ecclesiastical history and the other of natural history, whose peculiar origin we have mentioned above. It is not surprising, under these conditions, that Poleni obtained his laboratory of physics only with great difficulty, although since 1719 the secretary of the *Riformatori* warmly approved its foundation; nor that the new chair of experimental physics was appointed only after seven chairs had been dropped from the roll in 1738–1739.

This drastic reduction of chairs afforded such financial "relief" to the University that the *Riformatori* felt justified in gradually introducing new courses. Scientific research, which had already been enriched by the new courses held by Vallisneri junior and Poleni, was further strengthened in the early 1740s by the creation of courses in: Rudiments of geometry, Nautical theory, and naval architecture. The introduction of the last of these is of particular interest, in that it anticipated the spirit of reform which was to dominate the 1760s. Such a discipline could make no strong claim on academic grounds. It is clear that lawyers and physicians, the two main professions accredited by the University, would draw little advantage from becoming "apt at navigation". The Senate invoked the broad and effective patriotic argument: namely "the dignity of the public name and [......] the benefit of the state and subjects, in order to render them able to navigate and become expert in an art, whose cultivation is as necessary to our Republic, as to any other maritime power".

As a matter of fact, Venetian authorities expected mostly from the new professor, Gian Rinaldo Carli, a prominent figure, a contribution to the effort to improve the quality of the arts of navigation and of naval architecture of the Republic. He was expected to provide a successful shift from traditional and empirical knowledge, which was less and less adequate to equip the Venetian navy to face international competition, to a solid, scientifically based knowledge.[24]

It had long been the practice of the Venetian government to call on University professors as expert consultants. But, in the case of the chair of nautical theory there were two important innovations: first, its foundation was meant explicitly to develop one branch of the economy. Second, one of the charges to the professor was to examine candidates for the post of sea captain among the students from the Nautical school founded in Venice in 1739, and this was a sign of the shift of the University toward an "open" institution—in this case only indirectly—to the technical world; moreover by placing a specialized secondary school under the control of the University, it marked the first step toward a pyramidal education system like the one which could be created in the Napoleonic age.

Closer and closer attention was now being paid to economic progress (entailing a deeper link between science and technology or rather the requirement that traditional technical methods should give way to procedures based on scientific research). The University subjects most valid were those that could contribute to the needs of the State and society and to "public happiness". Thus it was a necessity, consistent with such progress, to transform those disciplines at the professors' disposal, into practical disciplines based, as far as possible, on the principle "everything was to be done according to experience". These are the more important characteristics of the reforms carried out in Padua from 1761 as regards scientific research. They did not have a very successful start with the foundation of the course in nautical science, for

Carli's contribution was poor and fleeting; he quickly gave up teaching and neglected scientific interests to devote himself to economy and management.

A letter, sent in late 1760 by Simone Stratico, who was at that time a professor of medicine, to a Venetian friend in close contact with the *Riformatori*, was the manifesto of the new project of reform. The "three principal objects" of the University were: to put at the disposal of the State an *élite* of "subjects expert at those sciences and arts which are useful for the benefit of society as a whole, and which can frequently be exploited for the benefit of the Serenissima in particular", "to spread by their efforts and study the achievements of sciences and arts throughout the Republic and to give people appointed to the use of arts, which human kind is in need of" and lastly, to attract "thanks to these hopes, a great number of young people to the city destined as the seat of the University", to ensure its prosperity—as Nicolò Donà stressed—as a "center of commerce". Given these aims, Stratico thought it necessary to reconsider—as Maffei did—the roll of subjects taught and to create two new educational courses: a degree in "philosophy" (that is sciences) and, a certificate in "mathematics" (that is, engineering).[25]

Although the new educational syllabus proposed by Stratico was not included in the reform passed in 1761, the necessity to cultivate not only the disciplines necessary for the professional education of men of law, physicians, surgeons, and theologians, but also all the other "arts, which humankind is in need of" was not put aside. Indeed, the *Riformatori* of 1761 could identify themselves with the principle of "the subjects intended to guarantee the greater utility and happiness of the peoples" that they made the Senate appoint the creation of a new course in agrarian science—the first in the world—and a course *de morbis mulierum, puerorum et artificum*, that is gynecology, pediatrics, and workers' illnesses.[26] Farmers and artisans, women and children, the poorest classes and conditions most ignored by the well-educated society of old régime were now to receive attention. Scientific progress thus—although obviously not on "populist" grounds—was meant to align itself with economical and social progress.

Despite conservatives' opposition, which at first led to a radical reduction of the 1761 reform, the "progressive" side of aristocracy ended by prevailing in the short term and was also able, as we have seen, to promote the creation of several new scientific "departments". In a very short time a chemical laboratory, an astronomical observatory, an experimental farming school, a school of obstetricians, medical and surgical clinics in San Francesco hospital, a school of practical architecture, and a college of veterinary provided with an Anatomy Theater. A doctor at the medicinal baths of Abano was also added to the roll, which allowed the University to acquire something approximating department, one specifically linked with the Venetian territory. That aim was also fulfilled, in different ways, by the experimental farm linked with an agriculture academy, by the school of obstetricians, which was intended to train rural practitioners and by the school of practical architecture mentioned above.

In addition, the *Riformatori* promoted the development of the University Library of Padua into an institutional instrument of research.[27] Finally, as a crowning achievement, they founded in Padua the *Accademia di Scienze, Lettere ed Arti*. It was the consecration of a "national" cultural strategy (the structure of the Academy consisted of 24 members *pensionari* living in Padua, most of them teachers of scientific subjects at the University, and of 16 "national" members drawn from the other cultural centers of the Republic). It was also designed to elicit from an academic body made up of teachers, who often did not show any interest in research, an *élite* of academic researchers who were to put their skills "in sciences and arts of use to the whole society" at the disposal of the state.[28]

If one assesses the actual impact of these reforms as a whole on the academic life of Padua, and even more on the Venetian economy and society in general, of course the wide gap between intention and limited achievement is noticeable. Nevertheless, one cannot forget that, although with limits, in the late 18th century Venetian policy toward the University of Padua manifested a belief in scientific progress that, compared to the contemporary experiences of other states, appears one of the most progressive.

NOTES

[1] P. D. Negro, "Appunti sul patriziato veneziano, la cultura e la politica della ricerca scientifica nel secondo Settecento", in G. Bozzolato, D. Negro, and C. Ghetti (eds.), *La Specola dell'Università di Padova* (Brugine: Edizioni 1 + 1, 1986), pp. 291–292.

[2] Ezio Vaccari, 1992–1993, "I Manoscritti di uno scienziato veneto del Settecento: notizie storiche e catalogo del fondo 'Giovanni Arduino' della Biblioteca Civica di Verona", *Atti dell'Istituto Veneto di scienze, lettere ed arti, Classe di scienze fisiche, matematiche e naturali*, tomo CLI, pp. 271–385; Vaccari, *Giovanni Arduino (1714–1795). Il contributo di uno scienziato veneto al dibattito settecentesco sulle scienze della Terra* (Firenze: Leo S. Olschki, 1993); D. Negro, "Giovanni Arduino e i Deputati all' agricoltura", in E. Curi (ed.), *Scienza tecnica e 'pubblico bene' nell'opera di Giovanni Arduino (1714–1795)* (Verona: Fondazione Cassa di Risparmio di Verona Vicenza Belluno e Ancona, 1999), pp. 145–192.

[3] D. Negro, "Proposte illuminate e conservazione nel dibattito sulla teoria e la prassi dello Stato", in G. Arnaldi and M. P. Stocchi (eds.), *Storia della cultura veneta,* V/2, *Il Settecento* (Vicenza: Neri Pozza, 1986), p. 144.

[4] *Materie attinenti al Magistrato de' Riformatori.* Venezia: Biblioteca del Museo Civico Correr, mss. Donà dalle Rose 335/III/33.

[5] G. Tabacco, *Andrea Tron (1712–1785) e la crisi dell'aristocrazia senatoria a Venezia* (Udine: Del Bianco, 1980), pp. 149–151 (but, in my opinion, Tabacco underestimates the importance of *nomenklatura* in Venetian policy).

[6] M. Cesarotti, 1797, *Introduzione generale* [to a reform plan of Padua University], Vicenza, Biblioteca Bertoliana, mss, 1223 (G.8.6.1), II/2, f. 5; D. Negro, " 'L'Università della ragione spregiudicata, della libertà e del patriotismo.' Melchiorre Cesarotti e il progetto di riforma dell'Università di Padova del 1797", in L. Rossetti (ed.), *Rapporti tra le Università di Padova e Bologna. Ricerche di filosofia medicina e scienza*, Centro per la storia dell'Università di Padova, Contributi 20 (Trieste: Lint, 1988), pp. 375–402.

[7] D. Negro, "Giacomo Nani e l'Università di Padova nel 1781. Per una storia delle relazioni culturali tra il patriziato veneziano e i professori dello Studio durante il XVIII secolo", *Quaderni per la storia dell'Università di Padova* 13: 77, 112 (1980).

[8] D. Negro, "L'Università", 1986, p. 69.

[9] M. Berengo, *La società veneta alla fine del Settecento. Ricerche storiche* (Firenze: Sansoni, 1956); G. Cozzi, "Politica e diritto nei tentativi di riforma del diritto penale veneto nel Settecento", in V. Branca (ed.), *Sensibilità e razionalità nel Settecento*, Vol. 1 (Firenze: Sansoni, 1967), pp. 373–421; Cozzi "Fortuna, o sfortuna, del diritto veneto nel Settecento", in Cozzi (ed.), *Repubblica di Venezia e Stati italiani. Politica e giustizia dal secolo XVI al secolo XVIII*

(Torino: Einaudi, 1982); G. Torcellan, *Una figura della Venezia settecentesca. Andrea Memmo. Ricerche sulla crisi dell'aristocrazia veneziana* (Venezia-Roma: Istituto per la collaborazione culturale, 1963); Torcellan, *Settecento veneto e altri scritti storici* (Torino: Giappichelli, 1969); F. Venturi, *Settecento Riformatore*, V, *L'Italia dei lumi*, tomo II, *La repubblica di Venezia (1761–1797)* (Torino: Einaudi, 1990); P. D. Negro and P. Preto (eds.), *Storia di Venezia*, VIII, *L'ultima fase della Serenissima* (Roma: Istituto della Enciclopedia italiana, 1998).

[10] A. Favaro, "Lo studio di Padova e la repubblica veneta", *Atti del R. Istituto Veneto di scienze, lettere e arti*, serie VI, tomo VI, p. 25.

[11] Attilio Simioni, 1941–1942, "Lineamenti di storia politica dell'Università di Padova", *Atti e memorie della R. Accademia di scienze lettere ed arti in Padova. Memorie della classe di scienze morali*, n.s. 58, p. 69.

[12] See the item "budget for the University of Padua" in F. Besta (ed.), *Bilanci generali*, III (1736–1755) (Venice: Tipografia Emiliana, 1903), and A. Ventura (ed.), *Bilanci generali*, IV (1756–1783) (Venice and Padua: Tipografia Antoniana, 1972), pp. 178–179 and 190–193.

[13] D. Negro, 1986, "Appunti sul patriziato veneziano", p. 265.

[14] G. A. Salandin and M. Pancino, *Il 'teatro' di filosofia sperimentale di Giovanni Poleni*, Centro per la storia dell'Università di Padova, Contributi 19 (Trieste: Lint, 1987), pp. 252–253; D. Negro, "Alcune note su Gianrinaldo Carli tra Padova e Venezia", in *Acta Histriae*, 1997, Contributi dal Convegno internazionale *Un gran riformatore del '700. Gian Rinaldo Carli tra l'Istria, Venezia e l'Impero*, Vol. 5, pp. 153–154.

[15] G. Brunetta, *Gli inizi dell'insegnamento pubblico dell'architettura a Padova e a Venezia. Cronaca e storia*, Centro per la storia dell'Università di Padova, Contributi 8 (Padova: Antenore, 1976).

[16] In 1759, Vallisneri's chair would receive the clearer and more dignified title of Natural History. On these distinctions, see Mocholi in this volume.

[17] Salandin and Pancino, *Il 'teatro' di filosofia sperimentale*, 1987.

[18] B. Dooley, "Science Teaching as a Career at Padua in the Early Eighteenth Century: the Case of Giovanni Poleni", *History of Universities*, Vol. 4, 1984, pp. 116–151; Dooley, "La scienza in aula nella rivoluzione scientifica: dallo Sbaraglia al Vallisneri", *Quaderni per la storia dell'Università di Padova* 21:23–41 (1988).

[19] A. Robinet, *L'empire leibnizien. La conquête de la chaire de mathématiques de l'Université de Padoue. Jacob Hermann et Nicholas Bernoulli (1707–1719)*, avec la collaboration de Maria-Vittoria Predaval et Nelly Bruyère, Centro per la storia dell'Università di Padova, Contributi 22 (Trieste: Lint, 1991).

[20] B. Brugi, "Un parere di Scipione Maffei intorno allo Studio di Padova sui principi del Settecento, Edizione del testo originale con introduzione e note", *Atti del R. Istituto Veneto di scienze, lettere e arti* 69:575–591 (1909–1910).

[21] "Informazione sopra lo Studio di Padova", anonymous but attributed to Francesco Grimani Calergi, in D. Negro, "L'Università", p. 59.

[22] D. Negro, "Bernardo Nani, Lorenzo Morosini e la riforma universitaria del 1761" *Quaderni per la storia dell'Università di Padova* 19:87 (1986).

[23] D. Negro, "L'Università di Padova negli anni 1730", in *Carlo Goldoni dottore 'in utroque iure' a Padova*, in *Quaderni per la storia dell'Università di Padova* 30:16–17 (1997).

[24] D. Negro, "Alcune note su Gianrinaldo Carli", 1997, pp. 141–150.

[25] D. Negro, "Pensieri di Simone Stratico sull'Università di Padova (1760)", *Quaderni per la storia dell'Università di Padova* 17:213, 221 (1984).

[26] "Scrittura dei Riformatori dello Studio di Padova del 24 aprile 1761", in D. Negro, "Bernardo Nani, Lorenzo Morosini e la riforma universitaria del 1761", *Quaderni per la storia dell'Università di Padova* 19:130 (1986).

[27] T. P. Marangon, *La Biblioteca Universitaria di Padova dalla sua istituzione alla fine della Repubblica veneta (1629–1797)*, Centro per la storia dell'Università di Padova, Contributi 11 (Padova: Antenore, 1979), pp. 143–156. See also, for this and the other Paduan scientific and

cultural institutions, L. S. Rea (ed.), *Istituzioni culturali, scienza, insegnamento nel Veneto dall'età delle riforme alla Restaurazione (1761–1818)*, Atti del Convegno di studi, Padova 28–29 maggio 1998, Centro per la storia dell'Università di Padova—*Contributi* 32 (Trieste: Lint, 2000); Università degli Studi di Padova—Centro Musei scientifici (ed.), *La curiosità e l'ingegno. Collezionismo scientifico e metodo sperimentale a Padova nel Settecento* (Padova: Centro interdipartimentale di servizi musei scientifici, 2000); *Dall'Accademia dei Ricovrati all'Accademia Galileiana*, Atti del Convegno storico per il IV centenario dalla fondazione (1599–1999), Padova, 11–12 aprile 2000 (Padova: Accademia Galileiana di scienze lettere ed arti in Padova, 2001); P. D. Negro and F. Piovan (eds.), *L'Università di Padova nei secoli (1601–1805). Documenti di storia dell'Ateneo*, Centro per la storia dell'Università di Padova (Treviso: Antilia).

[28] D. Negro, "Appunti sul patriziato veneziano", 1986, pp. 264–294. See the biographies of professors and scientists in S. Casellato and L. Sitran Rea (eds.), *Professori e scienziati a Padova nel Settecento*, Contributi alla storia dell'Università di Padova, *Profili biografici* 3 (Treviso: Antilia, 2002). See also L. Pigatto (ed.), *Giuseppe Toaldo e il suo tempo nel bicentenario della morte. Scienze e lumi tra Veneto e Europa*, Atti del Convegno, Padova, 10–13 novembre 1997, Centro per la storia dell'Università di Padova, Contributi, 33—Osservatorio astronomico di Padova (Cittadella: Bertoncello Artigrafiche, 2000).

PAUL WOOD

CANDIDE IN CALEDONIA: THE CULTURE OF SCIENCE IN THE SCOTTISH UNIVERSITIES, 1690–1805

Among the extant manuscripts of the Scottish polymath Thomas Reid (1710–1796), there is a single sheet taken up with an algebraic problem on the recto and a geometrical demonstration on the verso which also contains the note "April 24 1758 Planted the Northermost Row of my Pittatoes of Small Seed uncut the two next, of the largest seed Cut the fourth of Midling Seed Cut. and Same day planted 1700 Cabbage at 7 Score to the hundred".[1] In the spirit of Carlo Ginzburg, I want to take this brief entry as a "clue",[2] and use it to explore the place of natural knowledge in the culture of the Scottish universities during the Enlightenment. I do so in the belief that 18th-century Scotland constitutes one of the most important and revealing episodes in the history of academic science and medicine in early modern Europe. For it is worth remembering that during the so-called "long 18th century" Scotland emerged as one of the centers within the Atlantic world for teaching and research in the sciences, thanks to the dynamism of her universities and the dramatic rise of the Edinburgh medical school.[3] Although Scottish academe was riven by religious disputes through much of the second half of the 17th century, the establishment of a Commission of Visitation by the Scottish parliament in 1690 launched the universities on a course of reform and renewal which transformed Scottish higher education and nurtured the pursuit of the natural sciences and medicine. Such was the success of the Scottish universities that in the momentous year of 1789, Thomas Jefferson declared that as far as "Science" was concerned "no place in the world can pretend to a competition with Edinburgh".[4]

But the very success of university science and medicine in 18th-century Scotland poses a complex historical problem. For how was it that a nation which in the 1690s seemed to many of its own inhabitants to be impoverished and backward came to occupy such a prominent position in the republic of letters in the space of less than a hundred years? This is a question which has elicited a variety of responses since it was first posed by the Scottish moralist and mathematician Dugald Stewart in the early decades of the 19th century. For Stewart, part of the answer lay in the intrinsic "liberality" of the Scottish universities, which allowed them to respond creatively at the end of the 17th century to the new currents of thought flowing into Scotland from the continent and, more importantly, from England. Consequently, even though Stewart held that Scotland's "philosophical age" only really blossomed after the last Jacobite rebellion in 1745, he maintained that in the realm of "the mathematical sciences... Scotland, in proportion to the number of its inhabitants, was never, from

M. Feingold and V. Navarro-Brotóns (eds.), Universities and Science in the Early Modern Period, 183–199.

the time of Neper [sic], left behind by any country in Europe".[5] Aspects of Stewart's analysis have since been developed in very different ways by Lord Dacre and, more recently, by Roger Emerson, who has emphasized that the origins of the Scottish Enlightenment lay in the flourishing medical and scientific culture of late 17th century Scotland.[6] Taking Emerson's work as my starting point, in what follows I will sketch out one possible answer to Dugald Stewart's question as it relates to science and medicine in the Scottish universities. I will focus on the following themes and issues: (i) the role of the ethos of "improvement" in fostering the study of natural knowledge; (ii) the impact of the ideal of "politeness" on the transformation of the curricula of the Scottish universities during the long 18th century and the ways in which the related notion of "the gentleman" shaped the teaching of mathematics and the natural sciences; (iii) the interplay between religion and natural knowledge; (iv) the specific institutional features of the Scottish universities which affected the cultivation and transmission of natural knowledge; and (v) the part played by the universities in the creation of public science in the Scottish Enlightenment. By way of conclusion, I will also suggest that Scotland's "philosophical age" owed much to the enlightened patronage of political managers like the third Duke of Argyll.

I

Although it may seem unremarkable to us that Thomas Reid was planting potatoes and a large number of cabbages in his garden in lateApril 1758, his gardening activities are in fact of considerable historical interest, for it was only in the mid-18th century in Scotland that the potato began to be a staple part of the diet of most Scots rather than a luxury food of the landed classes.[7] Reid's cultivation of potatoes and the experimental investigation of the chemical properties of potatoes which he carried out earlier in the year alert us, therefore, to the ethos of improvement which pervaded the culture of the Scottish Enlightenment and which helped to sustain the growing importance of mathematics and the natural sciences within the universities from roughly the 1690s onwards.[8]

An early instance of the application of natural knowledge to the improvement of the Scottish economy is to be found in the work of the Glasgow regent and later Professor of Mathematics, George Sinclair. Like the Fellows of the Royal Society of London, with whom he had brief but unsatisfactory contact in 1662, Sinclair sought to satisfy economic need through the promotion and use of practical mathematics and experimental philosophy. Beginning in the 1660s, Sinclair advised various landed Scots about the draining of mines, and he was employed for a time by the Edinburgh Town Council to improve the town's water supply. Toward the end of his life, he also published an elementary textbook on astronomy and navigation which reflected the kind of teaching he did in Edinburgh during the years 1666–1689, when he was out of favor with the authorities in Glasgow because of his Presbyterian sympathies and had to support himself as a schoolmaster and public lecturer in the capital.[9] Sinclair's chequered career illustrates that at least some landowners in the Scottish lowlands and borders were looking to improve their estates in the period and that at least some academics were keen to promote economic advance through the application of mathematics and the natural sciences.

The desire for improvement became most acute, however, in the bleak years of the late 1690s, when successive crop failures resulted in serious famine and the dismal failure of the Scots colony on the Darien isthmus reinforced anxieties about the perilous state of the Scottish economy. An increasing number of Scots were also deeply troubled by the atmosphere of religious bigotry engendered by the long-standing schism between Episcopalians and Presbyterians, for such divisions seemed to compound Scotland's problems. Moreover, greater contact with men of science in England and on the continent meant that many Scottish savants familiar with the cabinets, botanic gardens, instruments, and institutions found abroad wanted to replicate them at home, as can be seen, for example, in Sir Robert Sibbald's abortive attempt to establish a Royal Society of Scotland in 1701.[10] Nor did the sense of shame and backwardness entirely dissipate as the century progressed. Returning to his native land after studies in London and Paris, the Aberdeen physician and naturalist David Skene despondently observed that "Scotland was a truly mortifying sight".[11] The desire for improvement was, therefore, both long-lasting and multifaceted, encompassing not only the economic but also the cultural and intellectual spheres.

The ethos of improvement which took shape in the late 17th century proved to be a potent resource for legitimating the promotion of mathematics and the natural sciences in the decades which followed. In our own day, many universities are starved for funding, and a similar situation existed in Scotland at the beginning of the 18th century. The shortage of cash meant that it was difficult for the Scottish universities in this period to acquire the scientific instruments needed for research, or for teaching courses of experimental philosophy featuring demonstration experiments patterned on those initially popularized in an academic setting by John Keill and J. T. Desaguliers. In order to raise the necessary funds, King's College Aberdeen, Glasgow, St. Andrews, and Marischal College Aberdeen all launched subscription schemes which targeted the nobility and landed gentlemen as potential donors. The earliest scheme was begun by the Professor of Mathematics at King's, Thomas Bower, in 1709. Bower shrewdly played the improvement card to attract financial support in his "Proposals For Bying Mathematical Instruments for the use of the Kings College of Aberdeen", arguing that benefactors should contribute money because:

> A stock of Mathematical Instruments about an University is of so publick advantage, as well for the many usefull observations and experiments which may be dayly made by the Professors themselves for the improvement of Learning as the instruction of the Youth under their care, by shewing them to their eys the Truths that are abstractly demonstrated in Mathematicks, and their great usefulness in all Arts and Sciences, which conduce to the happiness of human life and societies.[12]

The masters at Glasgow likewise emphasized the usefulness of experimental learning in their own appeal of 1710, adding that "Natural Philosophy as Founded on Demonstration and Experiment" was also "Pleasant and Entertaining". At St. Andrews nothing was said about utilitarian benefits to the public when a set of proposals was issued around 1724, but at Marischal College Aberdeen in 1726, the regents and professors were more explicit. Complaining of the "Meanness of their Funds", the members of Marischal urged "all Noblemen, Gentlemen and Lovers of Learning, particularly such as have been educated in this College" to contribute funds for the acquisition of experimental hardware, books, and "the newest *Machines* in Husbandry" because they hoped their proposed course of experimental philosophy would "very much tend to the Improvement and Advantage of these Northern Shires".[13] What these proposals show is that innovations in the teaching of mathematics and natural philosophy were typically justified in terms of material utility and improvement. And even if the proposals manifest an unstable mixture of financial self-interest and genuine concern for the public good, the professed aspirations of the universities suggest that the ethos of improvement did prompt the incorporation of courses of experimental philosophy into the curriculum, and hence contribute to the transformation of teaching methods which occurred in Scottish academe during the first half of the 18th century.[14]

The modalities of usefulness and improvement also shaped the systematic reform of the curricula at King's and Marischal Colleges which took place in the winter of 1752–1753. Whereas the curricula of the other Scottish universities changed in a piecemeal fashion during the course of the 18th century, the *cursus philosophicus* of the two Aberdeen colleges was restructured at one fell swoop and the remaining traces of scholasticism were eliminated. Although the teaching of natural philosophy was an integral part of the curriculum at both King's and Marischal prior to 1752, the reforms increased the amount of classroom time devoted to instruction in mathematics and the natural sciences. Natural history became a formal part of the philosophy course taught in the second year of study. The various branches of natural philosophy now occupied the whole of the third year, and pupils were given more training in mathematics so that they could cope with the technicalities of astronomy, optics, mechanics, and the other sciences they were introduced to. Thus over one-third of the new curricula at King's and Marischal would in future be given over to natural knowledge, and in 1754 King's resolved to establish a museum containing natural history specimens, instruments, and models, along with a chemical laboratory for the benefit of its students.

For their part, the masters and professors at the two colleges justified the sweeping changes they made in terms of the promotion of useful knowledge. One of the principal architects of the reforms at King's was Thomas Reid, and his committee stated that instead of dwelling on "the Logic and Metaphysic of the Schoolmen, which seem chiefly contrived to make Men subtle Disputants, a Profession justly of less Value in the present age", the college would in future devote itself to "teaching those parts of Philosophy, which may qualify Men for the more useful and important Offices of Society". Moreover, they noted that their projected museum was "intended, not for Shew, or for gratifying the Curiosity of idle People, the Use that is too commonly made of such Collections, but for the Improvement of the Students in the Knowledge of the Works of Nature, and the most useful Operations of Art".[15] Similar claims were made at Marischal. Reid's associate Alexander Gerard spoke for the college when he argued that the revised sequence of subjects they had adopted was "that in which the sciences will afford most light to one another, and in which they will have the most useful influence on life". Usefulness was also invoked in his defence of the inclusion of civil and natural history in the curriculum:

> it is ... a considerable advantage in their new plan of teaching, that by it these useful branches of study are introduced into the scheme of Education. Natural history, besides its advantages already mentioned [that is, it prepares the mind for the study of philosophy], is the immediate foundation of almost all the arts of life, agriculture, gardening, manufactures, medicines, &c.[16]

According to Gerard, Reid, and their Aberdeen colleagues, therefore, a university education ought to provide students with useful learning which would contribute to the improvement of the individual and of society more generally. In the expectation that natural knowledge could be used to realize their aspirations, the Aberdonians assigned mathematics and the natural sciences a much more prominent place in the curriculum than they had previously occupied.[17]

The emphasis on agriculture in the pamphlets and handbills issued by Marischal College illustrates just how important agricultural improvement was to many Scots during the long 18th century, and how the universities, albeit fitfully, responded to the perceived need for an institutional basis for research and teaching.[18] In 1743, the Secretary of the Honorable the Society of Improvers in the Knowledge of Agriculture in Scotland (founded in 1723), Robert Maxwell, called for the establishment by the Crown of a Professorship in Agriculture,[19] but it was not until 1790 that a chair was finally founded at the University of Edinburgh through the generosity of a private individual. In the interim William Cullen lectured on soils and the nourishment of plants in the context of his chemistry course at Glasgow beginning in 1748 and, after his move to the University of Edinburgh, Cullen delivered a series of private lectures on agriculture in 1768.[20] At Aberdeen, similar topics relevant to agricultural improvement were covered in the natural history lectures of Francis Skene and James Beattie the younger at Marischal College, and in those of Thomas Reid at King's College.[21] The most sustained academic engagement with agriculture began, however, in Edinburgh in 1779 with the appointment of the Rev. John Walker as Professor of Natural History. Walker discussed agricultural matters in his course and, in 1789, began giving an additional series of lectures devoted entirely to agriculture at the University. The following year, his students were so enthused that they launched the Edinburgh Agricultural Society under Walker's patronage.[22] Yet despite his obvious success, Walker was passed over when William Pulteney founded a Chair of Agriculture at Edinburgh in 1790. The position instead went to Dr Andrew Coventry, who thus became the first Professor of the subject in Britain thanks to Pulteney's endowment. Why Pulteney, who was himself a Scot who had married well and was living in England, decided to endow the chair is not known, but his decision marked the long-awaited consummation of the marriage between agricultural improvement and academe in the Scottish Enlightenment.[23]

II

I turn now to consider how the ideal of politeness and the codes of "the gentleman" were bound up with the teaching of mathematics and the natural sciences in the Scottish universities. From an historiographical point of view, the following discussion should be seen as a critical response to the influential writings of Nicholas Phillipson and Steven Shapin. Natural knowledge does not figure in Phillipson's interpretation of the culture of politeness in 18th-century Scotland; Shapin, in his account of the tension between the roles of the scholar and the gentleman in early modern England, claims that the attempted institutionalization of the refashioned identity of the gentleman-scholar was a "failure".[24] I want to suggest an alternative to these positions, and to argue that the value of politeness and the identity of the gentleman shaped the culture of science as it developed in the Scottish universities from the 1690s onwards.

Thomas Reid again provides a necessary clue. In his lectures on the "culture of the mind" delivered soon after he moved to the Chair of Moral Philosophy at the University of Glasgow in the autumn of 1764, Reid spent some time discussing the principles of education, and among the modern authorities he cited on the subject he included his teacher at Marischal College George Turnbull, his deceased friend David Fordyce, and John Locke.[25] Locke's inclusion in Reid's list is highly significant, for it points to the considerable impact of Locke's *Some Thoughts concerning Education* (1693) on the direction taken by curriculum reform in Scottish higher education during the course of the 18th century. Along with the writings of Lord Molesworth, Locke's work fueled a public debate regarding the nature and purpose of a university education which to some extent deflected public attention from the proceedings of the Visitation Commission established by the Scottish Parliament in 1690; indeed this debate arguably informed the reshaping of the curriculum in Scotland in the early decades of the 18th century. Given the influence of Locke and Molesworth on Scottish attitudes toward education, it is important to note that both of them maintained that the primary task of the educator was to produce gentlemen trained for a life of civic virtue. To that end, Locke and Molesworth sought to reconcile the worlds of the scholar and the gentleman by eliminating pedantry and by making learning both polite and useful. Echoing contemporary critics of Aristotelian scholasticism, the two of them attacked the pedantry and dogmatism which they believed were still characteristic of the scholarly realm; they urged that scholars become men of the world by cultivating the art of polite conversation rather than that of disputation.[26]

While Molesworth had nothing to say about natural knowledge, Locke was convinced that "[i]t is necessary for a Gentleman in this learned Age to look into some of [the systems of natural philosophy] to fit himself for Conversation". Although this passage may suggest that Locke saw natural philosophy as being of largely ornamental value, he elsewhere argued that a knowledge of nature was morally and materially useful:

> There are very many things in [natural philosophy] that are convenient and necessary to be known to a Gentleman: And a great many other, that will abundantly reward the Pains of the Curious with Delight and Advantage. But

> these, I think, are rather to be found amongst such Writers as have imploy'd themselves in making rational Experiments and Observations, than in starting barely speculative Systems.[27]

For Locke, the gentleman was better advised to consult the writings of Robert Boyle and Isaac Newton, rather than those of Descartes, Gassendi, and the other system builders of the 17th century: Boyle and Newton provided matters of fact which revealed Divine design or illustrated principles which could be applied to practical ends.[28] When applied to higher education, Locke's writings could be read as suggesting that polite and virtuous gentlemen could be formed through a course of study which emphasized usefulness as well as rational edification, and which highlighted the new experimental philosophy championed by Boyle instead of the traditional Aristotelian natural philosophy of the schools.[29]

Such a reading of Locke is to be found in Scotland in the pamphlet first published in 1704, *Proposals for the Reformation of Schools and Universities*, which is sometimes attributed to the noted patriot and political theorist, Andrew Fletcher of Saltoun. Like Locke and Molesworth, the pamphlet's author wanted to effect a reconciliation of the scholar and the gentleman. Worried by the apparent decline of learning in Scotland, the author observed that:

> the natural Tendency of our present Methods is to unfit a Scholar for a Gentleman, and to render a Gentleman ashamed of being a Scholar. And, till we reconcile the Gentleman with the Scholar, it is impossible Learning should ever flourish. But was this once done, was Learning taken out of the Hands of the Vulgar, and brought to be as honourable and fashionable among the Gentry, as it is now contemptible, I think it would be in a fair way of prospering.[30]

To a modern reader, the argument that schools and universities should become more socially exclusive in order to promote the interests of learning strikes a decidedly dissonant note, but the curriculum outlined in the pamphlet seems far less alien. For the author suggested that in addition to instruction in the standard subjects of Greek, Latin, rhetoric, logic, ethics, and metaphysics, students ought to be thoroughly trained in the various branches of mathematics (namely geography, chronology, arithmetic, arithmetic, the elements of geometry and algebra, and plain and spherical trigonometry) so that at the end of their course they could proceed to the study of "*mixt* Mathematicks, or *Natural* Philosophy, *viz.* the Laws of *Motion*, *Mechanicks*, *Hydrostaticks*; *Opticks*, *Astronomy*, *&c.* and *Experimental* Philosophy".[31] Following Locke, the anonymous author of the *Proposals* thus saw mathematics and the natural sciences as forming an integral part of a gentleman's education, and it was this perception which came to dominate educational theory (and practice) in 18th-century Scotland.

Confirmation of this point comes in the pedagogical writings of the two figures Thomas Reid encountered at Marischal College during his student days, George Turnbull and David Fordyce. From early in his career, Turnbull was deeply engaged with questions related to education, and his experience as a university teacher and private tutor was eventually distilled in his *Observations upon Liberal Education*

(1742). The details of Turnbull's educational theories need not concern us here; suffice it to say that he echoed the sentiments of Locke, Molesworth, and Lord Shaftesbury in his condemnation of the traditional scholastic curriculum, and in his recommendation of an alternative scheme of education which aimed to provide the young with the knowledge they needed to pursue virtuous and useful lives in both the public and private spheres. Consequently, Turnbull valued mathematics and natural philosophy as much for their moral as for their material utility, believing as he did that the study of nature served to illustrate the argument from design; much the same view of the moral utilities of natural knowledge can be found in Fordyce's *Dialogues concerning Education*.[32]

Moreover, for Turnbull, as for the author of the 1704 *Proposals*, politeness involved the rejection of the values and modes of behavior associated with scholasticism. This meant that he drew a close connection between being polite and acquiring a stock of experimental (or experiential) and useful learning. In the preface to *The Principles of Moral Philosophy*, he juxtaposed the "tedious ungainful manner" of treating natural and moral philosophy which "of late more generally prevailed in the schools", with the "free elegant and pleasing way" in which some moderns and many of the ancients had cultivated these subjects. Crucially, Turnbull defined this style of inquiry in terms of the empirical method championed by Bacon and Newton, and the Lockean programme of eliminating meaningless metaphysical jargon; it was only by adopting this style that he believed both branches of philosophy could be rescued by "people of a more liberal and polite, as well as [a] more useful and solid turn" from the pedants.[33] In Turnbull's eyes, then, politeness was as much about matters of fact as it was about moral instruction. His construal of what it meant to be polite thus throws into sharp relief the limited conceptual horizons of scholars like Nicholas Phillipson who equate politeness with the urbane practical moralizing of Addison and Steele.[34] Given that natural knowledge was intimately connected across Europe with the polite, gentlemanly worlds of the court and the cabinet in the 17th century, it is no surprise that this cultural configuration found a place in the universities of the Scottish Enlightenment, or that figures such as the Marischal Professor of Mathematics, John Stewart, and the Edinburgh Professor of Natural History, John Walker, should have claimed that their respective subjects formed a necessary part of the education of a gentleman.[35] Chemistry too was taught as a form of gentlemanly knowledge at Edinburgh by William Cullen and his friend Joseph Black.[36] While it may be true, as Shapin has argued, that in England the identity of the gentleman-scholar popularized in the works of Robert Boyle and the apologists for the early Royal Society had a chequered career, in Scotland that identity was institutionalized in Scottish academe from the 1690s onwards.

III

What does the career of Thomas Reid tell us about the relations between religion and the natural sciences in the Scottish universities during the long 18th century? One salient feature of this relationship to which it draws our attention is the number of men teaching mathematics and the natural sciences who were educated in divinity or who actually became ministers in the Church of Scotland. Reid, for example, followed in his father's footsteps and embarked on a career in the Kirk in 1731; other men of science who were also clergymen include John Walker, as well as the Edinburgh Professors of Mathematics Matthew Stewart and John Playfair. For these academics, the moral utilities of natural knowledge were just as important as the material ones, and they routinely deployed information about the natural order to strengthen the rational foundations of Christian belief. From the Boyle Lectures of Richard Bentley onwards, the Newtonian system was presented to the English public as the best defence of true religion, and this message was not lost on the Scots. In all of the Scottish universities, the natural theological implications of Newton's discoveries were repeatedly highlighted by his many disciples, and it is arguable that the perceived congruence between Newtonianism and Christianity partly explains why Newton's ideas were so rapidly assimilated in Scotland at the beginning of the 18th century.

But Newtonian natural philosophy was not the only support for the argument from design, since natural history was likewise mined for evidence of the existence and attributes of God. In his lectures on natural history given at King's College Aberdeen, for example, Thomas Reid affirmed that while natural historical knowledge satisfied our curiosity, served as the foundation for natural philosophy, and was of use in the arts, manufactures, trade, and commerce, it was "still more usefull as it tends to lead us to admire and love that Being who hath furnished this World with such a prodigious Variety of things for our Use and Conveniency".[37] After moving to Glasgow, Reid drew extensively on natural historical examples when rehearsing the argument from design in his lectures on natural theology. A similar use of natural history was made in public classes held at King's; a graduate of the college noted that,

> On Sunday, 'twixt Breakfast, and going to Church; the different Classes attend their Respective Masters, to hear Lectures, on natural History, general observations on the Structure of the Human Body, or other Subjects, calculated to establish the Belief of a God, to display the Wisdom and Goodness of Providence, and other Principles of natural Religion.[38]

Arguably, then, natural knowledge was more readily assimilated into the curricula of the Scottish universities because it could be used to reinforce the rational grounds for Christian belief. Although there are signs that the harmonious relations between science and religion were strained in the aftermath of the French Revolution, for most of the 18th century the argument from design served to integrate the natural sciences with moral philosophy and religion and thereby helped to legitimate the pursuit of natural knowledge within academe.

IV

In addition to these broader social and cultural factors, the cultivation of mathematics and the sciences in the 18th-century Scottish universities was shaped, and in some cases promoted, by a number of institutional features. Here too Thomas Reid's career is illuminating, for at King's College Aberdeen he taught within the traditional system of regenting, whereas at the University of Glasgow he lectured as a professor within a system of fixed chairs. In Scotland, regents were responsible for taking their students through the whole three years of their philosophy course, which meant that a regent like Reid typically had to cover a wide range of subjects from ethics and rhetoric to mathematics and natural philosophy. Needless to say, the kind of omnicompetence presupposed by this system was not always in evidence, and at the turn of the 18th century it seems that regenting adversely affected the teaching of the sciences because some regents could not cope with the mathematical or conceptual technicalities of Newton's *Principia*.[39] Regenting did have some benefits, however, not least in the conception of the unity of human knowledge which it assumed. The notion that all of the branches of the *cursus philosophicus* were parts of a whole contributed to the integration of the natural sciences with religion discussed above, and it also encouraged the view expounded by Reid and many of his fellow Scots that moralists should employ the same empirical methods in their study of the human world as natural philosophers and natural historians did in their investigation of the material creation.[40] The close relations between natural and moral philosophy fostered by the regenting system bore copious fruit in the work of leading figures of the Scottish Enlightenment such as Thomas Reid.

But regenting was gradually phased out through the course of the 18th century. Edinburgh was the first to switch to individual chairs in 1707, followed by Glasgow (1727), St. Andrews (1747), Marischal College Aberdeen (1753), and finally King's College (1800). The rise of the professor entailed a fragmentation of knowledge, but it did allow for the formation of new independent sciences such as chemistry, which was one of the most innovative areas of inquiry in 18th-century Scotland.[41] Moreover, a professor's income was derived primarily from class fees rather than his fixed salary and consequently, as Adam Smith noted,[42] financial self-interest acted as a spur to classroom performance. Because the salaries attached to the chairs and lectureships in mathematics and the sciences were almost invariably low (particularly in newly founded posts), incumbents had to become successful teachers in order to reap the potential financial rewards of their appointments. The demonstration experiment, therefore, became an art form which attracted large audiences and substantial class fees. For the most accomplished performers, like Joseph Black, their showmanship brought considerable wealth, but for others, like John Robison, their comparatively austere teaching style resulted in serious financial hardship. In order to increase their class sizes, instructors also opened their classroom doors to fee-paying townspeople, and in so doing helped to create the public science of the Scottish Enlightenment. Low salaries, furthermore, gave an impetus to research and consultation work because the publication of results and the ability to give expert advice could be lucrative. How university teachers were paid, then, had a significant effect on teaching and research, as well as on the place of natural knowledge in the public sphere.[43]

V

In order to map the broad contours of public science in 18th-century Scotland, I want to return to Reid and his garden. His cultivation of potatoes and cabbages did not take place in a social or intelectual vacuum; rather, his gardening activities need to be seen in the context of his affiliation with two local groups, the Gordon's Mill Farming Club (which flourished between 1758 and 1765) and the Aberdeen Philosophical Society (which met from 1758 to 1773). Reid discussed the distillation of potatoes with his fellow members of the Farming Club in March 1759, and they consulted him about the use of lime as a manure as well as the design of harnesses for horses. More importantly, he put his considerable mathematical skills to work for them in devising an improved method of keeping farm accounts; he did so because he believed that mathematics ought to be applied to "the affairs of Life".[44] Agricultural improvement was also discussed by Reid and his associates in the Aberdeen Philosophical Society, which served as a venue for the exploration of a wide range of topics, including mathematics, natural history, and astronomy.[45] Reid's membership and activities in these two groups, along with his later participation in the Glasgow Literary Society, are significant historically for they illustrate key features of the growth of the public sphere in Scotland during the long 18th century.

Recent scholarship has shown that natural knowledge was a constitutive element in the emergence of the public sphere in Europe and, as I have argued elsewhere, Scotland was no exception.[46] Yet the process by which the public sphere was formed in Scotland differed in important ways from that in England and France, because Scottish public culture was to a considerable extent an offshoot of academe.[47] The voluntary clubs and societies which Jürgen Habermas saw as primary sites of the public sphere were in Scotland at least, often founded by, and largely made up of, university men. The two universities of Aberdeen may have dominated public culture to a greater degree than was the case elsewhere, but the fact that the Farming Club and the Philosophical Society were primarily the creations of individuals who taught at Marischal and King's illustrates the broader point that academics were instrumental in the formation of an urban public culture in Scotland similar to that found elsewhere in enlightened Europe. Moreover, as the two Aberdeen groupings show, this culture encompassed agricultural improvement, experimental demonstration, and the natural history of God's creation, as well as music, literature and polite moralizing after the fashion of Addison and Steele. By making their lectures open to town and gown (for a fee), professors in medicine, the natural sciences and mathematics ensured that natural knowledge was at the same time public knowledge, and their participation in the many clubs which convened in 18th-century Scotland helped foster an audience which gave their activities social legitimation. The professors were thus able to bring about the reconciliation of the gentleman and the scholar called for in the *Proposals* of 1704.

Further research has, however, suggested some new avenues of enquiry concerning the Scottish example. First, it now seems that we can date the origins of the public sphere in Scotland to the 1670s rather than the 1690s. Evidence for this is to be found in George Sinclair's attempt to fashion a new career after being forced out of the University of Glasgow in 1666. The records of Edinburgh Town Council state that on 16 November 1670 Sinclair was licensed "to profess severall usefull sciences", including pneumatics, hydrostatics, mechanics, mathematics, geography, surveying, navigation, and gunnery. To illustrate his lectures, Sinclair proposed to replicate the Torricellian experiment and to employ an air pump as well as other instruments.[48] As yet, no information has come to light about the size or composition of Sinclair's audience, let alone the success of his course, but this episode is nevertheless of significance because it reinforces the argument that science was an integral part of Scottish public culture and it provides us with a clearer chronology for the genesis of the public sphere in Scotland. Furthermore, it suggests that at this early stage, the universities were not as closely involved in the formation of public culture as they were later to become.

Secondly, Sinclair's activities as a regent and an extramural lecturer underline the important role played by mathematical practitioners and professors of mathematics in the cultivation of the sciences in late 17th-century Scotland. His lecture course and his employment by the Edinburgh Town Council to improve the water supply bring to mind the range of activities engaged in by Galileo and other Italian mathematicians appointed by civic authorities to provide public instruction and practical advice.[49] Unfortunately, we know little as yet about the profile of the Scottish mathematical community in this period. The various mathematical members of the Gregory family, to whom Thomas Reid was related, have hitherto attracted the bulk of scholarly interest, but figures like the Edinburgh mathematician James Corss deserve wider attention. Mathematical practitioners continued to ply their trade in Scotland well into the 19th century, offering lectures on geography, practical mathematics, and natural philosophy.[50] Like the itinerant lecturers on the sciences,[51] the mathematical practitioners have their part in the story of public science in Scotland, a part that has thusfar been unduly neglected. Within the universities, the mathematics professors seem to have led the way in assimilating the new science into the curriculum. Prior to being outed, Sinclair seems to have been an innovator at Glasgow. Slightly later, the Gregories spearheaded the adoption of the Newtonian system in Scotland, and their lead was soon followed by men such as Thomas Bower, Robert Simson, Colin Maclaurin, and John Stewart. Thus it is arguable that the dynamism of the Scottish professors and mathematical practitioners fueled the construction of public science and the transformation of the curriculum at the dawn of the Enlightenment.[52]

VI

By way of conclusion, I want to consider briefly the ramifications of the fact that after the Union of 1707 all university appointments were controlled by powerful Scottish political magnates who managed the affairs of Scotland on behalf of the British crown. In order to obtain an academic post one needed a patron who could secure the requisite backing from those controlling an appointment, as can be seen in Thomas Reid's election to the Glasgow Chair of Moral Philosophy in 1764. Although Reid's first major biographer, Dugald Stewart, suggested that Reid gained his position solely on the basis of merit, his candidacy was supported by two highly influential figures in Scotland, the jurist Henry Home, Lord Kames, and the aristocrat, Lord Deskford, who were able to persuade the managers of the patronage system in London not to back a better placed candidate.[53] It is arguable that the success of the Scottish universities in the Enlightenment was due as much to the managers' vision of the world, as to the talents of the men they appointed. This point was already appreciated in the 18th century by the Italian man of letters Carlo Denina, who attributed the recent and dramatic revival of learning in Scotland in large part to Lord Ilay, the third Duke of Argyll, who "patronised the ingenious with a bounty worthy of himself, and paid particular attention to the university of GLASGOW, which has since become one of the most renowned in EUROPE".[54] Recent work by Roger Emerson has shown that Ilay was himself a serious man of science, with particular interests in botany, chemistry, and scientific instruments. Something of Ilay's conception of a proper education for a gentleman can be gleaned from a letter written to his nephew in 1738, in which he recommended that his young kinsman should study history, civil law, the classics, mathematics, and experimental philosophy (which Ilay thought was both "entertaining" and "very useful in many things").[55] It is no accident that the teaching of mathematics and the natural sciences benefited from the Duke's enlightened patronage in the period stretching from roughly the 1720s until his death in 1761. Similarly, his nephew, Lord Bute, who controlled appointments in the Scottish universities from 1761 until around 1780, was a keen naturalist and promoted the academic study of natural knowledge whenever he had the opportunity.[56] If Scotland became a leading centre for university science and medicine during the 18th century, it was partly because patrons like Argyll and Bute wanted it so. We should not neglect their contribution to the transformation of the ancient kingdom of Caledonia into a "hot-bed of genius", renowned in the Enlightenment not only for its moralists and historians, but also for its mathematicians, physicians, natural historians, astronomers, and natural philosophers.[57]

NOTES

[1] Aberdeen University Library (AUL) MS 2131/5/II/2. I am grateful to Dr Iain Beavan, Head of Historic Collections, Special Libraries and Archives, University of Aberdeen, for permission to quote from manuscript materials in his care.

[2] C. Ginzburg, "Clues: Roots of an Evidential Paradigm", in *Myths, Emblems, Clues*, trans. John and Anne C. Tedeschi (London, 1990), pp. 96–125.

[3] For a recent survey of science teaching in the Atlantic world during the 18th century see L. Brockliss, "Science, the Universities, and Other Public Spaces: Teaching Science in Europe and the Americas", in R. Porter (ed.), *The Cambridge History of Science, Volume 4: Eighteenth-Century Science* (Cambridge, 2003), pp. 44–86.

[4] Thomas Jefferson to Dugald Stewart, 21 June 1789, *The Papers of Thomas Jefferson*, J. P. Boyd et al. (ed.), 19 Vols to date (Princeton, 1950–), 15:204.

[5] D. Stewart, *Dissertation … exhibiting a General View of the Progress of Metaphysical and Ethical Philosophy, Since the Revival of Letters in Europe*, in *The Encyclopædia Britannica*, 8th ed., Vol. 1 (1840; reprint Farnborough: Gregg International Publishers, 1970), p. 249. But

compare Stewart's view with the claim made by the 17th-century Scottish mathematician James Corss that apart from Napier Scotland had produced few mathematicians of note; see Corss as quoted in *The Wealth of a Nation in the National Museums of Scotland*, J. Calder (ed.) (Edinburgh and Glasgow, 1989), p. 91.

[6] Hugh Trevor-Roper, "The Scottish Enlightenment", *Studies on Voltaire and the Eighteenth Century* 58:1635–1658 (1967); *idem.*, "The Scottish Enlightenment", *Blackwood's Magazine* 322:371–388 (1977); R. L. Emerson, "Natural Philosophy and the Problem of the Scottish Enlightenment", *Studies on Voltaire and the Eighteenth Century* 242:243–291 (1986); *idem.*, "Science and the Origins and Concerns of the Scottish Enlightenment", *History of Science* 26:333–366 (1988); *idem.*, "Sir Robert Sibbald, Kt, the Royal Society of Scotland and the Origins of the Scottish Enlightenment", *Annals of Science* 45:41–72 (1988); *idem.*, "Science and Moral Philosophy in the Scottish Enlightenment", in M. A. Stewart (ed.) *Studies in the Philosophy of the Scottish Enlightenment* (Oxford, 1990), pp. 11–36. I have discussed Stewart's historiographical legacy in "Dugald Stewart and the Invention of 'the Scottish Enlightenment' ", in P. Wood (ed.) *The Scottish Enlightenment: Essays in Reinterpretation* (Rochester, NY, 2000), pp. 1–35.

[7] T. C. Smout, *A History of the Scottish People, 1560–1830* (Glasgow, 1979), pp. 251–252.

[8] AUL MS 2131/6/V/11, fol. 2v.

[9] G. Sinclair, *The Hydrostaticks; or, The Weight, Force, and Pressure of Fluid Bodies, Made Evident by Physical, and Sensible Experiments. Together with some Miscellany Observations, the Last Whereof is a is a Short History of Coal, and of all the Common, and Proper Accidents thereof; a Subject never Treated of before* (Edinburgh, 1672); *idem.*, *The Principles of Astronomy and Navigation*... (Edinburgh, 1688). Sinclair's exchanges with Henry Oldenburg are discussed in A. Johns, *The Nature of the Book: Print and Knowledge in the Making* (Chicago and London, 1998), pp. 502–503.

[10] Emerson, "Sir Robert Sibbald".

[11] A. Thomson, *Biographical Sketch of David Skene, M.D., of Aberdeen; With Extracts from Correspondence Between Dr Skene and Linnæus and John Ellis, About the Year 1765* (Edinburgh, 1859), p. 6.

[12] National Archives of Scotland, Mar and Kellie MSS, GD124/15/966/2.

[13] "Proposals By the Faculty of the University of *Glasgow* for buying Instruments necessary for Experiments and Observations in Natural Philosophy", reproduced in P. Swinbank, "Experimental Science in the University of Glasgow at the Time of Joseph Black", in A. D. C. Simpson (ed.), *Joseph Black 1728–1799: A Commemorative Symposium* (Edinburgh, 1982), pp. 23–35, 31; "Proposals for An Annual Course of Experimental Philosophy, in St. Salvator's College of the University of St. Andrews" ([St. Andrews], n.d.); "Proposals for Setting on Foot a Compleat Course of *Experimental Philosophy* in the Marishal College of *Aberdeen*", AUL MS 3017/10/18/2.

[14] That is, the use of demonstration experiments in the classroom marked just as much of a break with traditional pedagogy as did the introduction of English as the language of instruction.

[15] *Abstract of Some Statutes and Orders of King's College in Old Aberdeen. M.DCC.LIII. With Additions M.DCC.LIV* [Aberdeen, 1754], pp. 13, 21.

[16] [A. Gerard], *Plan of Education in the Marischal College and University of Aberdeen, with the Reasons of it. Drawn up by the Order of the Faculty* (Aberdeen, 1755), pp. 30, 34; see also the announcement of the Marischal reforms in *The Scots Magazine* 14:606 (1752).

[17] I have discussed the curriculum reforms at King's and Marischal in greater detail in *The Aberdeen Enlightenment: The Arts Curriculum in the Eighteenth Century* (Aberdeen, 1993), chap. 3.

[18] For an insightful discussion of the role played by agricultural improvement in the formation of an audience for science in Edinburgh see S. Shapin, "The Audience for Science in Eighteenth Century Edinburgh", *History of Science* 12:95–121 (1974).

[19] *Select Transactions of the Honourable the Society of Improvers in the Knowledge of Agriculture in Scotland*, R. Maxwell (ed.) (Edinburgh, 1743), pp. x, xiii–iv.

[20] C. W. J. Withers, "William Cullen's Agricultural Lectures and Writings and the Development of Agricultural Science in Eighteenth-Century Scotland", *The Agricultural History Review* 37:144–156, 150–151 (1989).
[21] Wood, *Aberdeen Enlightenment*, pp. 92, 93–94; T. Reid, "Scheme of a Course of Philosophy", AUL MS 2131/8/V/1, fol. 1r-v; *idem*., "Natural History 1753", AUL MS 2131/6/V/10a, fol. 2v; J. Beattie, "Institutes of Natural History", AUL MS M. 189, 2:41.
[22] C. W. J. Withers, "A Neglected Scottish Agriculturalist: The 'Georgical Lectures' and Agricultural Writings of the Rev Dr John Walker (1731–1803)", *The Agricultural History Review* 33:132–146 (1985).
[23] I. J. Fleming and N. F. Robertson, *Britain's First Chair of Agriculture at the University of Edinburgh, 1790–1990: A History of the Chair founded by William Johnstone Pulteney* (Edinburgh, 1990), esp. chaps. 1–3.
[24] See in particular N. Phillipson, "Politics, Politeness and the Anglicisation of Early Eighteenth-Century Scottish Culture", in R. A. Mason (ed.), *Scotland and England, 1286–1815* (Edinburgh, 1987), pp. 226–247, and S. Shapin, " 'A Scholar and a Gentleman': The Problematic Identity of the Scientific Practitioner in Early Modern England", *History of Science* 29:279–327, 305 (1991).
[25] AUL MSS 2131/4/I/18, fol. 2v and 2131/4/I/31, 3; Reid's views on education are contextualized in J. C. Stewart-Robertson, "The Well-Principled Savage, or the Child of the Scottish Enlightenment", *Journal of the History of Ideas* 42:503–525 (1981).
[26] For their pointed attacks on scholasticism see J. Locke, *An Essay Concerning Human Understanding*, P. Nidditch (ed.) (Oxford, 1975), p. 10; [Robert Molesworth], *An Account of Denmark, As It was in the Year 1692* (London, 1694), C4r-v. For his part, Locke recognized that the needs of scholarship were somewhat different from those of public life, but he did not think that the roles of the scholar and the gentleman were necessarily antithetical. On the differing needs of the scholar and the gentleman see J. Locke, *Some Thoughts Concerning Education*, W. John and J. S. Yolton (ed.) (Oxford, 1989), p. 249.
[27] Locke, *Some Thoughts*, pp. 247, 248.
[28] Locke, *Some Thoughts*, pp. 248–249. Compare here Shapin's interpretation of these passages in "A Scholar and a Gentleman", pp. 304–305.
[29] Interestingly, in a manuscript dictated in 1703 and published posthumously in 1720, Locke said that the "proper calling" of a gentleman was "the service of his country", and therefore suggested that gentlemen need only cultivate those studies which "treat of virtues and vices, of civil society, and the arts of government, and so will take in also law and history"; see J. Locke, "Some Thoughts Concerning Reading and Study for a Gentleman", in M. Goldie (ed.), *Political Essays* (Cambridge, 1997), p. 350. There was, therefore, a degree of inconsistency in Locke's position.
[30] "Proposals for the Reformation of Schools and Universities, in order to the Better Education of Youth; Humbly Offered to the Serious Consideration of the High Court of Parliament", in *The Harleian Miscellany: Or, A Collection of Scarce, Curious, and Entertaining Pamphlets and Tracts*, 8 Vols (London, 1744–1746), pp. 1:485–490. On the question of attribution see A. Fletcher, *Political Works*, J. Robertson (ed.) (Cambridge, 1997), pp. xxxvi–xxxvii.
[31] "Proposals", p. 489.
[32] G. Turnbull, *Observations Upon Liberal Education*, T. O. Moore, Jr. (ed.) (Indianapolis, 2003), pp. 197–199, 319–327; D. Fordyce, *Dialogues Concerning Education*, 2nd ed. (London, 1745), esp. dialogue XI. Fordyce also discusses (pp. 23, 25) the teaching of mathematics and the sciences when describing his idealized academy (a description partly inspired by the academy run by Philip Doddridge). I have surveyed the educational ideals of Turnbull and Fordyce in Wood, *Aberdeen Enlightenment*, pp. 40–49, 50–55. For a slightly different interpretation of their significance see P. Jones, "The Polite Academy and the Presbyterians, 1720–1770", in J. Dwyer, R. A. Mason, and A. Murdoch (eds.), *New Perspectives on the Politics and Culture of Early Modern Scotland* (Edinburgh, [1982]), pp. 156–178; *idem*., "The Scottish Professoriate and the Polite Academy, 1720–1746", in I. Hont and M. Ignatieff (eds.), *Wealth and Virtue: The*

Shaping of Political Economy in the Scottish Enlightenment (Cambridge, 1983), pp. 89–117. On Turnbull see also M. A. Stewart, "George Turnbull and Educational Reform", in J. J. Carter and J. H. Pittock (eds.), *Aberdeen and the Enlightenment* (Aberdeen, 1987), pp. 95–103.

[33] G. Turnbull, *The Principles of Moral Philosophy*, 2 Vols (London, 1740), 1:i–iii.

[34] The same limitations are found in L. E. Klein, *Shaftesbury and the Culture of Politeness: Moral Discourse and Cultural Politics in Early Eighteenth-Century England* (Cambridge, 1994).

[35] J. Stewart, "Some Advantages of the Study of Mathematicks, with Directions for Prosecuting the Same", National Archives of Scotland, MS GD 248/616/2, 19; J. Walker, *Lectures on Geology*, H. W. Scott (ed.) (Chicago and London, 1966), p. 2. The perception of the relevance of mathematics to the education of gentlemen was, in fact, relatively widespread; see the comments made by the master of the Ayr burgh school in 1729 quoted in D. J. Withrington, "Education and Society in the Eighteenth Century", in N. T. Phillipson and R. Mitchison (eds.), *Scotland in the Age of Improvement* (Edinburgh, 1970), pp. 169–199, 170.

[36] P. Wood, "Science, the Universities, and the Public Sphere in Eighteenth-Century Scotland", *History of Universities* 13: pp. 99–135, 106–107 (1994).

[37] T. Reid, "Natural History 1753", AUL MS 2131/6/V/10a, fol. 1r.

[38] John Bethune to unidentified correspondent, 25 September 1787, Edinburgh University Library, MS La.III.379/42, fol. 1r-v, as quoted in Wood, *Aberdeen Enlightenment*, p. 94.

[39] On this point see Wood, *Aberdeen Enlightenment*, p. 2.

[40] I have developed this point in "Science and the Pursuit of Virtue in the Aberdeen Enlightenment", in Stewart, *Studies*, pp. 127–149, and "The Natural History of Man in the Scottish Enlightenment", *History of Science* 28:89–123 (1990). See also Emerson, "Science and Moral Philosophy".

[41] On the emergence of chemistry as a distinct science in Scotland see A. L. Donovan, *Philosophical Chemistry in the Scottish Enlightenment* (Edinburgh, 1975), and J. Golinski, *Science as Public Culture: Chemistry and Enlightenment in Britain, 1760–1820* (Cambridge, 1992), chap. 2. For a detailed study of the fragmentation of the curriculum at Glasgow see R. L. Emerson and P. Wood, "Science and Enlightenment in Glasgow, 1690–1802", in C. W. J. Withers and P. Wood (eds.), *Science and Medicine in the Scottish Enlightenment* (East Linton, UK, 2002), pp. 79–142.

[42] For Smith, the system of class fees was the key to the superiority of the Scottish universities when compared with those of Oxford and Cambridge; see A. Smith, *An Inquiry into the Nature and Causes of the Wealth of Nations*, R. H. Campbell, A. S. Skinner, and W. B. Todd (eds.), 2 Vols (Oxford, 1976), 2:759–760.

[43] On these points see the seminal article by J. B. Morrell, "The University of Edinburgh in the Late Eighteenth Century: Its Scientific Eminence and Academic Structure", *Isis* 62:158–171 (1970), along with Wood, "Public Sphere".

[44] "Minute Book of the Farming Club at Gordon's Mill 1758", AUL MS 49, 26, 52, 140, 231. Reid's method of keeping accounts is discussed in W. R. Humphries, "The Philosopher, The Farmer, and Commercial Education", *Scottish Educational Journal*, 14 May: 619–620 and 21 May: 661–663 (1937). On the practical application of mathematics see T. Reid to R. Price [1772/3], in P. Wood (ed.), *The Correspondence of Thomas Reid* (Edinburgh, 2002), p. 64.

[45] On the Society see *The Minutes of the Aberdeen Philosophical Society, 1758–1773*, H. L. Ulman (ed.) (Aberdeen, 1990).

[46] Wood, "Public Sphere"; see also the more recent analysis in C. W. J. Withers, "Towards a History of Geography in the Public Sphere", *History of Science* 36:45–78 (1998).

[47] Developments in Scotland were more like those in Germany, for which see T. Broman, "The Habermasian Public Sphere and 'Science *in* the Enlightenment' ", *History of Science* 36:123–149 (1998).

[48] *Extracts from the Records of the Burgh of Edinburgh 1665 to 1680*, M. Wood (ed.) (Edinburgh and London, 1950), pp. 92–93.
[49] On the Italian context see M. Biagioli, "The Social Status of Italian Mathematicians, 1450–1600", *History of Science* 27:41–95 (1989).
[50] Many of the mathematical practitioners active in the eighteenth century are listed in Withers, "Geography in the Public Sphere", table 1.
[51] Nothing of substance has been done on this subject since the appearance of two pioneering articles by J. A. Cable, "The Early History of Scottish Popular Science", *Studies in Adult Education* 4:34–45 (1972) and "Early Scottish Science: The Vocational Provision", *Annals of Science* 30:179–199 (1973).
[52] It should be noted that Scottish men of science benefited from the correspondence networks established by the mathematicians, for individuals like James and David Gregory as well as James Stirling provided contacts in London, Cambridge, Oxford, and on the continent. This subject deserves further research.
[53] On Reid's election see my " 'The Fittest Man in the Kingdom': Thomas Reid and the Glasgow Chair of Moral Philosophy", *Hume Studies* 23: 277–313, 289–291 (1997).
[54] C. Denina, *An Essay on the Revolutions of Literature*, trans. John Murdoch (London, [1771]), pp. 276–277.
[55] Wood, *Aberdeen Enlightenment*, p. 161. Ilay's scientific pursuits are described in R. L. Emerson, "The Scientific Interests of Archibald Campbell, 1st Earl of Ilay and 3rd Duke of Argyll (1682–1761)", *Annals of Science* 59: 21–56 (2002).
[56] On Bute see R. L. Emerson, "Lord Bute and the Scottish Universities, 1760–1792", and D. P. Miller, " 'My favourite studdys': Lord Bute as Naturalist", in K. W. Schweizer (ed.), *Lord Bute: Essays in Re-interpretation* (Leicester, 1988), pp. 147–179, 213–239.
[57] The phrase "hot-bed of genius" comes from T. Smollett, *The Expedition of Humphry Clinker*, L. M. Knapp (ed.), revised by Paul-Gabriel Boucé (Oxford, 1984), p. 233, where it is applied specifically to Edinburgh.

UGO BALDINI

THE SCIENCES AT THE UNIVERSITY OF ROME IN THE 18TH CENTURY

THE QUESTION

The role of the universities in the development of science in Italy, between the 16th and 18th centuries, has been evaluated in several different and contrasting ways, less because of the subjective tendencies of historians than from the evidence, which objectively is contradictory. On the one hand, it is true that many important scientists in the period were not teachers, while others were, but made their original contributions outside university institutions (and in contrast with its orientation and courses of study); it is true that the official programmes remained faithful to medieval models for a long time; it is true that the majority of professors remained untouched or even hostile to innovation, and resisted their inclusion in their teaching. On the other hand, the majority of leading scientists taught in the universities (Cardano and Galileo, Morgagni and Spallanzani and the greatest mathematicians of the second half of the 18 century),[1] and this idea is strengthened if the term "university" is held to include some of the higher religious schools (which had quite different origins and organization, but were also on a university level because they awarded doctoral degrees, not only in theology but also in philosophy).[2] Naturally some of these contrasts disappear if the evaluation is narrowed chronologically or geographically, by discipline or by research question: universities were more remote from the new science before 1650 than afterwards (and even less so after 1750); for epistemological-scientific, as well as professional reasons, the resistance was greater in natural philosophy than in mathematical physics, in medicine than in normal anatomy or pathology, in philology than in descriptive botany, in cosmology and physical astronomy than in the astronomy of positions; *overall*, the renewal was greater and faster in the universities between Padua and Pisa (and, from midway through the 18th century in those of the north-west, that is, Pavia and Turin) than in those south of Tuscany. Furthermore, it should be noted that not all the facts that denote (or seem to denote) a remoteness of the universities from scientific progress are interpretable as effects of cultural hostility. A good example comes from pure mathematics. The *algebra cossica*, and later that of Viète, analytic geometry, and the developments that finally led to calculus remained outside university teaching until the beginning of the 18th century at least, not because of an official ban, but only because the programmes were designed for simple applications, and the society at large did not explicitly demand their improvement.[3]

M. Feingold and V. Navarro-Brotóns (eds.), Universities and Science in the Early Modern Period, 201–230.

But this general question contains a more limited, but important, one, in which the studies are still less numerous and less conclusive. The writings on Italian science during the centuries mentioned consider the role of the universities by looking only at a few of them, mostly in the North: Pisa, Bologna, and Padua from the 16th century; Pavia and (to a lesser degree) Turin from the middle of the 18th. This corresponds to a different quantity and quality of scientific work also outside the universities, which can be traced back to more general historical circumstances.[4] In some ways this has seemed predictable and natural, and so it has not stimulated more specific analysis of the situation in the scientific chairs of the other universities, of their institutional and organizational features, or of what they actually taught. In this way, however, important facts about the overall dynamics of Italian intellectual life are still known only superficially and incompletely, and the role of universities in the cultural reality of cities like Rome, Naples, Messina, Catania, and Palermo is underestimated, or evaluated too generally or in a biased manner. In recent years, there has been some progress in our knowledge of the scientific culture of university circles of Naples; as for the University of the Viceroyalty of Sicily (which was nearly always in Catania), it was second rate, so that the lack of studies is consequently less serious in making a general evaluation.[5] The history of science at the University of Rome (commonly called "la Sapienza", as is written in the title in its most extended history)[6] is almost unknown for entire periods and for specific disciplines, and its secondary role presents an interesting problem. Rome was the capital of one of the largest Italian states; its role as the center of the Church certainly led a large number of scholars, comparable with that of any European city, to reside temporarily or permanently in the city; it was the home of a large number of higher institutions of teaching (even if almost all of them were religious) that should have produced a positive exchange of ideas, which seems not to have existed. This justifies the following discussion, that cannot be exhaustive, but is limited to establishing the fundamental facts and circumstances.

THE INSTITUTIONAL HISTORY OF THE ROMAN SAPIENZA AND THE REFORMS OF THE 18TH CENTURY

The University of Rome had been created by Boniface VIII (1303) as a state university, like those of Naples and Turin, not as a city university like all the others in Italy.[7] In the 14th century, before the transfer of the Papal Seat to Avignon and the crisis of the Western schism, its work was precarious; it began to function continuously only in 1431, when Eugene IV renewed its bylaws.[8] In the modern age, it was based on the Statute of Leo X, issued in 1513, which remained essentially unchanged until the reforms of the 18th and early 19th century.[9] The faculties (which more precisely should be called "universities")[10] were, as almost always in Italy, the three traditional ones: Arts (literary and philosophical disciplines, including mathematics) and Medicine; Law; and Theology.[11] In Rome, as in other places, the first, the only important one for this essay, included very different disciplines, which in the following centuries where arranged very differently. As is well known this was not an original arrangement, but the result of one originated in the Middle Ages and due, principally, to the need to supply a basis for the instruction of physicians.[12] In reality, the faculty was a heterogeneous container whose contents only shared their lack of pertinence to the other two faculties.[13]

Consequently, the organizational model and the chairs of the Roman University were practically identical to that of other Italian universities. Until the 15th century it was never on the level of Padua and Bologna (and, probably, not even on that of Pavia and Pisa). The number of students was always smaller; despite the historic and religious role of the city, that attracted many foreigners, the percentage of non-Italian students was always small (they were not organized in *nationes* of foreign students, as in the North); few professors of the philosophical and scientific disciplines were first rate. The number of specialized publications of professors of these disciplines (except for opening lectures and other circumstantial writings) was smaller than elsewhere, and neither in philosophy nor in medicine does a Roman school appear that is recognizable for any methodological or scientific character. Until 1600 the chair of mathematics—decisive for the quality of modern science in universities—was discontinuous and was often entrusted to persons of a low level; even later (with few exceptions) it was not a center of advanced research and did not form a school (until the 18th century it was rare for professors to be students of a former professor, and the great majority were not educated in Rome). Complete analyses are lacking, but some significant factors are certain; it is useful to list them distinctly.

The best high schools of the religious Orders had courses of philosophy (as an introduction to theology) open to lay students, which rivaled with the university; a competition also existed between the theology courses of these schools and the theological faculty of the *Sapienza* (some of the professors were shared, but those of the schools of the Orders were often better, and the number of the disciplines taught—and thus their specialization—was higher);[14] the need that philosophical teaching be strictly in line with Christian doctrine excluded the teaching of Arab and Alexandrine Aristotelianism (the more "naturalistic" ones) and assured the predominance of scholastic philosophy, the more so because the professors were almost always clerics, and the great majority were members of the regular Orders.[15] This was a major difference in respect to the universities (Pisa, Bologna, Padua) where professors of philosophy were laymen and often physicians, that considered the chair as a step on the road to one in medicine. Even if faithfulness to scholasticism would not make the teaching necessarily more backward and less qualified, it made it more attentive to the needs of the theology course than to those of medicine, and this tended to exclude the more concrete and factual contents of the Aristotelian works and a tendency to observation and experimentation.[16]

But the effect of these cultural factors was increased by other regulatory and administrative ones, even more typical of the *Sapienza*. A fundamental one was the control structure of the university, that can be represented schematically in the following manner:

The Pope
The Cardinal Secretary of State
The Congregation of Cardinals "pro Universitate Romana"
The Cardinal Camerlengo
The College of Concistorial Advocates
The Rector.[17]

The greater part of the governing hierarchy was thus formed by individuals and offices that, in part because of their religious status, but also because of their cultural background and—that is to say—professional specialization, were not always able to perceive the social needs that would tend to strengthen and modernize scientific teaching.

Besides the Pope and the top of the Curia bodies, whose actions were sporadic, and usually general and indirect in nature, it should be pointed out the particular nature of the College of Concistorial Advocates, of which the Rector, responsible for the daily administration of the university, was a product and, substantially, only an executor. It was supposed to communicate problems and needs for reforms to the Curia;[18] but, as a body of functionaries highly bureaucratized and sensitive to formal questions, it was largely foreign not only to scientific questions, but also to sectors of society interested in the technical questions that could have promoted the development of the scientific components of the chairs and programmes. If to this is added the fact that in the Papal State strict feudalism lived longer than in central and northern Italy, and that it was no incentive to develop technical public offices with a permanent and qualified staff, it is possible to understand why in Rome's Faculty of Arts scientific members continued to be fewer, and generally less qualified and modern, than in those of other universities.[19] The lesser attention of Papal government, and of Roman society in general, to the university in respect to what occurred in other areas of Italy is demonstrated by the salaries given to professors, which were among the lowest in Italy (and Europe).[20] This was both a cause and an effect. As a cause, it dissuaded many able people from pursuing an academic career, or it made them consider it only as a source of prestige to be exploited in private professional work. This lowered the number of famous professors summoned from outside the Papal State, compromised the effectiveness of courses, and left chairs to unqualified individuals, or to clergy, who had fewer economic problems and accepted lower salaries; the religious peculiarity of Rome was not, therefore, the only reason why in Rome some chairs, elsewhere held by laymen, were held for centuries by clergy.[21] As an effect, it shows the lesser importance given to university lectures, and the fact of the abundance of ecclesiastic scholars in the city made candidates with modest economic demands available.

Finally, some specific characteristics of the Roman environment should be considered. The creation of chairs of mathematics in the Faculty of Arts, which began in Italy in the 14th century, was due to the need to offer students the astronomical basis of astrology, for its use in medicine. So a need that today does not appear scientific was decisive in the permanent introduction of mathematics in the university *curricula*, and contributed to providing a social and professional basis to groups of specialists.

For this reason, from the 15th to the 17th century, important Italian university centers of mathematical studies (in the quadrilateral Pavia–Padua–Bologna–Pisa) were also the main centers of theoretical and practical writing in astrology.[22] In Rome, these circumstances occurred partially. The old condemnation of astrology, even if it tended to refer judicial and not natural astrology, was applied rather blandly and discontinuously, and was taken up again for the use made of astrology in the 16th century non-scholastic Aristotelians and by natural philosophers;[23] so it was of quite common interest and commonly taught privately in Rome, but its teaching in public schools was minimal, and this discouraged students from attending mathematics courses and discouraged professional scholars.[24]

A second factor was the Roman schools of the teaching orders (the Jesuits, but also the Barnabites, Theatines, and Piarists). After modest beginnings in the first half of the 17th century, some of them took on university *status*, with teaching that was often of a superior quality to that of the university and, unlike the university, attracting a large number of young Italians and foreigners, guests of numerous colleges (both private and supported by the religious orders themselves). So, if the courses of arts, philosophy, and theology are considered (Loyola had excluded medicine from Jesuit teaching), it may be said that for the number of courses and the students enrolled the Roman College of the Jesuits was, in the 17th and the beginning of the 18th century, the real University of Rome.[25] As for the other courses of study, some contemporaries were of the opinion that Rome did not have a true university, or that it was so backward that it was not reformable.[26] The situation was so bad that Innocent XII (1691–1700) seriously considered suppressing the university and giving the palace to the Piarists, who could have developed the higher school they already had in Rome (the Nazareno College), bringing it up to the level of the Jesuits; more than the resistance of the university bodies, what blocked the proposal was that this would have deprived Rome of a university school of medicine.[27]

Medicine was the only area of scientific study that was present in the university (with associated fields such as anatomy and, to a lesser extent, botany and natural history).[28] Within it, already in the 16th and 17th centuries, the average level of teaching and professorship was not far behind that of the other universities; there were original scholars, or at least open to new areas of study (Realdo Colombo, Andrea Cesalpino, Johann Faber, Pietro Castelli, Giovanni Battista Trionfetti, Lucantonio Porzio, Giovanni Maria Lancisi, Giorgio Baglivi).[29] But a large part of the teaching was based only on reading (enhancing a characteristic shared by many universities); in addition the chairs in medicine had to face real competition from the medical schools of several hospitals, that seem to have trained the majority of physicians, especially in the oldest and most famous of them, Santo Spirito.[30] Finally—and this should be extended to all disciplines—it is essential not to assess the real work of the university courses simply on the basis of their programmes and the quality of the professors. In Rome (and, it seems, also in Naples) a phenomenon that occurred elsewhere was more pronounced: the low level of effort of professors in teaching, with frequent interruptions and the use to entrust teaching to inexperienced assistants not selected by the university. For this reason, a history of the Faculty of Medicine in Rome (as of others) that evaluates the quality and adequacy of teaching on the basis of the works of professors risks overestimation or real distortion.[31]

One circumstance demonstrates the decisive value of these facts. In the 18th century, all the major Italian universities underwent reforms, that transformed—in certain cases drastically—their structure, the level of their teaching and their position in respect to the others in the peninsula. From the second half of that century the University of Turin, which had always been second rate, became one of the best for science; Pavia, illustrious in the past but which had declined progressively during the Spanish dominion in Lombardy, after 1760 was perhaps the most advanced in Italy; Padua and Pisa, historically the heart of the Galilean tradition, eliminated some of their conservative tendencies in certain areas and (especially the former) were centers of research on an European level. In Bologna, in the first years of the century, the introduction of young philosophy and mathematics lecturers who were in favor of modern approaches and experimentation, and the creation of new research organizations such as the Institute of Sciences and the associated Academy, revived the ancient prestige of the university and gave it importance through the Napoleonic years. Even the University of Naples, despite the somewhat backward socio-political context and the burden of academic practices mentioned above, experienced significant improvements.[32] In Rome, however, the changes were not so extensive or effective, and at least until the 1780s, the distance between the *Sapienza* and other Italian universities increased. This did not occur in the absence of reforms, which were attempted on several occasions of the same kind and scale as those introduced elsewhere. Rather, as revealed by a brief outline of them, few were applied thoroughly and persistently; none of them eliminated all the abuses or defects that they intended to combat, and at times Popes (or Secretaries of State, or their agents) who ordered them also maintained, and even reinforced, ancient practices that had been factors in the university's decline.[33]

1660 Alexander VII established the university Botanical Garden on the slopes of Mount Gianicolo; its establishment was contemporary, having moved the university to a new site, designed by F. Borromini.[34]

1667 Again Alexander VII established, with the Bull *In suprema pastoralis officii specula*, the University Library (which was called Alexandrine after him), one of the oldest and largest Italian university libraries.[35]

1695–1700 After abandoning the idea of transforming the university into a school of the Piarists, Innocent XII took measures. In addition to calling Giorgio Baglivi (1696) as the successor of Lancisi to the chair of anatomy and Luca Tozzi to that of practical medicine, during these years, an anatomical theater was built (beforehand there was no permanent place for anatomy). Furthermore, the professors of law were prohibited from giving private lessons.[36] Even if the decision did not apply to the Faculty of Arts, it concerned a general problem, that is the absenteeism of lecturers and their external activities, not satisfactorily resolved for most of the 18th century.

1701 A congregation appointed by Clement XI divided the chairs (except the theological ones) into three classes, Law, Medicine, and Arts (beforehand the second and third formed one). Without effecting the traditional medical chairs, it fixed the number of Arts chairs: logic, philosophy, metaphysics, ethics, rhetoric (Latin language, stylistics, and literature), Greek language, Hebrew language, Arabic language, Syriac language. In addition, it fixed the stipend of each chair, proposed that the lessons of theoretical and practical medicine were daily, against the abuse of lecturers who held lessons only occasionally, and set a minimum of seven students for a course to be held.[37]

1703 Clement XI enlarged the Botanical Garden.[38]

1713 The same Pope introduced the practice of assigning the chair of philosophy to the Secretary of the Congregation of the Index.

1730 A commission established by Clement XII formulated some reforms, but they were blocked by the resistance of university bodies.

1744 In the Bull *Inter conspicuos ordines* Benedict XIV ordered that university chairs, up until then filled by appointment (by the Pope, the Secretary of State or the College of Concistorial Advocates), would be filled by public examination, and established the procedure.[39]

1748 With the Bull *Quanta Reipublicae commoda*[40] Benedict XIV: transformed one of the two lectureships in elementary mathematics into one in higher mathematics (in reality, one of advanced applications); established the teaching of Chemistry, and with it a chemistry laboratory;[41] suppressed (implicitly) the practice of assigning the lectureship in philosophy to the Secretary of the Congregation of the Index, giving it to F. Jacquier, and he gave it a more experimental character, establishing a laboratory under a technician.

Furthermore, with a chirograph promulgated on 14 October by the Rector C. Argenvilliers (which he had in part suggested),[42] the Pope ordered that: all the lectureships which became vacant were reassigned with an obligation of daily lessons (earlier 60 lessons were required per year, but as has been mentioned, this number was rarely reached); for violators there was to be a proportional reduction in their stipend; the ordinary chairs of medicine were six, each with a ordinary lectureship and no more than one extraordinary, while earlier the number of extraordinary positions was often high even if lecturers did not teach, which created both financial and teaching difficulties;[43] the "ostensore" (exhibitor) of the simples (manager of the Botanical Garden and of the exhibitions that were held there) would be distinct from the lecturer of botany, and he could not pass to a medical lectureship (the exhibitors of the simples had often been a temporary role for physicians who aspired to a more important medical chair, and thus often did not dedicate themselves to the job and research).[44]

Finally, the Pope: regulated the stipends and the award of pensions (earlier handled case by case); required that every professor presented in advance a detailed annual programme, whose performance the academic authorities were to verify; and finally required every lecturer to have a set timetable over the year.

But, again, the application of these orders was only partial, and almost never long-lasting, in part due to the opposition of the professors, from whom the reform required much more work.[45] The work of the chemistry and physics laboratories, even if some of the technicians who worked in them were competent, was limited;[46] degrees continued to be awarded without any real control of the preparation of the candidates.[47]

1773 After the suppression of the Society of Jesus, the university asked Clement XIV to amalgamate its building with that of the Roman College, in order to increase the number of lectureships by adding that of "commerce", agriculture, natural history, astronomy, diplomacy.[48]

This all seems to be at least a partial failure.[49] As mentioned above, if around 1780 the university showed some signs of improvement compared to the beginning of the century, the changes occurred in the other Italian universities had increased the backwardness of la Sapienza in respect to them. However, from the middle of the century, even within limits and in ways caused by the peculiar state of the city and its government, and the form of control exercised by it, the new science and, in part, the needs of reform and rationalization of the Enlightenment had progressively penetrated Rome, and in part the clergy and the Curia itself.[50]

Between 1750 and 1790, interest in science had increased in the academies, not only superficially, and the offices and individuals in the government who had been promoted, began important initiatives. In these new circumstances, the last Pope of the *ancien régime*, and some of his collaborators, made a more complete and decided attempt than earlier efforts, whose effects were proportionally greater, even if after a few years the revolutionary and Napoleonic reform gave the Papal government different priorities. In 1786 and 1788, Pius VI dedicated to the University of Rome two Bulls, *Ad supremum Apostolatus fastigium* and *Postquam Divinae Sapientiae*.[51] The second sanctioned several innovations already part of a *Regolamento dell'Archiginnasio Romano*, published by the Concistorial Advocates also in 1788. A simple list of the provisions of the three documents demonstrates that the academic authorities were finally aware that the situation required energetic action.[52] The fundamental provisions of Benedict XIV (including the requirement of daily lessons) were carried out, or made effective; they, furthermore, were applied to a new organization of chairs in five classes, far removed from the old tradition of the three classes of Clement XI: religious subjects (six lectureships); law (six lectureships); medicine and surgery (nine lectureships);[53] philosophy and arts (five lectureships);[54] and languages (five lectureships).[55] Two new lectureships in medicine were established: one in surgical operations (with the associated elements of legal medicine), one of obstetrics;[56] the library, the chemistry library, and the anatomical theater were reorganized; the lectureship in botany was amalgamated with the work of "exhibition" of the simples and the direction of the Garden, for greater unity between theory and practice.

Until 1798, the university operated under this new organization, with significant signs of renewal and growth in scientific work by the professors. In that year, however, the French occupation of Rome, the declaration of the Roman Republic (1799), and the deportation of the Pope led to the suspension of university courses until the 1800–1801 academic year, which began after the return of the new Pope, Pius the VII. Despite the serious political situation and the fact that—even if normally with caution and indirectly—several important professors had collaborated with the French and the Republican government, Pius VII undertook almost immediately the stabilization and the completion of the reforms of his predecessor. The constitution *Uberes dum menti nostrae* (1804)[57] established the chair of mineralogy and natural history, to which was added a museum for natural history, that gathered together the collections of various professors from the preceding century; the following *Inter multiplices curas* (1806) established the chair of veterinary medicine and assigned to it a professor that the government had sent to Paris for specialist training.[58] To these provisions, as will now be seen, there was added the selection of professors who were more modern and better connected in the Italian and European scientific community, so that they may be considered the completion of the reforms of Pius VI and the true conclusion of the institutional history of the Sapienza in the 18th century.

For the second time, however, the political situation blocked the implementation of the reforms. With the new occupation of Rome (1808) and its annexation to the Empire (1809), the regulations of the imperial universities were extended to Rome, and this meant substantial change; their actual application, however, began only in 1812, and lasted for only 2 years for in May 1914, Pope Pius VII returned to Rome.[59] The intellectual climate and the state of the scientific disciplines had by then been changed too greatly to re-establish the university organization outlined in 1786–1788, so that Pius VII formed a reform commission. This commission had a much more arduous task than the former, because the updating of the chairs and scientific programmes did not raise solely technical questions: in the 17th century, the biggest problem that had to be faced by reform was heliocentrism (with its related questions), but after 1815 a good part of the accepted theories (in astronomy, biology, and geology) seemed to have anti-Christian or more generally anti-religious implications, and in particular contradicted the traditional literal interpretation of the Genesis cosmogony. The commission proposed a reform of the chairs and programmes that accounted—for most if not all—the contemporary teaching and research, without however accepting the sections that seemed to have materialistic implications, or contradicted the literal meaning of passages in the Bible. For this and other reasons, the task was a long one (from 1816 to 1823) and the reform was introduced in 1824 when Pius the VII had already been succeeded by Leo XII. It remained in effect in Rome until the Papal State was annexed by the kingdom of Italy, and thus characterized the situation of the university in the period following that considered in this essay.[60]

SCIENCE TEACHING THROUGHOUT THE CENTURY

The weaknesses of a teaching institution effect the average level of teaching, the preparation of students, and the quality of professors, but only statistically, because they are compatible with significant exceptions. Already in the first half of the 17th century, in a period of stagnation at the Roman University, the presence of Benedetto Castelli in the chair of mathematics (from 1626 to 1642) had allowed the formation of three of the most important Italian scientists in the middle of the century: E. Torricelli, G. A. Borelli, M. Ricci.[61] All the more so in the next century, when the above reforms, even if they were only partially effective, changed some aspects of the situation. As will be shown, this was not only true about the education of students, but also about the quality of professors. The indifference to teaching of many of them, and their inadequate training, more bookish than experimental-observational, made their research and publications usually inferior—both for quantity and quality—to those of professors in the other major universities. In the 18th century, however, there were some figures who produced some significant works. The coexistence of these two aspects makes it difficult to describe the scientific life of the Sapienza in a complete and balanced manner; the most natural and least biased way is perhaps to consider the succession of professors and the evolution of theory in relation to individual chairs or groups of chairs separately.

Philosophy

Traditionally this chair, in Rome as in all the Italian universities and generally in European ones, regarded only natural philosophy (with the theory of the mind and its operations), because logic (not only its inferential concerns, but also the ontological and psychological basis which Aristotle attributed to them) and metaphysics were subjects of separate chairs. Its function in the structure of university courses and the religious schools was essential, not only for the importance of its content, but because it defined the cosmological principles, the limits of the knowable and its regions; so it indicated the architecture of knowledge, giving a place to each specific scientific discipline and fixing the limits of each. In other words, it was at a same time a theory and a metatheory.[62] In addition to this double function, something other made it strategic in "cultural policy", so the discipline most subject to doctrinal control. Some of its concerns (the creation–eternity and finiteness–infiniteness of the world, the causation–finality and contingency–necessity of events, the substantialness of the mind and spirit, their mortality–immortality) were connected directly to religion; the Church fought Aristotelian schools such as Averroism and Alexandrinism, that promulgated solutions that were different from Scholasticism, and criticized the universities (like Padua) where they prevailed, by submitting some professors to the Inquisition and the Index.[63] In Rome, therefore, the teaching of philosophy followed strict Scholastic lines; this did not mean that interest in natural things themselves was necessarily less than in other universities, but teachers certainly had to follow predefined channels, which guaranteed the congruence of cosmological-physical doctrines with the needs of natural Scholastic theology.[64] It would seem that at the end of the 17th century, a novelty emerged with the teaching of Francesco Nazzari (1670–1700), founder and director of the *Giornale de' letterati di Roma*, the first journal that spread knowledge of the authors and works of the new philosophy and the new science in the city.[65] However, even if the contents of his teaching are unknown, it has not been established that they strayed from the usual canons; in addition, all his successors were regular clergy and strict Scholastics, few of them published philosophical works and none of them were original.[66] This situation was sanctioned institutionally after 1713, as mentioned, by the practice of assigning the chair to the secretary of the Congregation of the Index.[67] So, whatever the updating of the programmes of the individual chairs over the century, they could not introduce as proven, facts or theories that contradicted, or seemed to contradict, the basic features of Scholastic or Biblical cosmology (if not all of their traditional content). Given the theoretical progress in the period, in certain sectors of research, this was constricting, and reinforced by the links of the philosophical programme to the teaching of metaphysics (held constantly by regular clergy), which in Rome did not suffer the progressive isolation that occurred in the other Italian universities from the middle of the 18th century.

This places the changes produced by the reform of Benedict XIV in a realistic historical context, which by its content and chronology seems—at a first sight—analogous to those of the same years in other Italian universities. In some of them the nature of the chair in philosophy had changed significantly; in others, such as Padua, also a chair in experimental philosophy (physics) had been established (1739).[68] In Rome, nevertheless, cultural, as well as environmental factors made the changes less radical in their premises, and probably even less so in their effects. The abandoning of the use of assigning this chair to the Secretary of the Index, the appointment of a professor of the accomplishment of François Jacquier, the creation of a physics library, that of a lectureship of higher and applied mathematics, a natural complement to that of experimental physics, assigned to T. Le Seur, colleague and close friend of Jacquier, who had written with him a large commentary to Newton's *Principia*, were facts that, seen from today, seemed to be destined to cancel out the deformities of the Roman situation. Nevertheless, the restrictions of the environment continued to be influential, if to a more limited degree. Elsewhere the changes in the programmes of philosophy were proceeded by a modification (at times a near disappearance) of the Aristotelian character of the discipline, with the influence (beginning at the end of the 17th century) of Cartesian, atomistic and empiricist approaches, which allowed for an increase in the scientific content of teaching. In Rome, on the contrary, the approach remained basically Scholastic, with all the ontological, gnoseological, and cosmological restrictions implied by this. In addition Jacquier (like Le Seur) did not give much energy to his new position, and the physics laboratory was used much more for teaching demonstrations than research.[69] It is uncertain if the approach of the two Minim friars was more a cause or effect of the situation; in any case, at least until after 1780, the content of philosophy teaching changed much less than elsewhere at least in one central aspect—the relationship between the metaphysical-cosmological and the concrete scientific elements.[70] So, without an updated general framework in which they could take on a new general importance, most of the novelties introduced in individual disciplines remained circumscribed technical details.

Mathematics

This chair is perhaps that which best demonstrates the scanty stimulus to develop technically and, as a consequence, to provide Rome and the Papal State with high level specialists. The fact that best characterizes this is the tardiness with which the new mathematics influenced teaching. In some Italian universities, at least analytic geometry has been taught since midway through the 17th century, and calculus began to be taught between 1710 and 1730; in Rome, on the other hand, synthetic geometry, together with the Cavalieri method, dominated the programme until the middle of the 18th century. Only in part did this come from the professors' inadequacy. Vitale Giordano, who taught from 1685 to 1711, was a good mathematician who played a role in the discussion of Euclid's fifth postulate, and his work in synthetic geometry was among the best of the period.[71] But his education had occurred in an environment isolated from European progress, and this backwardness allowed him to remain for decades at the top of mathematics of the city, even though he did little to modify his technical standard. This is proven by a document of great interest, that does not seem to have equivalents in other Italian universities. In 1685, the chair being vacant, the academic authorities evaluated a group of candidates, and chose Giordani, by an unusual procedure for the time, that is by examination.[72] Nevertheless, it is even more interesting how the examination was carried out: the candidates had to demonstrate some propositions from the *Elements* of Euclid, chosen from a manuscript volume that contained only the terms and the figures of all the propositions of some books (I–VI, XI, XII) and a few of the others, arranged in a different order than the original to eliminate the aid of memory.[73] The manuscript was kept so that it could be used on future occasions; it is not clear if this occurred, but what is known about the level of professors, the content of their works, and their teaching, at least until 1748, when the second chair was created, does not strictly exclude this possibility.[74] If this was the case, then until middle 18th century professors were chosen on the basis of their knowledge of some propositions of only one classic of the discipline (to the exclusion—partial or total—not only of the other classics, but of all recent mathematics).[75]

So 1748 seems to be a turning point. But perhaps, as in the case of philosophy, it was more on the level of regulation than on the level of actual teaching. The mathematical culture of Jacquier and Le Seur was certainly higher than that of their predecessors, and the chair of the first implied an updating in content, but it is not known to what extent this happened; he does not seem to have prepared notable students, and his effort was judged to be insufficient.[76] An updating of the reality to fit the intentions may have occurred in 1768, when Jacquier and the professor of elementary mathematics, C. Pozzi, left their chairs and were replaced, respectively, by the Piarist Francesco Maria Gaudio and another regular cleric, Carlo Maria Quarantotti. While the second was an obscure personality, who does not seem to have published anything, Gaudio (1726–1793) was competent in recent mathematical physics, but the treatise of mathematics he wrote for the schools of the Piarists was a mere handbook, although a good one, and his more original work was on hydraulic reclamation, so that it is not certain that his teaching was on the level of the best Italian universities.[77] A substantial transformation had to wait until 1787, when Gaudio was replaced by Gioacchino Pessuti, the best mathematician born in Rome in the 18th century, whose education, however, had been much more than Roman.[78]

Medicine, physiology, anatomy

The highest point in the Roman university school of medicine, on the level of research, was in the final years of the 17th century, with the teaching of Lancisi and Baglivi, who combined normal anatomy and pathology with nosology, medical theory, and clinical practice.[79] In the work of their successors, medical research and anatomical-physiological research were less connected (even if some of them held teaching positions in both areas), and none of them obtained—even in restricted areas—comparable achievements. Some of them (Alessandro Pascoli, Antonio Celestino Cocchi, Giovanni Antonio Volpi, Giorgio Bonelli) had great reputations, but produced few truly original written works or results, apart from their noteworthy efforts in teaching.[80] This is true also for two professors that held the chair of anatomy the longest period of time (sometimes amalgamated with that of surgery), that is Natale Saliceti (from 1750 to 1769 or 1770) and Leopoldo Micheli (from 1771 until after 1786), although they were esteemed for their commitment and if their demonstrations were an attraction for other professors and people from outside the university.[81] The situation can be summarized as follows: until the reforms of Pius VI at least, a traditional tone prevailed and the courses of medicine in Rome did not usually train physicians on a comparable level to those of the north Italian universities (perhaps not even to those of Naples, where by 1760/70 the Bourbon reforms had notable effects).[82]

Chemistry

Among the reforms of Benedict XIV, the establishment of the chair in chemistry and the associated laboratory was perhaps the most timely (it preceded similar facts in other Italian universities). As elsewhere, the chair was established within the faculty of Arts and Medicine not only because it gathered together all the scientific subjects, but because chemistry was introduced less as an area of study in itself than as a subsidiary course for natural philosophy and, in particular, for medicine (in this way its position was analogous to the courses of simples). Nevertheless, in Italian universities in which organizational progress, ties to local industrial realities and knowledge of the European developments in the discipline were greater, the laboratory of chemistry quickly obtained some degree of autonomy, preparing the scientific codification of the discipline in the years between Priestley and Lavoisier and its separation from medicine later. Even in this case, in Rome this occurred slowly, and to a degree that is difficult to document.[83] The laboratory seems to have been used exclusively for teaching;[84] lecturers do not seem to have published significant work or have contributed to the development of the discipline in the decisive years between 1760 and 1790; their names are practically absent from the academic collections and specialized journals appearing in the second half of the century, even in the Italian ones.[85] The significance of the chair should be seen mostly in the medical field, in having contributed to discrediting the use of charlatan chemical remedies, that in Rome had an established tradition, and also to the spread of a naturalistic approach and of elementary techniques of analysis used especially in geological and mineralogical investigations that, as will be described, soon developed in the city.[86] Moreover, in propagating among students an interest for the chemistry and knowledge of its principal schools, a fact which favored sound training for keen students with a scientific background.[87]

Botany and natural history

Unlike chemistry and experimental physics, in Rome as in almost all the universities, botany had two centuries of precedents. Its institutional position, the lectureship of the simples, had a good local tradition because of the specialists trained in the monasteries, to which could be added, for the demonstrations, the Botanical Garden of the Gianicolo.[88] The lectureship, however, was not the equivalent of one of botany in the modern sense, not only because of its inadequate programme but also because it did not use a complete and trustworthy classification of plants, but one that only included those considered useful in pharmacology, and almost exclusively in terms of external description. In addition, the demonstrations in the Roman Garden, besides these technical and programmatic limits, were neglected for long periods, as was the Garden itself.[89] The permanent division introduced by Benedict XIV between the professor appointed to the theoretical lectureship at the Sapienza, which by 1751 was called botany,[90] and the position assigned to the "exhibitions" of the Garden was aimed at the former concentrating on the conceptual progress of the discipline and to liberate the latter from its traditional procedures, so that it would organize the cultivation of plants and demonstrations on an updated scientific basis. From 1747, the Garden was no longer left to a head gardener ("Custodian") with experience in the medical use of plants, who was only formally under the direction of the lecturer of simples (who seldom oversaw his work); it became directed by a professor (the exhibitor), who the custodian answered to.[91] As always the aims were only partially achieved: the lecturers in the simples, generally physicians who were more interested in the therapeutic use of plants rather than an independent study of them, contributed less to botanical research than the professor assigned to exhibitions and the custodians themselves, and probably delayed the acceptance of theoretical development in the Roman scientific environment.[92] Despite the limits of the reform it was a first step towards decisive change both culturally and institutionally: the creation of an official position and a career for naturalists, distinct from the medical one. This process, however, was completed in Rome only in the first decade of the 19th century, whilst in other universities it had begun in the first decades of the 18th.[93] In addition, in the Pontifical capital the progress not only occurred later, but in some sense it was less open to the future. In Padua, and later elsewhere, the lectureship in the simples was divided into one for botany and one for natural history, and this gave some institutional support to teaching and research in geology, mineralogy, paleontology, zoology, vegetable and comparative anatomy. In Rome, on the contrary, a chair in natural history was created only in 1804 by Pius VII; until then, those new disciplines were not present—at least officially—in the university.[94]

All this provides only the organizational context of a wide range of activities, research, and interests that, with a small number (except in botany) of professors and assistants, was carried on outside the university (especially in the academies); only slowly did the university become a reference point for research other than botany and a center of co-ordination, beginning in the years of Pius VI. A general study of scientific work in Rome between 1750 and 1800 does not exist, and any summary description will be inadequate.[95] As for the situation in the university, it will be enough to outline generally the development of ideas in botany, separating it from the writings and evidence of those years related to natural history generally.[96] In the Sapienza midway through the century, the affirmation of botany as an autonomous discipline from medicine was due to G. Bonelli and F. Maratti. Neither of the two was innovative, but both were deeply interested in the subject and maintained contact with the better centers of research in Italy; the second carried out promotion of herbalism in many parts of Latium.[97] However, they did not accept the Linnaean reform, even if they had contact with some of its supporters, and the system of Tournefort remained the basis of their major works: Bonelli's first volume of *Hortus Romanus*, and Maratti's *Flora romana*.[98] It has been pointed out that this approach certainly corresponded to their convictions, and also expresses the required conformity with the professors of botanical theory until the end of the century; this seems confirmed by the fact that the most important collaborator of Bonelli and Maratti, Nicolò Martelli, whose convictions were more modern, expressed them only partially and gradually, while in writings produced in the Roman scientific environment by other scholars, no more notable nor more informed, the ideas of Linnaeus were adopted without hesitation around 1780.[99]

But the history of botanical ideas cannot be reduced to debates between the supporters of two systems of classification, even in the second half of the 18th century, when that debate was in the forefront. The work of the Roman botanical school in those years has to be looked for in contributions more limited in scope but not without importance, regarding the description of species, their place in the classification system and questions of plant chemistry and physiology. They became more numerous and more advanced after the work of Maratti, but have been recognized less than they merited, for a number of reasons: they often remained unpublished; scientists in Rome were often poorly connected with the scientific circles in northern Italy; until the early years of the 19th century (when the publication of the acts of the new Lincei academy began) the city had no journal dedicated exclusively to scientific works.[100] The study of those contributions began with Pirotta and Chiovenda,[101] but their work has not been followed up; new research will not discover decisive facts, but it certainly will give a place in the history of "normal" science (in the Kuhnian sense) to authors who up to now have not had one. In particular, the numerous work of the two Sabbati (especially the first) offers enough data to enrich the usual image of the role of Rome in the practice of natural sciences at the turn of the century.[102]

A BRIEF CONCLUSION

It has been shown that in the 18th century the Sapienza underwent various reforms, the chronology of which substantially coincides with that of others elsewhere, but that until the 1780s were put into effect more slowly and less completely. In those years some earlier reforms were resumed and new ones came, together with the assigning of important chairs to teachers of a more modern preparation; some of them, for partly scientific and partly ideological reasons (as in the case of Pessuti), tended to change the cultural environment of the Sapienza and created a new situation. So the practice of the sciences in Rome approached the level that for several decades had consolidated itself in the universities of Padua, Pavia, Turin, Bologna, Pisa, and (to some extent) Naples. It should be noted that this development did not occur against, or independently, from the intentions of the Papal government, but because of initiatives of at least part of it (as earlier the lack of development or the failure of some reforms did not depend only on the hostility or insensitivity of part of the Curia and the academic offices but also on the resistance of professional corporations and the insensitivity of the professors). In this phase, Pius VI and, perhaps principally, his Secretary of State, Cardinal Francesco Saverio de Zelada, had a decisive role. The latter, both a lover of science and a private patron, between 1780 and 1787 acted to give the city some important scientific structures, both inside and outside the university.[103] Not only did he inspire the two Bulls of 1786 and 1788 and gave Pessuti the chair in applied mathematics, but in 1787 he also had the astronomical observatory of the Roman College built (the first public one in the city) and supported the appointment of other important professors and the building of laboratories and collections.[104] So it is not an exaggeration to talk of a period of scientific "Papal reform" before the revolutionary age: somewhat tardy, and with intentions and results less complete than those that occurred in other universities, but nonetheless, real. However, environmental limits, philosophical-theological obstacles mentioned at the beginning, and soon afterwards revolutionary events—that first absorbed the attention of Papal government and then overwhelmed it—caused this phase not to fully express its potential. In part, it remained unexpressed, in part it was interrupted, in part—having shown some effects during the French occupation of Rome—it produced something that has been (and still is) attributed entirely to the institutional and cultural changes about by the new political situation. When, with the restoration, the Papal government re-established its control of the University, the socio-political and cultural conditions were so changed that it could not simply restore the rules created in the past century, but had to create a new organization. These reasons make Rome something of a special case in the social history of science: a case that is still waiting for careful analysis.

NOTES

[1] Before leaving Italy, Lagrange taught in the School of Artillery in Turin, outside the university, but substantially a specialized school on the university level.

[2] At least three cases can be distinguished: the schools of the religious orders open to laymen (especially the Jesuits, Barnabites, and Piarists), some of which were authorized to give doctoral degrees; other universities governed by states or by local magistrates, but entrusted to the Company of Jesus (as were, at different times, Parma, Mantua, Macerata); universities that were formally "secular", but organized wholly or in part according to the Jesuit model, and that used some Jesuits as teachers. See U. Baldini, *Die Philosophie an den Universitäten*, in *Grundriss der Geschichte der Philosophie. Die Philosophie des 17. Jahrhunderts. Band 1. Allgemeine Themen. Iberiche Halbinsel. Italien*, Herausgegeben von Jan-Pierre Schöbinger (Basel, 1998) pp. 621–668.

[3] This is illustrated by something to which we will return below: until the beginning of the 18th century (and perhaps until nearly half way through it) at the University of Rome the professors of mathematics were selected through an examination based on a few of the propositions of Euclid's *Elements*. Even in the most advanced universities the gap between the mathematics programmes and avant-garde research remained clear until the beginning of the 18th century. This does not mean that research was carried out by people outside the university: the greatest mathematicians of the 17th and 18th century (Torricelli, J. Riccati, and G. C. Fagnano being the only important exceptions) were professors. It means that, as in the case of Galileo in astronomy, there was a clear separation between the content and methods of teaching and that of research.

[4] Naturally individual scientists or academic groups (Maurolico, G. C. Gloriosi, Borelli, the Investiganti and their successors in Naples, some of the professors in the university and in the specialized schools opened by the Bourbons in the 18th century) carried out significant research and attempted to introduce some new elements into the Southern scientific culture, at times even before they were introduced in the North. But the work in nearly all of these disciplines exhibits a clear difference between the two regions, both quantitatively and qualitatively, that naturally should not be attributed to individual incapacity, but to complex and long-term historical conditions.

[5] Even in the case of Naples, however, many gaps persist and, especially, the studies reflect much more the intellectual climate that existed in certain academic groups than their actual teaching activities and research (as we will see in Rome, very significant differences may exist between these two realities, and judgments about the latter derived from the former may be very misleading).

[6] F. M. Renazzi, *Storia dell'Università degli Studi di Roma detta comunemente la Sapienza*, 4 volumes, Roma, 1803–1806. This work for certain questions is still valid. The only comprehensive history which is more recent (N. Spano, *L'Università di Roma*, Roma, 1935) completes it for the 19th century and the beginning of the 20th, but it is quite summary and for the years before the Napoleonic epoch it adds nothing new. The few other useful works for the period examined will be cited below.

[7] Others like those of Padua, Pavia, and Parma, had become state universities when the city in which they are located lost their communal autonomy to become part of the state (the Republic of Venice, the Dukedoms of Milan and Parma); except for Parma (that, as has been said above, was intermediate between a civil and ecclesiastical university), until the end of the 18th century they maintained some characteristics of city universities (such as the setting aside of some chairs for persons born in the city, which often had direct consequences on the quality of the faculty).

[8] The founding Bull of Boniface VIII, *In supremae praeminentia dignitatis* (20 April 1303) is published in N. Spano, *L'Università di Roma*, p. 15. The new Bull of Eugene IV, with the same title, is dated 10 October 1431.

[9] A brief exposition is in Renazzi, *Storia dell'Università*, II, p. 26.

[10] As is commonly known, according to the model of medieval origins, followed in Rome as elsewhere, a single administrative structure did not exist, but several distinct *universitates*. As will be seen below, the only unifying element was the administrative bureaucracy, because the overall management of the Sapienza was entrusted to the College of Concistorial Advocates, who named the Rector.

[11] Elsewhere in Italy, even if the course of theology was formally within the universities (in the sense that one of its subjects, metaphysics, was taught within them, and the theology degrees were conferred in the same place), its fundamental courses (scholastic theology, Holy Scripture) were entrusted to readers of these disciplines in *Studia generalia* (theological universities) of the existing religious orders in the same city, and the students attended the lessons in those *Studia* (for documentation, see Baldini, *Die Philosophie*). This does not seem to have occurred in Rome, in which professors often taught also in the schools of the religious orders, but held distinct courses in the university. This circumstance was not directly relevant to scientific

teaching, but as will be seen below the relationship between the Sapienza and the religious schools is one of the decisive factors in understanding its history before the revolutionary period.

[12] While, for example, the teaching of logic and philosophy naturally originated with the faculties (at the end of the 13th and the beginning of the 14th century), because they were considered essential for the preparation of physicians, that of mathematics—also introduced for the sake of medicine, being the basis of medical astrology—was begun after the mid-14th century. Metaphysics, uniting historically and conceptually the faculties of Arts and Theology, existed stably in Italian universities only from the 15th century. Literary subjects (Latin and Greek languages and literatures, and more rarely Hebrew and oriental languages) likewise were begun late, and were justified by the need to study the classics of philosophy and medicine from the original texts or (for the theologians) as necessary in Biblical exegesis.

[13] Latin and—in part—Greek could be necessary in the study of theology and law. The course of theology, like that of medicine, required some knowledge of philosophy; thus, even if it did not form a single *universitas* with philosophy, as occurred in medicine (because of its different organization as mentioned above and its earlier and independent origin), the presence of classical languages in the faculty of arts could have also served its needs. But this did not hold for the study of law, whose *curriculum* was entirely separate. A student of law that wanted to understand the classics of the discipline and who had not studied the classical languages at secondary school then had to attend the faculty of Arts, which thus functioned both as a container of specialized disciplines and of other "instrumental" common disciplines.

[14] According to Renazzi the course of theology of the Sapienza for this reason remained almost without students (*Storia dell'Università*, IV, p. 2).

[15] The predominance of regular clerics over laymen was due to economic as well as doctrinal reasons (the former could accept much lower compensation).

[16] There is no study of the annual programmes of the courses of philosophy at the Sapienza, but what is known of the interests of professors and their writings seems to suggest this conclusion.

[17] The "*pro Universitate*" congregation was established by Sixtus V in his Bull *Immensa aeterni Dei* (January 1587), which defined the structure and the tasks of the 15 congregations that, from then to the fall of the Papal State in 1870, formed the Curia, which handled both the technical-political questions of the Papal State (public works, finance, the fleet and army, etc.) and questions related to the organization and doctrine of the Church. It was co-ordinated by the Cardinal Camerlingo (one of traditional positions of the Curia), to whom the administrative body of the university, the college of advocates, answered. The college named the Rector, responsible for its ordinary operations and who was head of administrative and teaching personnel.

[18] The college, that brought together the advocates authorized to plead cases before the highest courts of the Papal State, had its seat within the university; it was assigned this task by the Bull *Sacri Apostolatus ministerio* of Sixtus V (1587: see the text in G. Carafa, *De Gymnasio Romano et de eius professoribus*, Romae, 1751, II, pp. 595–599). In the 16th and 17th century the Rector was always named from its own members, and its judgment was imperative for the Rector in all questions of importance. One can speak of a "cultural policy" of individual Rectors, distinct and in contrast to the college, only when they acted on the orders, or with the support, of the Cardinal Camerlingo, the Secretary of State of the Pope.

[19] In the State there was a Congregation (equivalent functionally to a modern ministry, but with a different structure because it was formed by a group of cardinals, one of whom presided with the title of Prefect) for rivers and roads. This was never on the same level of the analogous offices along the Po and in the Republic of Venice, and was almost the only permanent one, because for the other technical questions (including military engineering) there were few offices with a defined structure but they were assigned on a case by case basis. The job given by Urban VIII to Bendetto Castelli (1626) to study the floods of the river Reno is the first known case of a professor of mathematics from the Sapienza assigned an engineering question, and similar cases were

rare in the future. As has been said above, the permanent Congregations of the Roman Curia were 15 in the Bull *Immensa aeterni Dei* of Sixtus V (1588), that established the framework of the Curia until the revolutionary period; others were formed for transitory questions (such as, the one formed by Gregory XIII to reform the calendar, in which, significantly, there were no professors from the Sapienza). On the structure of the Curia: N. Del Re, *La Curia romana: lineamenti storico-giuridici*, IV ed., Roma, 1998. On the technical offices of state, the preparation of specialists and the reforms that occurred between the end of the 18th and the beginning of the 19th century, see O. Verdi, *Agrimensori, architetti ed ingegneri nello Stato Pontificio del primo Ottocento: dalla professione privata all'impiego pubblico*, in "Roma moderna e contemporanea", VI, 1998, 3, pp. 367–396, and the bibliography that is cited in it. This article demonstrates that until the period of the restoration the engineers and architects of the state were not trained in the university or the related schools (except a course of architecture in the Roman Academy of St. Luke, that was primarily artistic in character), but were formed mainly in the field.

[20] M. R. Di Simone, *La "Sapienza" romana nel Settecento. Organizzazione universitaria e insegnamento del diritto*, Roma, 1980, pp. 153–154.

[21] Besides the clear case of theology (whose lecturers were systematically members of regular orders), this was true especially for chairs in logic and philosophy. While metaphysics, a conceptual bond between natural philosophy and theology, from the Middle Ages until the mid-18th century was assigned to regular clergy in nearly all Italian universities, logic and philosophy tended to be assigned to laymen (often physicians, who began their teaching careers to go on to lectureships in medicine, which was of greater interest to them and which were, on the average, better paid). In Rome, on the contrary, from 1690 to 1748 all the readers of logic and philosophy (except two in logic) were regular clergy. Naturally what is said here holds over the long run and for most cases, not for every particular case. Occasionally Rome called professors with great reputations from the outside (like Francesco Patrizi in philosophy at the end of the 16th century); these cases, however, were not the product of a preconceived policy, but were the personal initiatives of popes or cardinals.

[22] Astrology remained among the subjects of the chair of mathematics until the end of the 17th century, and in various universities until the beginning of the 18th century the professor was required to write the yearly prediction about the foreseeable climatic-medical events of the coming year (an almanac called *taccuino*).

[23] On the history of the Church's prohibition of astrology, its new forms in the 16th century and its possible historic effects, see U. Baldini, "The Roman Inquisition's Condemnation of Astrology: Antecedents, Reasons and Consequences", in G. Fragnito (ed.), *Church, Censorship and Culture in Early Modern Italy* (Cambridge, 2001), pp. 79–110.

[24] The case of Christoph Clavius is interesting. In his first teaching years in the Collegio Romano he dealt with astrology in his private courses, but not in the official ones or in published works (see the essay mentioned in the previous note). One of the most famous Italian mathematician-astrologers of the first half of the 17th century, Andrea Argoli, who was educated and began his university teaching of mathematics in Rome (1621–1624), but after a short stay he moved on to the University of Padua and published almost all of his works outside the Papal State.

[25] There was a course in higher mathematics in the College (the "academy of mathematics" or "academy of Clavius") that over about 50 years between the 16th and 17th centuries (to some extent even later) trained many specialists, while at the same time in 20 years of teaching Luca Valerio, a mathematician of great talent, did not produce even one significant pupil (U. Baldini, *Saggi sulla cultura della Compagnia di Gesù*, Padua 2000, Chap. II; U. Baldini and P. D. Napoletani, *Per una biografia di Luca Valerio. Fonti edite e inedite per una ricostruzione della sua carriera scientifica*, in "Bollettino di storia delle scienze matematiche", XI, 1991, n. 1, pp. 3–157). The existence of a crisis in the relationship between the Sapienza and the Roman College beginning in the 16th century with the development of the latter has been often noted (see, e.g., Renazzi, *Storia dell'Università*, III, pp. 173–174; the same historian wrote

that, because of the competition of the religious schools, in the years between the 17th and 18th centuries the courses of theology and philosophy were almost without students: IV, pp. 2–3).

[26] Mons. G. Giustino Ciampini, who in the last decades of the 17th century was of the people most interested in the new science and founded an experimental academy, left money for the establishment of a "Roman University" on a new model, in which science, figurative arts, music, and modern languages (in the Sapienza only the classical languages and Hebrew were taught, and until the end of the 18th century the lessons were in Latin) were to be taught. The project was never carried out but its existence is revealing. See H. Gross, *Rome in the Age of Enlightenment* (Cambridge, 1990), pp. 236–237; on Ciampini's Academy: W. E. Knowles Middleton, "Science in Rome, 1675–1700", *The British Journal for the History of Science* VIII (2):138–154 (1975).

[27] Renazzi, *Storia dell'Università*, IV, pp. 10–12.

[28] See the historical overview and the bibliography in A. L. Bonella, "La professione medica a Roma tra Sei e Settecento", in *Roma moderna e contemporanea* VI (1998), 3, pp. 349–366.

[29] Colombo taught anatomy around 1559; Cisalpino practical medicine, from 1592 to 1601; Faber the "simples" (botany) from 1601 to 1628; Castelli held the same position from 1629 to 1634, and Trionfetti from 1678 to 1708; Porzio taught practical medicine from 1672 to 1681, and surgery and anatomy in 1682 and 1683; Lancisi taught surgery and anatomy from 1682 to 1695, extraordinary theoretical medicine from 1696 to 1701 and practical medicine from 1702 to 1719; Baglivi succeeded him in 1696 until 1701 in the lectureship of anatomy, and from 1702 to 1706 in extraordinary theoretical medicine (see *I Maestri della Sapienza di Roma dal 1514 al 1787: i Rotuli e altre fonti*, edited by Emanuele Conte, Roma, 1991).

[30] On this school see: Bonella, *La professione medica*, pp. 362–365. Paolo Zacchia (1584–1659), one of the founders of social and legal medicine, and Giovanni Guglielmo Riva (1627–ca. 1677), who carried out important research in anatomy, only taught in the hospitals where they practiced. Generally, the pontifical archiaters (both Zaccaria and Riva were archiaters) were physicians in the hospitals rather than academics. Lancisi, a pupil of Riva, in addition to the university taught at Santo Spirito, where he had practiced since his youth, and set up a medical academy that was a sort of graduate course compared to the university course. Antonio Leprotti (1687–1746), archiater to both Clement XII and Benedict XIV, taught only in the hospitals and tried to energize Lancisi's Academy. Among Roman doctors were born several naturalistic-medical academies as early as the 17th century; beginning in 1715 Lancisi's Academy went much further with an ambitious modern programme, but in the 18th century it ceased to function in several periods, and even when it did, it did not live up to original intentions (Renazzi, *Storia dell'Università*, IV, 165 ss.; P. De Angelis, *Giovanni Maria Lancisi, la Biblioteca Lancisiana, l'Accademia Lancisiana nel 250. anno di fondazione*, Roma, 1965). In theory the hospital schools were not supposed to train physicians, but to increase their practical experience after their academic training; in fact, however, it seems that a good number of physicians were trained almost exclusively in these schools. The control of medical education was so lax that Clement X (1670–1676) introduced a "matricola" (a register) in which they had to be enrolled to practice, and subordinated enrolment to a prior examination; but it does not seem that the situation improved significantly.

[31] Naturally this behavior is only rarely documented in official records, but it is found unequivocally in contemporary memoirs. In the most important work of this kind, *Memorie della Sapienza* di Pantaleo Balsarini, lecturer in logic from 1727 to 1746 and librarian, written between 1740 and 1770 (Rome, Bibl. Universitaria Alessandrina, mss. 60–64), chapter 12 on the chairs of the Sapienza begins with this phrase: "This chapter must be described with tears" (ms. 60, p. 298). Balsarini wrote (p. 301) that many professors (especially, but not only, in medicine) seriously neglected their teaching, and in his consideration of individual professors (particularly in ms. 62) blamed some of the best known (like F. Jacquier) for this. In the chirograph of 1748 with which some reforms were introduced, and of which more will be said below, Benedict XIV wrote that the Rector of the University, C. Argenvilliers, had informed him that

in the Sapienza "good studies . . . day after day are nearly abandoned" (text in G. Carafa, *De Gymnasio Romano et de eius professoribus*, Romae, 1751, II, p. 643 and ff., and Renazzi, *Storia dell'Università*, IV, p. 453 and ff.). Even after these partial reforms were introduced later in the century the Rector C. L. Costantini wrote that after 1780 it was common to not hold any or only a part of the lessons (*Relazione dell'Archiginnasio Romano nel Rettorato di Monsignore Carlo Luigi Costantini*, Rome, Bibl. Univ. Alessandrina, ms. 117, f. 16r). For Naples, Filippo Caravita's report of 1714 is important, in which we read that two of the most prominent professors of medicine, Luca Tozzi and Luca Antonio Porzio (both associated with the Neapolitan circles most open to the new science and corpuscularism, and because of this considered in all the histories of scientific and philosophical thought of the Kingdom of Naples between the 17th and 18th centuries), had for many years dedicated themselves exclusively to their medical practices, delegating their lessons to obscure substitutes who were unpaid (G. De Blasiis, "L'Università di Napoli nel 1714", in *Archivio storico per le province napoletane*, I, 1876, pp. 142–166).

[32] A discussion or list of the more recent works on the history of these universities, well-known and easily found, may be omitted.

[33] These limits are perhaps the reason why the reforms of University of Rome, unlike those which occurred in Turin, Pavia, Padua and elsewhere, are considered only in a few specialized works, and have had little influence on the study of the development of the Papal State in the 18th century (they are not considered in F. Venturi, "Elementi e tentativi di riforma nello Stato Pontificio del Settecento" in *Rivista storica italiana*, LXXV (1963), pp. 778–817).

[34] On the creation of the Botanical Garden (preceded in the century before by some private gardens and that of the Vatican): Renazzi, *Storia dell'Università*, III, pp. 158–159; R. Pirotta and E. Chiovenda, *Flora romana* (Roma, 1900), pp. 112–113.

[35] Text in Carafa, *De Gymnasio*, pp. 600–607.

[36] Which often were substituted for public lessons, which were not offered. See the text of the prohibition in Renazzi, *Storia dell'Università*, IV, pp. 424–425.

[37] See the commission report in Renazzi, *Storia dell'Università*, IV, pp. 429–430. More radical proposals had come from Gian Vincenzo Gravina, professor of civil law beginning 1699 (after 1703 he also taught canon law). He had sent several memoranda to the Pope, describing the state of the university very negatively, and had indicated as a priority taking its direction away from the Concistorial Advocates. Nevertheless, despite its passivity and traditionalism, this body, considered by many as a cause of the problems, kept its role until Napoleonic years. The final two recommendations of the congregation, as appear below, were not permanently applied. The obstacles to change were not only the Concistorial Advocates, but also the professors, who propagated, at times publicly, their objections, or proposals that did not produce any real innovation. Gravina's memoirs were perhaps produced at several different times rather than as a single text: *Sbozzo di supplica al Papa a pro dell'Università degli Studi, contra gli Avvocati Concistoriali* (Naples, Biblioteca Nazionale, ms. XIII B 44, ff. 22v–23r), and *Per l'Università della Sapienza contro il Collegio degli Avocati Concistoriali alla Santità di Nostro Signore Papa Clemente XI* (Biblioteca Apostolica Vaticana, ms. Vat. lat. 9790, ff. 1–41; another copy in ms. Ottob. lat. 3137, ff. 132v–145v). A long memoir addressed by a group of professors to Clement XI was published in Rome in 1705 (*Memoriale alla Santità del Sommo Pontefice Clemente XI Nostro Signore intorno allo Stato antico, e Moderno dello Studio Generale della Sapienza di Roma*; a sample was included in Balsarini's *Memorie*, ms. 61, f. 122 and ff.).

[38] However, it was neglected to the point that in 1742 it had to be reorganized because "it had gone to ruin": Balsarini, *Memorie*, ms. 60, p. 86 and ff. (in the name of the Rector T. Antamori).

[39] Text in Carafa, *De Gymnasio*, pp. 608–635, and in Renazzi, *Storia dell'Università*, IV, pp. 450–452. In Italy this regulation was not adapted for the entire 18th century. At the same time, however, the Pope confirmed the prerogatives of the Concistorial Advocates, including the appointment of the Rector. Before the decisions of 1744 and 1748 Benedict XIV asked a commission of professors (formed in 1741) to make proposals. The commission used principally

an external model, the reforms introduced at the University of Turin between 1720 and 1727 (Balsarini, *Memorie*, ms. 63, ff. 19r–20r). In 1743, as the first provision, the anatomical theater was improved.

[40] Text in Carafa, *De Gymnasio*, pp. 636–642, and in Renazzi, *Storia dell'Università*, IV, pp. 459–460.

[41] The proposal to establish the two chairs of advanced mathematics and chemistry was made by the Secretary of State, Cardinal Valenti Gonzaga (Renazzi, *Storia dell'Università*, IV, pp. 221–222). In his youth he studied with Celestino Galiani, a student of Newtonian physics and a promoter of the reforms in Naples, and he was a personal friend and protector of Boscovich; so it cannot be excluded that the proposal came from the mathematician of the Roman College (on the Cardinal and his relationship with Boscovich: U. Baldini, *Saggi sulla cultura della Compagnia di Gesù*, cap. IX).

[42] Text in Carafa, *De Gymnasio*, pp. 643–655, and in Renazzi, *Storia dell'Università*, IV, pp. 453–458.

[43] Some of the new chairs were: Fundamentals (*Institutiones*) of medical theory; Fundamentals of medical practice; Fundamentals of surgery and anatomy; Fundamentals and experiments in chemistry. Earlier lectureships were often more numerous, despite the low number of students, because for each ordinary professor their was a variable number of extraordinary ones, and because occasionally lectureships on special topics, or *ad personam*, were established.

[44] G. Bonelli, one of the professors of medicine who was most interested in botany and the Botanical Garden, complained that after 1720 this neglect had become habitual (Pirotta-Chiovenda, *Flora romana*, p. 217).

[45] The *Memoirs* of Balsarini point out that at least a few of the professors of medicine continued to neglect their lessons and anatomical demonstrations; a short time later, the physicians reactivated some of the extraordinary chairs and lectureships that had been suspended. See A. Pazzini, *La storia della Facoltà Medica di Roma*, Roma, 1961, I, p. 46.

[46] Student attendance was not required, and remained the lowest of the universities of the other major Italian cities; moreover, the direction of botanical garden continued to be considered only as an initial step in the *curriculum* of professors of medicine (this was claimed by Bonelli: see note 44).

[47] Renazzi wrote that "once the first enthusiasm passed", the university returned "back to its original state of languor and disorder" (*Storia dell'Università*, IV, pp. 218–219); the same judgment is found in the *Relazione* of Costantini (f. 4r), who added that even the archive "was the most perfect picture of the confusion" (f. 5r). The lack of real improvement is reflected in the low number of enrolments; in 1753 the students enrolled in all the courses were 140, and 180 in 1773 (Di Simone, *La "Sapienza" romana*, p. 146). The significance of these numbers is clear if one considers that the attendance was lower than enrolment, that the University of Padua had more than a thousand students and that in Rome most of the enrolled studied law: 101 in 1753, 125 in 1773. Between 1748 and 1773 the only regulatory change was the Bull *Splendor paternae gloriae* of Clement XIII, of 1759 (text in Renazzi, *Storia dell'Università*, IV, pp. 462–464), which however only concerned the course in theology: it established moral theology (cases of conscience), granted in perpetuity to the Carmelites, who already held other theological chairs.

[48] The request, however, was not accepted, and the building of the College became the new home of the Roman Seminary.

[49] C. L. Costantini, who became Rector in 1787, wrote that he had to "pass on to his successors a Sapienza not reduced to such a state that he would have to blush", and much of the work he accomplished was merely applying the reforms of Benedict XIV (*Relazione*, f. 7v). This confirms the fact that they had not been put into effect.

[50] It should be noted that, while the Inquisition had been abolished in nearly all the Italian states, which took over both the preventive and *a posteriori* censorship of the press, so decreasing

also the role of the Index, the two Congregations were still active in the Papal State. So, while elsewhere the decree of 1616 that prohibited works in favor of heliocentrism stopped being observed between the end of the 17th and half way through the 18th century, in Papal territory it remained formally in effect until 1820; and even after 1758, when the general formula of condemnation of heliocentricism stopped being published in the *Index librorum prohibitorum*, scientists working in Rome avoided open declarations of the subject (see Baldini, *Saggi sulla cultura della Compagnia di Gesù*, cap. IX).

[51] Reproduced in Renazzi, *Storia dell'Università*, IV, pp. 471–475, and in Pazzini, *La storia della Facoltà Medica,* pp. 636–637 and 639–641.

[52] From 1781, when Lorenzo Prospero Bottini became Rector (1737–1818, then a Cardinal: see *Dizionario Biografico degli Italiani*, 13, Roma 1971, pp. 471–472), at the top of the bureaucracy of the Sapienza there were a group of men who, even though they came from the College of Consistorial Advocates or were appointed by them, judged the old situation to be intolerable and tried to change it. One of them was Costantini, already cited as the author of the *Relazione*, librarian beginning in 1782 and Rector in 1787, the real implementer of the reforms of Pius VI; another was Renazzi, the future author of the *Storia dell'Università.* The latter, in his opening address in 1781 (*Oratio de studiis litterarum ad bonum Reipublicae referendis habita in Romano Archigymnasio VI. Kal. Decembr. An. MDCCCLXXXI in solemni studiorum instauratione*, Romae [1781]), announced the Pope's intention to restructure the university.

[53] Anatomy; Fundamentals of theoretical medicine; Fundamentals of practical medicine; *Materia medica* (2); chemistry; botany; surgical operations; obstetrics.

[54] Ethics; logic and metaphysics; philosophy (natural philosophy and experimental physics); geometry (elementary mathematics); mixed mathematics (higher mathematics with applications).

[55] Only classical languages: Latin, Greek, Arabic, Syro-Chaldean, Hebrew.

[56] In Rome, as in other Italian universities, the teaching of surgery had existed since the 16th century, but it was often combined with anatomy, given irregularly and without a practical section. With the reforms of 1786–1788 it took on a modern character. Also the teaching of obstetrics had a precedent in the lectureship *de morbis mulierum*, which had existed since the 17th century in various Italian universities (in Rome from 1615 to 1738); but in this case as well it changed radically in the breadth of its anatomical basis and had a much less theoretical nature.

[57] Text in Renazzi, *Storia dell'Università*, IV, pp. 476–477.

[58] Text in Renazzi, *Storia dell'Università*, IV, pp. 477–478.

[59] An Italian translation of the regulations of the Napoleonic universities, to be used in Rome, was printed only in 1812 (*Decreti, statuti e regolamenti principali dell'Università imperiale*, Roma 1812). In 1813 professors were required to swear loyalty to the Empire, and those who refused had to transfer to the school of the Roman College. On this period: R. Boudard, *Experiences françaises de l'Italie napoleonienne: Rome dans le systeme universitaire napoleonien et l'organization des academies et universités de Pise, Parme et Turin, 1806–1814*, Rome, 1988.

[60] On the reform of 1824: A. Gemelli and S. Vismara, *La riforma degli studi universitari negli Stati pontifici, 1816–1824*, Milan 1933. See also A. P. Bidolli, "Contributi alla storia dell'Università degli Studi di Roma—La Sapienza durante la Restaurazione", in *Annali della Scuola Speciale per gli Archivisti e Bibliotecari dell'Università di Roma*, XIX–XX (1979–1980), pp. 71–110. Here it is enough to mention one of the more unusual aspects of the reform: the establishment in 1816 of a chair of "holy physics" separated from ordinary physics, intended to demonstrate the compatibility of cosmology (as well as of biology, zoology, and botany) with the Bible, understood mostly in its literal meaning. The chair remained active for about 40 years, and about it there is only one brief work (S. Proja, *Cenni intorno la cattedra di fisica sacra nell'Archiginnasio Romano*, Roma 1838).

[61] While it is certain that they studied with Castelli in Rome, it is not known if they were his students in the university or in private courses, but this is irrelevant for the considerations

made in this text. Torricelli and Borelli are known generally, while Ricci is known only to specialists; nevertheless his *Exercitatio geometrica* (1666) is one of the most important achievements of Italian mathematicians in the years between Cavalieri and Torricelli and the diffusion of analytical geometry and later calculus (J. E. Hoffmann, "Ueber die Exercitatio geometrica des M.A. Ricci", in *Centaurus* (1964), pp. 139–193). Another important pupil of Castelli in Rome, even less known because of his early death, was R. Magiotti.

[62] On this, and the programme of the discipline, see Baldini, *Die Philosophie*.

[63] The most well-known facts, such as the initiatives against Pomponazzi's writings and the inquisitorial proceedings against Cremonini, are only a few of those regarding philosophers of these schools. Furthermore, there were the proceedings against the books and persons of naturalistic or Platonic philosophers (from Cardano to Patrizi), partly for the same reasons, partly for others. It suffices to mention that Patrizi was investigated by the Index for his *Nova de universis philosophia* (later banned) while he held the chair of philosophy in Rome, which was offered to him by Clement VIII in 1592.

[64] The Holy Office suspected Giulio Cesare Lagalla, a professor at the Sapienza in the first decades of the 17th century, an acquaintance and correspondent of Galileo who provided Aristotelian explanations of the observations made possible by the telescope, to be inclined to Averroism. His private position, however, did not influence his teaching, and after him nearly all the lecturers were strictly orthodox regular clerics. On Lagalla see the relevant passages in the National Edition of the *Opere* of Galileo and, more recently, I. Gallo, Ancora su G. C. La galla, in *Rassegna storica salernitana*, 8 (1987), pp. 17–29.

[65] On Nazzari and the journal: J.-M. Gardair, *Le "Giornale de' letterati" de Rome, 1668–1681*, Florence 1984.

[66] The list of the philosophy professors in Conte, *I maestri della Sapienza*, p. 1081 and ff., has gaps after 1678; a more complete one is found in Renazzi, *Storia dell'Università*, III, pp. 181–184.

[67] In his *Memorie* Balsarini denied that it was based on an official document, and he added: "the Dominicans maintain that [the chair] is assigned to the Secretary of the Index to indemnify him of the many expenses he has to meet because of this post" (ms. 62, p. 112). In fact, while there is no evidence that this use derived from a precise intention of control, there is little doubt that it had also this effect.

[68] Beginning in 1744, when the first chair of ordinary philosophy in Padua was given to Giuseppe Suzzi (who was mainly a mathematician) it approached becoming one of mathematical physics, because the professors, even though they followed the formal order of the Aristotelian texts imposed by the programmes, dealt with each point in its modern form, giving little importance to ontology and metaphysics. The lectureship of experimental philosophy, given to G. Poleni, who held the chair of mathematics, reinforced this tendency.

[69] From 1765 to 1768 Jacquier and Le Seur taught in the Court of Parma; later the former seems to have interrupted his teaching several times, even if his name appeared constantly on the university roll until 1787. Balsarini, who knew the internal happenings of the university, wrote that he "over time lost the concourse [that is the attendance of the students], and credit" (*Memorie*, ms. 62, pp. 115–116). One of the reasons may be the fact that the "minimum" French father also taught in the Roman College *De Propaganda Fide*, where he probably had a larger number of students and more impelling teaching obligations. As for the laboratory, nothing proves that it was used for anything other than teaching; in 1789 the Rector C. L. Costantini wrote that until 1780 in the laboratory of physics and that of chemistry there were 15 demonstrations per year, Wednesdays and Saturdays, and that the professors suspended lessons the day before (Tuesdays and Fridays) to prepare the lessons; so the students did not go to the university on those days, deserting the lessons of medicine (*Relazione dell'Archiginnasio Romano*, f. 16v). No contemporary descriptions of the laboratory apparatus exist, and there has been no attempt to reconstruct its organization.

[70] Jacquier taught until 1787; from 1767 to 1792 also taught philosophy as an extraordinary lecturer the Piarist Girolamo Maria Fonda (1732–1800), who, during the long absence of Jacquier, was the only professor. He was more scrupulous, but not a scientist of the highest level: his notable scientific works are *Elementi di architettura civile e militare*, for the School of the Piarists (Roma, 1764), and an essay *Sopra la maniera di preservare gli edifici dal fulmine* (Roma, 1770). On Fonda: P. Stancovich, *Biografia degli uomini distinti dell'Istria*, Capodistria 1888, pp. 229–230. On Jacquier and Le Seur, so united in life and in their scientific work that it is difficult to discuss them separately, there is no adequate work, despite the fact that their long commentary on the *Principia* is mentioned in all the histories of physics of the 18th century. The fullest biographical treatment is still G. Ceruti, *Elogio funebre del Padre Jacquier detto in Arcadia*, Roma 1788; on both of them see also F. Bonnard, *Histoire du couvent royal de la Trinità du mont Pincio à Rome* (Rome, Paris, 1933), pp. 178–186. Perhaps Jacquier's university lessons were more technical than the manual he wrote for his courses in the College of Propaganda Fide (*Institutiones philosophicae ad studia teologica potissimum accommodatae. ... Physicae pars I*, Venetiis 1785), but it is improbable that they were much different.

[71] See the summary of C. Preti in *Dizionario Biografico degli Italiani*, 55, Roma, 2000, pp. 289–291.

[72] As has been said, competitive examination for chairs was introduced on a stable basis in Rome only by Benedict XIV, and it became common in other Italian universities even later. In 1685 the probable reason for it was that the academic authorities did not have enough information to evaluate the candidates or, perhaps, that no one was supported more authoritatively than the others: so the use of an objective procedure was imposed by uncertainty.

[73] This is now ms. 109 of the Alexandrine University Library. In the first folio it functions to explain in terms worth quoting: "Hoc opus, quod complectitur Propositiones omnes, et Figuras tam priorum sex librorum Euclidis, quam etiam XI, et XII, singulas necessarias, permixtas singulas, et in fine operis ad Euclideam methodum redactas per Indicem; hoc ordine constructum est iussu Ill<ustrissi>mi et R<everendissi>mi Praesulis D. Marci Antonii Buratti Sac. Consistorii Advocati, et Almae Urbis Gymnasii Rectoris Dep<utati>, ut eo ad explicandum proposito in Concursu ad Cathedram Matheseos, tuto possit iudicari, quis explicantium Matematica fondamenta percalleat.
Ideoque ab eodem Ill<ustrissi>mo ac R<everendissi>mo D. Buratto Bibliothecae donatum fuit, ut nunquam inde, nisi similium Concursuum causa, possit extrahi. Die IV Februarii MDCLXXXV, praecedente publico Edicto, in Palatio Quirinali in Aedibus Em<inentissi>mi ac R<everendissi>mi Cardinalis Cybi [that is Alderano Cybo] hora XXI coram eodem Em<inentissi>mo ac Sacri Collegii Advocatis, cum quatuor Examinatoribus, uti magis idoneus fuit electus Vitalis Iordanus Bitontinus".

[74] After Giordano, who died in 1711, Domenico Quarteroni taught (first as Giordano's assistant, then as an extraordinary professor after 1699), whose scientific publications deal only with calendar theory and land reclamation (P. Riccardi, *Biblioteca matematica italiana*, republished Milan 1952, I, columns 326–327). In 1727 Diego Revillas, an abbot of the order of St. Jerome, joined him; Revillas published on astronomy, historical metrology (his study on the Roman foot is important), and meteorological instruments (Riccardi, I, columns 551–552). He was also a notable personality in the academic circles of Parma, where he worked before Rome (A. Vallisneri, *Epistolario*, I, edited by D. Generali (Milan, 1991), pp. 62, 65–67). In Rome he had a role in the canonization of St. Juan de Avila. According to Balsarini he was highly esteemed in the university (*Memorie*, ms. 60, p. 306), but his teaching remained traditional. In 1742 the Benedict (Olivetan) monk, Cesareo Pozzi (1718–1782), joined him and from 1745 to 1769 succeeded him (in 1748 he joined Le Seur in the new chair of "mixed" mathematics); however, he does not seem to have published scientific works. The views on him are varied, but they agree on the ineffectiveness of his teaching: for Balsarini it contained "little theory" and was "neglected" (*Memorie*, ms. 60, p. 306); Renazzi saw him as a man of

talent, who dedicated to other activities (*Storia dell'Università*, IV, p. 101). For information on him see *Carteggio di Pietro e Alessandro Verri* (Milan, 1923) and ff., IV, pp. 239–240.

[75] This does not mean that the professors did not know the basic elements of modern mathematics (with the possible exception of calculus), but this characterizes the basis of their preparation and their teaching. Renazzi, who had personal experience with the university of the second half of the century, wrote that until Benedict the XIV the programme was limited to the elements of geometry (*Storia dell'Università*, IV, p. 221).

[76] As mentioned, in 1765 Le Seur and Jacquier went to the Court of Parma; when returning in 1768, the first rejected the chair due to its limited retribution (Balsarini, *Memorie*, ms. 62, p. 117). He did not publish a textbook for his courses, and no manuscript notes of them are know to exist.

[77] Riccardi (*Biblioteca matematica*, I, col. 576) lists only a part of his writings. Three of them [*De natura extensionis*, Naples 1760; *De naturae vi et lege generali*, Romae 1756 (reviewed in *Acta eruditorum*, 1757, p. 648); *De rectilinea lucis propagatione*, Romae 1760] are of interest, being critiques of Boscovich's theory of force. From 1754 to 1761, when he was still teaching in his Order's Collegio Calasanzio, Gaudio was in Rome the more decided opposer of the Jesuit's theory; he criticized it not only in the works mentioned, but also in at least nine dissertations discussed in public by his students during those years. His *Institutiones mathematicae ad usum Scholarum Piarum*, in 3 volumes, were published in Rome between 1772 and 1779. The only work specifically on him is still A. Amoretti, *Elogio di P. Gaudio* (Nice 1848).

[78] Pessuti (1743–1814) had studied in the Piarist college Calasanzio, in Rome, where mathematics was taught by Gaudio; in 1760 he published his philosophical and mathematical theses. Before 1770, in an unknown way, he became a teacher in the military academy of St. Petersburg, where he met Euler, then returned in Rome before 1780. A first rate analyst and a member of important scientific academies, he was unable to create a significant and stable school (the best Roman mathematicians of early 19th century were not his pupils). He is not of interest only to historians of science, but also to cultural and political-ideological historians of the city, because he directed and collaborated for a long time with the most important local journals, and he was the most eminent academic who collaborated with the French during their occupation of Rome. Despite this no specific study of him exists. See: E. De Tipaldo, *Biografia degli Italiani illustri*, III (Venice, 1836), pp. 266–269; G. Ferretto, *Note storico-biografiche di archeologia cristiana* (Città del Vaticano, 1942), *ad ind.*; M. Caffiero, *Le* "Efemeridi letterarie di Roma (1772–1798)", in *Dimensioni e problemi della ricerca storica* (1997), 1, pp. 77–78. Some unpublished papers of his, entitled *Schede di Gioacchino Pessuti*, are in the codex Vat. lat. 9828 of the Vatican Library; interesting information on his mathematical work and his debates with G. Calandrelli (n. 104) may be found in a work by A. Eximeno, a Jesuit from Valencia who lived in Rome for some years: *De studiis philosophicis et mathematicis instituendis* (Madrid, 1789), part II, *passim*.

[79] As for Lancisi the first aspect prevailed in his university teaching and his most original research, such as *De subitaneis mortibus*, while the second was obviously central in his work in the hospital of the Holy Ghost. On him there is no more adequate monograph than De Angelis, *Giovanni Maria Lancisi*. The work of Baglivi, as is commonly known, includes many subjects from medical practice to nosologic systems and physiology (he was one of the most rigorous theoreticians of iatromechanics), but his most original work was in pathological anatomy. For a summary of his work see: *Dizionario Biografico degli Italiani*, 5 (Roma, 1963), pp. 250–252.

[80] Pascoli (1669–1757) taught anatomy and surgery (1702–1708), theoretical medicine (1709–1719) and then, after Lancisi, practical medicine (1720–1749). For the essential information see E. De Tipaldo, *Biografia degli Italiani illustri*, II (Venice, 1835), pp. 209–210; L. Castaldi, "A. Pascoli filosofo e anatomista perugino", in *Rivista di storia delle scienze mediche e naturali*, XV (1924), pp. 173–180. Cocchi (1685–1747)—not to be confused with a contemporary Antonio Cocchi, a better known physician in Florence—was the lecturer of simples in 1726 and 1727; then ha taught anatomy and surgery from 1728 and 1740, and theoretical medicine after 1744. A

pupil of Baglivi and Lancisi and a correspondent of Morgagni, he published several writings on medicine which had a considerable reputation (in one of which he proposed a correct interpretation of the mechanism of cataracts), but his teaching does not seem to have been very effective: Balsarini thought he was a "great talent", but neglected (*Memorie*, ms. 60, p. 140, and ms. 62, p. 84). See also *Dizionario Biografico degli Italiani*, 26 (Roma, 1982), pp. 461–463. Volpi, earlier an assistant of Lancisi, taught theoretical medicine from 1726 to 1740; afterwards he held the lectureship of the simples from 1719 to 1725, and later taught chemistry (from 1760 to 1773); see Pirotta-Chiovenda, *Flora romana*, pp. 219–220. Bonelli taught chemistry in 1759 and practical medicine from 1763 to 1793; on him see P. Casini in *Dizionario Biografico degli Italiani*, 11 (Roma, 1969), pp. 757–758. As said below, his most significant work—not truly original—was not in medicine, but in botany, and its origins were in Turin and Padua, not in Rome.

[81] Balsarini attributed to Saliceti (1714–1789) "ultimate diligence" (*Memorie*, ms. 62, p. 261). Even if in the university he only taught anatomy, Saliceti was known as a physician (later he was also an archiater of Pius VI), member of various Italian academies and correspondent of famous scientists of the time; he was not, however, an original researcher. For his biography see A. Fabroni, *Elogi di uomini illustri*, II (Pisa, 1789), pp. 269–282. In 1767 the demonstrations were usually held on holidays (to not interfere with the lessons) and Thursdays, especially concentrated in the period of Advent (Balsarini, *ibid.*; on the attraction of the demonstrations see ms. 63, f. 140r–140v).

[82] In the current state of research, nevertheless, any precise conclusions are premature. The courses of medicine in the hospitals also had their own scientific equipment (the Academy of Lancisi had chemistry and physics laboratories), in which university professors often taught, so that to some extent these courses were continuations of university teaching (with the same function of university clinics today). Usually they were usually attended by people who were already physicians which dealt with specific problems raised by clinical practice and were less restricted by rigid and antiquated programmes; so it is plausible that these courses not only touched on more defined questions, but were more in touch with contemporary research. The best example is Santo Spirito, where around 1770 the school was restructured with the building of an anatomical theater. Demonstrations were given regularly, anatomical and obstetric models in wax were purchased, practical exercises for students went on, with awards for the best of them (Renazzi, IV, pp. 295–296). On medicine as profession in XVIIIth century Rome see Bonella, *La professione medica*.

[83] Since its origins there was resistance to the creation of a chair; this happened elsewhere as well, but in Rome it seemed greater: "in those times many said that Chemistry was not profitable for Rome, but harmful" (Balsarini, *Memorie*, ms. 62, p. 116).

[84] On this subject there is the same lack of information as for physics (see n. 68).

[85] Up to the pontificate of Pius VI they were: Luigi Filippo Giraldi (1748–1759); G. Bonelli e G. M. Volpi, already mentioned as a medical lecturer (1759–1760 and 1760–1773); Pasquale Adinolfi (1774–post 1787). Evidence of theoretical backwardness of teaching there is the faithfulness of Adinolfi to the chemistry of the phlogist (Renazzi, *Storia dell'Università*, IV, pp. 266–267).

[86] The works on chemistry published in Rome in the second half of the century are few, so that most of the evidence about knowledge and research in the discipline comes from writings on botany and geology, which used it. For example, Liberato Sabbati, discussed below as one of the "custodians" of the Botanical Garden, had a good background in chemistry, which he used in his studies on the medicinal properties of plants.

[87] While an approximation of the programme of physics courses, after the reform of 1748, can be drawn from the textbook written by Jacquier and by others for other Roman schools in those year (the Roman College, the Nazareno College, the college of Propaganda Fide and others), the texts used for chemistry are unknown.

[88] Already at the end of the 17th century a transition from monastic herbalism and the medieval and renaissance tradition of the "simples" to a broader conception of botanical

research was begun by Giovanni Battista Trionfetti (1656–ca. 1708), professor from 1682 to his death, educated in Bologna in the naturalistic tradition begun by U. Aldrovandi. See Pirotta-Chiovenda, *Flora romana*, pp. 123–142.

[89] This situation was denounced by Giuseppe De Panicis, lecturer in the simples from 1744 to 1746, in the inaugural address of 1745 (*Oratio pro studiis botanicis abita in Horto Accademico MDCCXLV* (Romae, 1745)); see Pirotta-Chiovenda, *Flora romana*, pp. 226–227; U. Baldini, "De Panicis, Giuseppe", in *Dizionario Biografico degli Italiani*, 39 (Roma, 1991), pp. 8–9.

[90] E. Conte, *I Maestri della Sapienza*, p. 1099.

[91] Conte, *I Maestri*, pp. 1099–1102, does not correctly differentiate between the professors of the two categories, putting some of the first in the second. A better classification is found in Pirotta-Chiovenda, *Flora romana*, pp. 223–247: After De Panicis the lecturers of the Sapienza were Francesco Aurelio Ginanneschi (from 1747 to 1768) and Giuseppe Micciari (from 1768 to at least 1787); the "ostensori" were Vallombrosian Benedictine priest Francesco Maratti (from 1746 or 1747 to 1777) and Nicolò Martelli (from 1777 to 1794). The custodians were Francesco Bertoldi (from 1744 to 1747), Liberato Sabbati (from about 1747 to 1779) and his son Costantino (from 1780 to after 1809). The new organization implied a higher status for the custodians: until Bertoldi, they were merely gardeners, chosen for having worked in other botanical gardens, while beginning with L. Sabbati (surgeon and expert in chemistry) they were university educated, and often physicians. The higher academic qualifications of individuals who worked in the botanical garden were such that, even when Pius VI renewed Benedict's XIV prohibition for individuals to move from the "ostensione" to medical lecturers, it was not followed: in 1794 Martelli became the lecturer of anatomy.

[92] The investigation of Pirotta-Chiovenda (*ibid.*) demonstrates that most of the important botanical writings by authors working in Rome came from the personnel of the Botanical Garden (like G. Bonelli), and almost none by lecturers in theoretical botany. However, given that the expression of an official position on innovative theoretical proposals—like Linnean classification—was an attribution of the latter, the former could not declare themselves without violating their area of academic competence. Even if there does not seem to be explicit testimony of this kind, this could be one of the reasons for the delay of the acceptance of the new botany in Rome. In fact, the programmes of theoretic botany were faithful to the Tournefort system until 1809 (Pirotta-Chiovenda, *Flora romana*, pp. 253–254), while as will be seen writings produced by the staff of the Botanical Garden raised doubts about it 25 years earlier.

[93] Principally in Padua, with the assignment of the direction of the botanical garden to Giulio Pontedera (beginning in 1719) and that of the lectureship of the simples, transformed into one in natural history and botany, to Antonio Vallisnieri Jr. (beginning in 1734). Other factors in dividing medical teaching in distinct autonomous disciplines were the renewal of the lectureship of philosophy (even if, for reasons already mentioned, in Rome it was more separated than elsewhere from medical disciplines), and the creation of one in chemistry. For the latter, however, the formation of a professional role was slower because there was only one lectureship, and at least in the first decades the laboratory technician was—unlike the custodian of the botanical garden—merely an operative, without academic training and unable to carry out autonomous research.

[94] Between 1750 and 1800 the work in this fields was normally outside the Sapienza and divorced from it, and only near the end of the century they gained some ground, not in teaching but in the personal interests of the professors of anatomy and botany. In 1804 the direction of a naturalistic museum, formed out of biological and mineralogical collections made by professors over the previous decades, was entrusted to the professor of natural history: see the Bull that established it in Renazzi, *Storia dell'Università*, IV, pp. 476–477.

[95] A scientific academy came into existence only in 1799–1800, when the Academy of the Lincei was re-established, but in the preceding decades other bodies, mainly erudite, literary, and historical (the best known academies being the Arcadia, the Tiberina, the Quirina, and the Academy of the Occulti), held scientific discussions that were not superficial or merely

popularizing. A good example was the report of the physician G. G. Lapi to the Quirina Academy (September 1758, but published in 1759) on the volcanic origins of the lakes Albano and Nemi, which began the study of the volcanic origins of part of Latium. Another, perhaps more important place for this were the schools of the religious orders, which traditionally had mathematical chairs, and in which from the mid-1700s the chairs of philosophy were places of naturalistic and physical research and had important collections as well (if the Kircherian Museum of the Roman College, formed in the 17th century, was inspired by the old *wunderkammer*, the mineralogical collection of the Nazareno College of the Piarists was of the highest level: Renazzi, IV, pp. 297–298). Some colleges and convents had astronomical observatories (in the second half of the century there were at least five observatories in Rome), and a center of culture and scientific research were the palaces of some of the high nobility (such as the Caetani) and some cardinals, in particular F. S. Zelada (who had a collection of wax anatomical models, later donated to the hospital of Santo Spirito, an important library, a private observatory and a collection of physical apparatus, later donated to the Roman College). On all of this the studies (except for articles on individuals or generic summaries of single academies) are few and not up to date. See Renazzi, IV, pp. 288–304; Pirotta-Chiovenda, "Flora romana", in F. Millosevich (eds.), *Le scienze fisiche e biologiche in Roma e nel Lazio* (Roma, 1933) (uneven essays on single disciplines); *Arte, scienza e cultura in Roma cristiana* (Bologna, 1971) (very general); M. Calisi, *Le Specole romane nel Settecento*, in L. Pigatto (ed.), *Giuseppe Toaldo e il suo tempo. Nel bicentenario della morte. Scienza e lumi tra Veneto e Europa* (Padova, 2000), pp. 423–445.

[96] For botany the widest study (and in certain ways one of a kind) remains Pirotta-Chiovenda, *Flora romana*. Among the fields omitted in this article the most developed (discussed mainly in the academies because of its relationship to archaeological research, then in great expansion in Rome), is geo-paleontology. For some basic information see E. Clerici, "La geologia e la paleontologia in Roma e nel Lazio", in F. Millosevich (ed.), *Le scienze fisiche e biologiche*, pp. 79–111.

[97] On the first see Casini, *Bonelli, Giorgio*. He was trained as a physician at the University of Turin, where he was a pupil of the noted botanist, C. Allioni, and he was also in contact with the curator of the Botanical Garden of Padua, G. Pontedera. Both, as members of the preceding generation, followed the Tournefort system, and Bonelli was never able to disassociate himself. As mentioned, his interest in botany was not professional, because he taught practical medicine. On Maratti see Pirotta-Chiovenda, *Flora romana*, pp. 229–236.

[98] The work of Maratti (in reality only a collection of note cards) was only published in 1822 in Rome by M. B. Olivieri in an inexact edition, which has reduced its importance; in any case, he remained pre-Linnean, even if in 1756 the author met M. Koehler, pupil of the Swedish naturalist. The first volume of *Hortus* (a description of the plants in the University Botanical Garden), published in Rome in 1782, declares his dedication to Tournefort in its title, and its arrangement faithfully followed that of the Botany Garden, which Maratti kept strictly according to Tournefort. In 1772 Bonelli showed the provisional text to J. J. Ferber, another pupil of Linneaus, but this did not produce any change in convictions.

[99] Martelli was a pupil of Maratti, whom he succeeded as *ostensore* in 1777. He replaced Bonelli in 1784 to edit the remaining volumes (II–VIII) of the *Hortus romanus*. In the preface to Vol. II he disassociated himself from the convictions of his predecessor, perhaps as an effect of a visit of M. Vahl, a third pupil of Linnaeus (1783–1784). Nevertheless Maratti only introduced Linnean nomenclature in Vol. VI and (perhaps to be coherent, and because the Garden still maintained its old organization) maintained until the last volume the Tournefort division of plants. That this may have derived from institutional restrictions is supported by the fact that in Rome the Linnean classification was publicly supported by two non-university botanists, Francesco (Cesare) Majoli (1746–1823) and Luigi Filippo Gilii (1756–1821), in the same years. The 27 volumes unpublished *Plantarum collectio* by Majoli, written "juxta Linnaeanum sistema", is conserved in the City Library of Forlì; in 1785 Gilii published the first Linnean

work to appear in Rome, a *Delineazione dei generi naturali divisi in VI classi a norma del <<Sistema naturae>> di Linneo* (on both see: Pirotta-Chiovenda, *Flora romana*, pp. 254–261).

[100] The burden of these factors was only partially attenuated by an extra-institutional fact, the personal relationships begun during the visits of Italian and foreign scientists to Rome. The succession of these visits, obviously more common in Rome than other Italian cities, has not been reassembled for scientific history; above the three disciples of Linnaeus many personalities are involved (including Goethe who during his travels in Italy also made some naturalistic observations).

[101] Their *Flora romana* should have been in several volumes, but only the first was published; it, in addition, was not dedicated to a technical analysis of the works, but to a general description of the development of the discipline in Rome.

[102] See the basic information in Pirotta-Chiovenda, pp. 235–247.

[103] On Zelada there is no recent or adequate study. For the essential facts, see G. Moroni, *Dizionario di erudizione storico-ecclesiastica*, CIII (Venice 1861), pp. 460–469.

[104] The initiatives of the Cardinal were often strictly tied to the most important Roman scientists of those years (such as the astronomers Giuseppe Calandrelli and Andrea Conti), and to notable private initiatives, such as the construction of the observatory of the Caetani dukes, perhaps the most significant private initiative of its kind in Italy in the late 18th century (for the basic information, see: L. Fiorani, *Onorato Caetani* (Roma, 1969)).

JOSÉ LUIS PESET

ENLIGHTENMENT AND RENOVATION IN THE SPANISH UNIVERSITY

The Spanish monarchy drew its scientific needs from three sources: the universities, the church, and the army. The first was the essential fount of knowledge, but this came from a pact between the monarch and the pope, who needed jurists and clerics in particular. As scientists, there were the physicians who treated the noble and ecclesiastical courts and the cities and who required suitable scientific training. Accordingly, the Faculty of Arts had some scientific chairs, particularly Physics and Mathematics: Botany and Chemistry were introduced as time went on. But the church also concerned itself with scientific knowledge: many scientific scholars were churchmen, and the orders had teaching institutions. In particular, the Jesuit colleges and, within them, the Imperial College and the Nobles Seminaries cultivated science. The military, for its part, always had academies with good scientific and technical training, in Flanders or later in Cádiz.

Two reforms were first necessary for the improvement in university education. The first was the expulsion of the Jesuits, who occupied chairs in Grammar, Philosophy, and Theology. The universities had to be released from the order's control, but the path chosen was to entrust it to others, particularly the Dominicans and their allies. The other reform was of the "colegios mayores" which were, in principle, hostels for poor students. However, over time, the collegiate places had been converted into privileged scholarships for future administrators of the Crown. The agreement between collegians and public officials meant that the University and Councils were dominated by a handful of families and their allies. Moreover, in some, such as Alcalá de Henares, the Colegio Mayor de San Ildefonso dominated the University, its rector controlling rents, grants, studies, students, and appointments. The reform saw the beginning of the end of the colleges and the freedom of the collegiate-style universities.

The Spanish university modernized through multiple channels, the King's endorsement of the new study courses being one of the most significant, introducing up-to-date textbooks which taught modern science. The idea was for teachers to draft the study courses but in most cases, that was not possible. As a result, there was frequent recourse to foreign texts and the universities even had their own printing works or agreements with printers. The Royal Printing House also collaborated in this task. Some university scholars, such as the physician Andrés Piquer, were extremely successful with their texts; at other times, persons outside the universities were involved, such as the mathematician Benito Bails. Nonetheless, it was very common to continue

M. Feingold and V. Navarro-Brotóns (eds.), Universities and Science in the Early Modern Period, 231–239.

publishing foreign authors in Latin and, occasionally, as with William Cullen, backing their translation.

Another channel was the emphasis on practical training, particularly for science and medicine, always directed at the training of medical professionals. Experimental physics laboratories, botanical gardens, natural history collections and, chemistry laboratories were ordered in all universities. In the medical field, good training required dissection rooms and hospitals. Libraries, non-existent in some universities, were no less important, particularly for science teaching. The expulsion of the Jesuits was an essential step in the appearance of these facilities in the universities: the Company had, since its inception, been concerned with science and, when it was expelled, it left books, instruments, collections, and buildings, most of which were passed to the universities. The installation of astronomical observatories was also important to knowledge of astronomy: but they were also useful to physicians. With the revival of Hippocrates—and no less of Sydenham—consideration was given to the effect of climate and the heavenly bodies on illness. While the astral role—the moon, critical day, bleeding dates—slipped into oblivion, clearly climatic aspects were increasingly emphasized. As a result, the daily charts—for example, of the Vienna School, given its influence—became not just splendid clinical records but also early meteorological observations.

There was also an increase in the importance of Arts or Philosophy Faculties. Always considered minor faculties, their teachers—as heard in the complaints of the mathematician Diego Torres de Villarroel—were at the end of the pecking order whether in terms of seats, salaries, or votes. Thus it was sought to promote an increase in the importance of this faculty. Of note was the attempt to create a College of Philosophy in the University of Salamanca: this concealed both a wish to increase power and salaries as well as to improve teaching. The end of the century saw a search for education which would be more modern, Castilian, practical, with new chairs and facilities. To some degree, this reflects the impetus which Immanuel Kant saw in the Faculties of Philosophy in *The Polemic of the Faculties*. There is no doubt that scientists were seeking a high academic and teaching standard, but in the Spanish universities they had to rely on the backing of physicians, who were quickly joined by the jurists, announcing a professional, non-scientific university, the university of the 19th century.

It became possible to introduce modern knowledge, putting an end to a university which was living in former centuries. Galen was ousted from the Faculty of Medicine, and the possibility of clinical medicine was restored in the name of Hippocrates. Anatomy and surgery got under way. Natural history first followed the system of Tournefort and, soon, Carl Linné. Chemistry saw the introduction, through Fourcroy, of the new terminology of Lavoisier. In Physics, Copernicus and Newton became known, and infinitesimal calculus was introduced in Mathematics. Other institutions were more advanced with these novelties, such as the Colleges of Surgery and the Military Academies, whose teachers, textbooks, knowledge, and installations influenced universities.

Centralization was another major tool of modernization: the aim was for there to be a centre in Madrid which would be a model and guideline for other universities. The Jesuit's long-standing attempt to establish a University of Madrid was taken up again in the 18th century and it became a reality in the 19th century. This had begun in the 18th-century thanks to many institutions—old and new—in the city. The Jesuits had the Imperial College and were to create the College of Nobles. Their books and some facilities were transferred to the University of Madrid. Madrid also had the Royal Academies, and professional associations began to appear. The Natural History Museum and Botanical Garden were set up, followed by a chemistry laboratory and the Royal Clinical Studies, through which medical Bachelors had to pass to obtain licence from the *Protomedicato* or the *Juntas* or Boards for practice. This was an old tradition, because the *Protomedicato* resided in the Court, with the royal physicians. The establishment of the San Carlos College of Surgery in the General Hospital and the San Fernando College of Pharmacy were advances along this path. The attempt to create the *Facultad Reunida*, to unify all health education, made Madrid the leader in medicine. This was where the *Protomedicato* or the *Juntas* were. Other universities were relegated to positions of secondary importance, and the hegemony of Salamanca and Valladolid came to an end. Others, as in Barcelona, took a special route: it was re-established by the liberals and was able to take advantage of resources in institutions, libraries, and among teachers in Catalonia. However, it was in permanent battle with Madrid, in its wish to be recognized as an equal. Here, ultimately, the conversion of the University of Alcalá de Henares into the Central or Madrid University was the decisive step.

MEDICAL TEXTS IN THE UNIVERSITY OF VALENCIA

From among the lines along which Spanish universities modernized, we shall select some. In the first place, the use of modern textbooks for teaching. This was not entirely novel in the 18th century since modern books had been known in universities since the Renaissance. But university requirements were for teaching using the treatises of the classics, particularly Galen and Hippocrates. This was the rule from the foundation of the universities until the mid-18th century. The teacher read a fragment from the classic text in Latin and then discussed, explained and enlarged on it. It was in this enlargement where there was space for modern texts. However, even these were selected from authors with a great respect for tradition, e.g., Galenism in medicine. This was the nature of the books used in this Faculty and which, given my specialization, I will speak about today. We may mention the texts of Mercado and Segarra, written in the late Renaissance, for the universities of Castile or Valencia.

To concentrate on the University which has received us here today, the 1733 Valencia Constitutions still recommended classical authors, although these were to be explained using modern texts, to facilitate their study. In fact, Renaissance works following Galen allowed for a more systematic examination, more rational explanation and easier memorization. Thus, the medical teachers who wrote the Constitutions wished for the works of Jaime Segarra to be republished, since some were no longer available.

However, it was from this wish to modernize that the attempt arose to have this University's teachers write their own texts to facilitate modern knowledge. It was natural for Mariano Seguer to recommend that good texts be obtained for education and to begin with the translation of a follower of chemical medicine, which was so accepted at the beginning of the century. This was Joseph Jackson's *Enchiridion medicum theorico-practicum* (1734), published for the *medicinae Tyrones*. This is a comprehensive view of human pathology, arranged in the ancient manner from head to foot. It ends with a treatise on syphilis and is accompanied by a letter on epidemics written by Hecquet as a consequence of the Marseille's plague. These pages make it clear that Seguer was an excellent clinical physician, in the line of Hippocrates and Sydenham. He was also an eclectic: when Jackson asserts that melancholia is caused by acidity of the animal spirits, he adds that the cause is in solids, which obstruct the movement of fluids.

This eclecticism, which allowed the introduction of the main medical doctrines in university classes is also seen in *Institutiones Medicae ad usum Scholae Valentinae* (1762), by Andrés Piquer. The professor from Valencia was to abandon his initial enthusiasm for mechanical explanations and moved toward vitalist theory. However, this tendency was compensated with those more inspired by solids theory. He maintains that living matter is not explained by physics or by chemistry, but by the laws of life—*ex distinctis vitae legibus*. Thus he offers a broad explanation of the *res naturales* and the structure of the parts, fibres, humours, and spirits. He has the corruption of the humours co-exist with the admission of blood cells and blood circulation. He takes his backing from microscope readings, injections, and vivisection. In general, he respects the classics such as Hippocrates and Galen, the great clinicians like Sydenham or Baglivi and the magnificent teachers like Boerhaave or van Swieten. His clinical skill is seen in the fact that he was one of the first writers to defend the link between mania and melancholia.

With the teaching of Pathology included in the Chair of Clinic Medicine introduced in the 1786 study course, good teachers were soon in the offing. We may recall Félix Miquel, so well studied by Jorge Navarro. The novelties arising in the teaching of clinical medicine can be summarized under three headings: its practical nature, eclecticism, and the desire to renew. There is emphasis on the relation between student and patient and between them and the teacher. Clinical sessions and records were thus introduced, together with meteorological and medical day charts. Good books were written and perused attentively by students. Here we may point to the edition by Ángel Sanz Muñoz of *Compendio de medicina practica arreglado a las explicaciones del Doctor Don Félix Miquel* (1820). This modern study of human clinics was influenced by the 18th-century classification of diseases, by anatomical-clinical thinking, and by the breadth of the Vienna School and the narrowness of Brown's system. However, once more, Hippocrates and clinical observation sought to be the guideline, on the way to a classification system which would link the relationships of diseases with the lesions they produce.

To return to the matter of pathological melancholia, while Seguer explained it in terms of science (physics and chemistry), and Piquer described it accurately, Miquel analyzed it from the standpoint of temperament and inheritance and of passions and excesses. "Hypochondria—and melancholia—are inherent to severe, dry and arid subjects whose nervous system is excessively mobile: this may be hereditary or the result of an indolent lifestyle characterised by excess". It was not found among the poor or workers, but rather among the wealthy, leisured, and disordered, separated from the divine nature of the classical authors. This was due among other things to "passions of the spirit", "profound meditations", "sensual delights", and the "abuse of narcotics". A long tradition of psychiatry, from Huarte de San Juan through Tissot to Pinel and Esquirol, is seen in these pages.

Two notable changes therefore took place. On the one hand, as we have seen, modern knowledge replaced Galenism, using the name of the master of Cos against the master of Pergamum. On the other hand, the reading of the ancient classical treatises was abandoned which, in Renaissance editions in Latin, had been the duty of all scholars. The old task of listening and copying was also ended, the aim being to read and memorize the modern text. It was also intended for the texts to be by the teachers themselves so that, with the Spanish university reforms begun in 1771 in Salamanca, an attempt was made to have the teachers write their own courses. This was however to fail, since Gregorio Mayans proposed a number of courses by various foreign authors: this was to be the trend in the future—to import, publish or translate foreign writers. This was clearly seen in the imposition in the 1786 study *curriculum* in Valencia of the work of Professor William Cullen, of Edinburgh, so that Vitalism was also well represented in the classrooms of this institution.

However, the future of medical education lay in the observation of the patient or corpse and in the modernity represented by anatomy and medical clinics. Although, as pointed out by López Piñero, the teaching of anatomy was not a great success in Valencia, since the future belonged to the Colleges of Surgery, clinical education had a great future. The role of the student was regained, albeit with differences of meaning. Hence the requirement to observe the patient and then the corpse. And to listen to the teacher and to take notes. In short, to write the clinical history and the daily meteorological and medical records. Writing by students was not lost and, despite the prohibition on dictating in class, lessons of masters from throughout the 18th century have been preserved: indeed some seem to be by the teachers themselves, such as the anatomy lessons of Andrés Piquer. However, now, the student's activity was more intense so that it is not surprising that books by teachers appeared copied or compiled by their students, and which were subsequently published. Latin had of course by now fallen by the wayside. In the next two centuries, excellent textbooks were written, which students had to learn by rote. Hence the outstanding treaties on histology and pathology of the last 100 years of the Spanish university. They have nonetheless also always been accompanied by many translations, first French texts, then German and, nowadays, from the United States.

The textbooks were of a three-way significance. On the one hand, they represented a major teaching reform in the 18th-century university. Then too they were good business for the authors and indeed for some universities and publishers. Thirdly, they were a means of control of primary importance, against doctrines threatening to the monarchy or the church, such as astral revolution, or blood circulation, as well as a control on students' familiarity with scientific knowledge. These books, which synthesized everything known to the modern world in a discipline, giving meaning and structure to knowledge, were fundamental to the initiation of the modern university.

THE COLLEGE OF PHILOSOPHY OF THE UNIVERSITY OF SALAMANCA

In 1798, Immanuel Kant published *Der Streit der Fakultaten*, in which, as the result of interference in universities from authority and religion, he sought to recapture the importance of the Faculty of Philosophy. From the time of the establishment of the universities, this had been a minor school, of little importance in comparison with the Major Faculties, particularly Theology and Canonical Law. Torres Villarroel, a renowned professor of mathematics in the University of Salamanca, complained that he was despised among the senate. That is hardly surprising since these professors had secondary seats and votes and poor remuneration and category. This battle between faculties was genuinely being waged at the same time in universities in the rest of Europe, and this is the aspect we are going to consider in relation to the University of Salamanca. Our approach will of course be quite different from that of the philosopher Kant, who asks who is responsible for the decisions in each faculty—the Church (Theology and Canonical Studies), the King (in Civil Law), Nature (in Medicine), or Reason (in Philosophy). My standpoint is rather more sociological or, if you will, institutional since what happened in our faculties was really an attempt to enhance the social standing of philosophers and scientists, something which did not exclude improvement to teaching and knowledge. This is the second channel of modernization of the universities to which I shall refer.

There were many facets to the improvements, such as changes in the selection of teaching staff. This was the case with the admission of Juan Justo García as teacher, first as assistant and subsequently to the Chair. This well-known late 18th-century author raised the teaching of mathematics to a modern standard after the unfortunate period of the Torres family, particularly Diego de Torres who was, on the other hand, an excellent writer. García's texts on mathematics, geography, and even theory allowed an unquestionable improvement to the Faculty of Philosophy. After his competitive examinations, the Consejo de Castilla suggested reform of the selection system, consulting a variety of scientists and sages, who recommended examination using modern texts, the posing of questions, written responses, and marking and grading. It was also recommended that the tribunal be made up of experts. These novelties reveal how, through science, a new form of knowledge was appearing.

In 1792, in the University of Salamanca, the Consejo de Castilla accepted the College of Philosophy, with the aim of providing independence and dignity to philosophers and scientists. The attempts of Isidoro Ortíz, nephew of Torres Villarroel and more notable scientist than his uncle, were furthered by Juan Justo García, who headed the petition for the improvement of the Faculty of Philosophy. This had been requested since 1779: they wished to upgrade mathematics, banish Aristotelianism and teach modern science. They wished to be independent of Theology—which considered itself master and lord—and demanded professional improvements. They also sought the permanence of chairs, and their improvement and control: they should be for life, well paid, and granted by philosophers or scientists.

Teaching should continue to be according to François Jacquier—introduced in 1788—and the preparatory studies for senior degrees should not be changed, though innovations could be introduced to philosophical study. Benito Bails' *Principios de Matemáticas*, written for the Academy of Fine Arts, was introduced, on Jorge Juan's recommendation. According to the 1771 plan, the course was to be designed by scholars, removing useless items and scholasticisms. It was drafted, but Madrid did not approve it. At the beginning of the 19th century, there was debate as to who would be professors, judges of competitive examinations, who might attend senates, and in what order of seat and intervention.

Theologians and canonists attacked the College, which was defended by jurists and physicians. They were accused of not accepting university teaching methods, disputes, syllogisms, and Latin. Preliminary studies for Law had changed and teaching had used the text of Juan Justo García. It was claimed that they had not run experiments or used scientific instruments, thus repudiating their scientific interest. They were accused of impiety and of having books of atheists and libertines, of defending the death of the soul or attacking expenditure on funeral ceremonies. We should not forget the vast luxury displayed by the universities, whose income went in ceremonies. This luxury came under fire but was nevertheless replaced by personal luxury: money was spent ostentatiously on dress or adornment and on consumption, while attacking alms as penitence in the confessional or those informing on this modernity. Criticism of absolutist monarchy or the catholic religion of course filtered into the classroom, new authors were read, from erotic literature to philosophers of the quality of Bentham. We should remember that it was during these years that an illustrious jurist of Salamanca, Ramón de Salas, was gaoled. Scientific modernity was introduced together with the new liberal opinions, and with the new way of life.

The battles continued. However there came times of sickness, famine, poverty and war and everything was reduced to fights over salaries and subsistence. The 1807 plan partly reflected this modernization, but it added little more. The solution was to be quite different: the Spanish universities of the 19th century were not significant because of their Science or Philosophy Faculties. They were in fact set up, having been established in the 1857 Act, but they served only to train teachers for secondary education. The important faculties of this time—and until the middle of the 20th century—were those of Law and Medicine, which had more teachers, more funds, and more students. They were to have greater social repercussion, since the contemporary Spanish university was to be essentially professional in nature, more concerned to award degrees to professionals than for knowledge and research. The ignorance of the university in Germany or the English-speaking countries and the choice of the French model had serious consequences. Teaching was given priority over research and, as in France, other institutions had to appear to initiate research. The creation in 1907 of the Higher Education Board made it possible to provide grants to graduates overseas and then create laboratories for the furtherance of their research.

BIBLIOGRAPHY

G. M. Addy, *The Enlightenment in the University of Salamanca* (Durham: Duke University, 1966).

S. Albiñana, *Universidad e Ilustración. Valencia en la época de Carlos III* (Valencia: IVEI, Universidad, 1988).

M. Baldó, *Profesores y estudiantes en la época romántica. La Universidad de Valencia en la crisis del Antiguo Régimen (1786–1843)* (Valencia: Ayuntamiento, 1984).

N. Cuesta Dutari, *Filosofía natural y pugna de facultades en la Universidad de Salamanca (1779–1796)* (Salamanca: Universidad, 1971).

N. Cuesta Dutari, *El maestro Juan Justo García*, 2 Vols (Salamanca: Universidad, 1974).

N. Cuesta Dutari, *Historia de la invención del análisis infinitesimal y de su introducción en España* (Salamanca: Universidad, 1985).

J. J. García, *Elementos de Aritmética, Geometría y Álgebra* (Madrid: Joaquín Ibarra, 1782).

J. J. García, *Elementos de verdadera Lógica* (Madrid: Mateo Repullés, 1821).

S. Garma, "Juan Justo García", in J. M. López Piñero, Th. F. Glick, V. Navarro and E. Portela, *Diccionario histórico de la ciencia moderna en España* (Barcelona: Península, 1983), I, pp. 368–369.

E. Hernández Sandoica and J. L. Peset, *Universidad, poder académico y cambio social (Alcalá de Henares 1508–Madrid 1874)* (Madrid: Consejo de Universidades, 1990).

J. Jackson, *Enchiridion medicum theorico-practicum, sive tractatus de morborum theoria, et praxi* (Antonio Marín: Madrid, 1734).

J. M. López Piñero and V. Navarro Brotóns (eds.), *Historia de la ciencia al País Valencià*, (Alfons el Magnànim: Valencia, 1995).

J. M. López Piñero et al., *La actividad científica valenciana de la Ilustración*, 2 Vols (Valencia: Diputación and Ayuntamiento, 1998).

M. L. López Terrada, *Libros y folletos científicos en la Valencia de la Ilustración (1700 –1808)* (Valencia: IVEI, Institut d'Estudis Gil Albert, 1987).

J. A. Micó Navarro, "Los médicos en la Valencia del siglo XVIII, tras la supresión de la organización foral", *Estudios sobre la profesión médica en la sociedad valenciana* (Valencia: Ayuntamiento, 1998), pp. 215–222.

J. Navarro, *La introducción de la Clínica en Valencia. Félix Miquel y Micó, 1754–1824* (Valencia: Ayuntamiento, 1998).

V. Navarro, *Tradició i canvi científic al País Valencià modern (1660–1720): Les ciències fisico-matemátiques* (Valencia: Tres i Quatre, 1985).

T. Núñez, *Sistema de la ciencia social* (Salamanca: Bernardo Martín, 1820).

J. L. Peset, "El plan de estudios médicos de la Universidad de Salamanca de 18 de enero de 1804", *Asclepio* 21:305–317 (1969).

J. L. Peset, "Reforma de los estudios médicos en la Universidad de Valencia. El plan del rector Blasco de 1786", *Cuadernos de Historia de la Medicina Española* 12: 213–264.

J. L. Peset, "Andrés Piquer y la enseñanza de la medicina", *Primer Congreso de Historia del País Valenciano* (Valencia: Universidad, 1976) III, pp. 725–729.

J. L. Peset, "Los caminos de la ciencia. 2. El siglo XVIII", in M. Fernández Álvarez, L. Robles Carcedo, and L. E. Rodríguez San Pedro (eds.), *La Universidad de Salamanca. II. Docencia e Investigación* (Salamanca: Universidad, 1990) pp. 137–149.

J. L. Peset and M. Peset, *Carlos IV y la Universidad de Salamanca* (Madrid: CSIC, 1983).

M. Peset and J. L. Peset, *El reformismo de Carlos III y la Universidad de Salamanca. Plan general de estudios dirigido a la Universidad de Salamanca por el Real y Supremo Consejo de Castilla en 1771* (Salamanca: Universidad, 1969).

V. Peset, "Andrés Piquer y la Psiquiatría de la Ilustración", *Clínica y laboratorio* 63:153–160 (1957).

V. Peset, "El informe del claustro de Medicina de Valencia sobre renovación de estudios (1721)", *Archivo Iberoamericano de Historia de la Medicina* 13: 143–155 (1961).

V. Peset, "El Dr. Seguer (1702?–1759) y la moderna historiografía médica española", *Asclepio*18–19:261–268 (1966–1967).

V. Peset, *Gregori Mayans i la cultura de la Illustració*, Barcelona, Valencia, Curial, Tres i Quatre.

A. Piquer (1762). *Institutiones medicae ad usum Scholae Valentinae*, in J. Ibarra (ed.) (Madrid: Third edition in 1790).

S. Rodríguez Domínguez, *Renacimiento universitario salmantino a finales del siglo XVIII*, (Salamanca: Universidad, 1979).

A. Sanz Muñoz, *Compendio de medicina practica arreglado a las explicaciones del Doctor Don Félix Miquel*, 2nd ed. (Valencia, Imprenta de Estévan, 1820).

Ten, (ed.), *Plan de estudios por S. M. y mandado observar en la Universidad de Valencia* (Valencia: Ayuntamiento, 1984).

ANTONIO GARCÍA BELMAR AND JOSÉ RAMÓN BERTOMEU SÁNCHEZ

SPANISH CHEMISTRY TEXTBOOKS DURING LATE 18TH CENTURY: BUILDING UP A NEW GENRE OF SCIENTIFIC LITERATURE

Science teaching, as historiographic subject, has until recently fallen between specialties which, for different reasons, have had scant interest in addressing it directly. Historians of education have focused their interest on the history of institutions and have paid little attention to scientific disciplines. Historians of science, at least those who pursue what used to be called internal history of science, have usually regarded science teaching as a secondary theme, marginal to the story of scientific creativity, which is carried forward in laboratories and academic institutions. In part, that view was due to a widespread image of teaching as intrinsically a matter of passive, routine transmission accepted scientific knowledge, of historical interest only as it relates to the training of an individual scientist, or his or her teaching activities. Otherwise, teaching is thought of as little more than a useful clue, which can indicate when and where a given idea was introduced into a country, or accepted by a scientific community.

Institutional studies of science, it is true, have scrutinized other aspects of the teaching of sciences: the structure, the regulations, the economic and social support of scientific and educational institutions, as well as their function inside the general educational system. Less interest has been paid to students than to teachers and to the research schools associated to these institutions; and very scarce attention has been paid to the explicit and tacit knowledge transmitted through teaching practices. What is transmitted is assumed to be similar to what has been created in the laboratories. Accordingly, historians have generally been concerned only to identify the presence or the absence of particular theories or scientific laws in textbooks or notebooks, to study their diffusion from laboratories to classrooms, from center to scientific periphery.

There are other possible approaches, as chapters discussed in this volume show. Studies on the transmission of science have remarked that scientific knowledge hardly ever travels unchanged from one context to another. Peripheries are now regarded as active actors in the appropriation of scientific knowledge.[1] Similarly, studies on science teaching have begun to take into account the practices of all the participants in teaching activities. These new historiographical trends have created a renewed interest in some sources related with teaching practices: Chemistry textbooks are good examples.

Several studies about this subject have been published in the last decades. Owen Hannaway's studies of the first chemistry textbooks have showed that these books

M. Feingold and V. Navarro-Brotóns (eds.), Universities and Science in the Early Modern Period, 241–257.

played a major role in defining the boundaries of chemistry as an independent discipline. By a careful analysis of Texts written by Oswald Croll and Andreas Libavius, he showed major differences in attitudes concerning communication in chemistry and in alchemy. In contrast with alchemical literature, chemistry texts, exemplified in Andreas Libavius's *Alchemia* (1597), were aimed at transmitting theories and practices as clearer as possible. Therefore, Hannaway equated the birth of chemistry with the publication of Andreas Libavius's textbook, which presented chemical operations in a new order that was inspired in Petrus Ramus pedagogical ideas. In this case, then, teaching was regarded as a leading force in the constitution of chemistry as a scientific discipline; and Hannaway has highlighted what he called the "self-generating didactic power".[2]

The relationship between teaching and research is another issue that has been revisited. Instead of assuming a fixed hierarchical relationship between research and teaching, scholars have made that relationship a topic of research.[3] From another point of view, studies on rhetorics of science have shed light on the creative practices related to scientific language in academical and teaching contexts. An increasing number of studies have analyzed the rhetorical dimension of different kinds of scientific writing.[4] In some sense, communication is no longer regarded as a passive transmission of knowledge but as one of the chief spaces in which scientific knowledge is constructed.

In such new perspectives, teachers and students are considered as active actors in creating scientific knowledge, and teaching as a multidirectional activity that implies a strong interaction of all participants. We agree with John Christie and Jan Golinski when they said that there is not a linear relationship between theoretical and experimental research and teaching. They regard it as a sort of non-Euclidean relation in which the trajectories of interaction were far more complex.[5] Scientific textbooks offer hints about this complicate space because they are in a crucial place among the multiple and diverse factors and actors that shape educational practices.[6] They are written by concrete authors (with particular backgrounds and goals), but they are produced by printers (with different technological means) and sold by publishers and booksellers in specific technical, economic, and commercial contexts. Moreover, textbooks are read and used by a great variety of audiences with different aims and reading practices. From this point of view, scientific textbooks are at the crossroad between disciplines such as history of science, history of education, and history of books and reading.[7]

In this chapter, we analyze some Spanish chemistry textbooks of the late 18th century and the early 19th century. In our study, we regarded chemistry textbooks not only as a source to chemistry teaching, but also as a valuable subject of research. Chemistry textbooks are not only windows into the classrooms, but also material and intellectual objects that are constructed under specific circumstances. Using some Spanish and French examples, we illustrate some of the main issues that govern our current research. First of all, we consider audiences; second, we turn into publishers and authors; lastly, we briefly discuss the structure and contents of the chemistry textbooks themselves.

THE ACTIVE ROLE OF AUDIENCES

Chemistry textbooks are good sources to study the main audiences of chemistry and their changes. During the late 18th century and the early 19th century, many forewords contained substantial information, both explicit and implicit, about their intended audiences, those whom authors and publishers bore in mind when writing and producing their books.

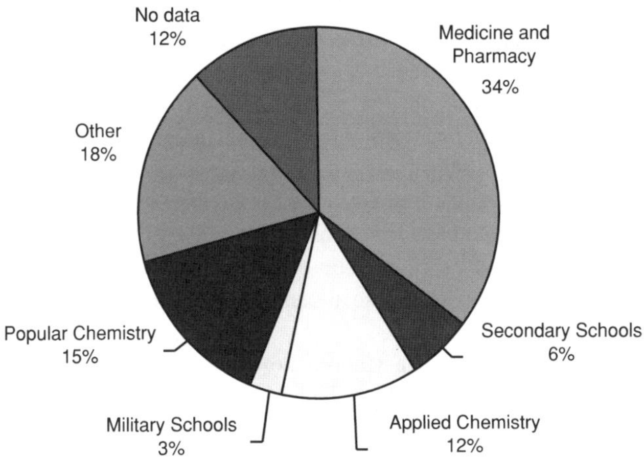

Graphic 1. Spanish Chemistry Textbooks, Audiences, 1780–1830. Source: Bertomeu (2000).

These data offer some hints about the publics of chemistry during this period: pharmacists and physicians, military students, craftsmen, and secondary school students. As already mentioned, forewords offer substantial information about the social origin, background, institutional situation, formative necessities, or professional perspectives of their readers. That allows us to begin to reconstruct the readers' universe of expectations: that is, what they expected to find inside these books.

In some textbooks, this information is very specific—particular institutions or examinations, for instance, are mentioned—while in other textbooks there are only general references to different levels of teaching. More rarely, and more often in the 18th than in the 19th century, information of this sort is not given, and we must look to complementary sources. In general, 18th century books were not destined to specific publics and they were often employed in very diverse contexts. This is a prominent characteristic of the early generations of chemistry textbooks, both Spanish and French.

For example, the famous *Traité élémentaire de chimie* by Antoine Lavoisier (Paris, 1789) was translated into Spanish, first in the Mexican School of Mines (1797) and, again, in the following years, in the Segovia School of Artillery by Juan Manuel Munárriz.[8] A few years later, Munarriz' translation was recommended in the Schools of Pharmacy created in 1804. In France, it went through several editions and was widely used in secondary schools (*écoles centrales*).[9] In short, the same book—Lavoisier's *Traité*—appears to have been used in very different contexts: in the teaching of mining and mineralogy, of military arts and pharmacy, and in secondary education. It is hard to name a 19th century chemistry textbook with such a wide use in so many different educational contexts.[10]

Secondary chemistry textbooks were the most important group that contributed to this change. In France, these books appeared during the first years of the 19th century as a result of the development of the system of secondary schools during the revolutionary years.[11] In Spain, few similar textbooks were published until the second third of the 19th century, along with the development of a new Spanish secondary educational system.

After the years of the Napoleonic wars (1808–1814), most new textbooks in chemistry were aimed at students of medicine and pharmacy while a small group was intended for students in military institutions and Royal Laboratories. The other important group of books was addressed to students who attended lectures on chemistry applied to arts and manufactures such as those established by the "Sociedades Económicas de Amigos del País", a group of societies that encouraged the development of agriculture and industry. These audiences were, in fact, mixed in classrooms as shown in the following examples.

In Valencia, Tomás de Villanova Muñoz i Poyanos (1737–1802), teacher of chemistry in the Medical Faculty, was obliged by university regulations to teach chemistry for medical students in the morning and for craftsmen during the afternoon; the "Sociedad Económica de Amigos del País de Valencia" had supported the establishment of his chair of chemistry during the last years of the 18th century.[12] In Barcelona, Francesc Carbonell i Bravo (1768–1837) was appointed professor of chemistry of the Junta de Comerç, an institution mainly interested in agriculture and industry. His teaching was thus directed to workers and employers in those fields but, in fact, an important part of his audience was made up of pharmacists and surgeons.[13] In Madrid, the pharmacist Pedro Gutiérrez Bueno (1745–1822) was the teacher of chemistry in the Royal Laboratory. He was the most important writer of chemistry textbooks during late 18th century.[14] His audience was mostly made up of pharmacists but he also mentioned the presence of "enlightened gentlemen", a new audience for chemistry lectures that emerged during 18th century in Spain as well as in other European countries.[15]

Medical and pharmaceutical students, craftsmen, industrialists and agriculturalists, and enlightened gentlemen formed the most important public for chemistry at the end of 18th century in Spain. That situation changed during the early 19th century. In Spain as in France, most of the chemistry textbooks were mainly directed to students of medicine and pharmacy as a result of the diffusion of chemistry lectures in Medical Faculties and Pharmaceutical Schools. It was only after the 30s and 40s that secondary education became a very important vector in transforming chemistry textbooks.[16]

That conclusion is at odds with a very common image of the chemical revolution. According to the image constructed by the 19th century chemist-historians, chemistry became an independent discipline by cutting its ties with pharmacy and medicine. The fact that the common hero in these narratives, Antoine Lavoisier, did not have a medical background has contributed to reinforce this image. Current historical studies

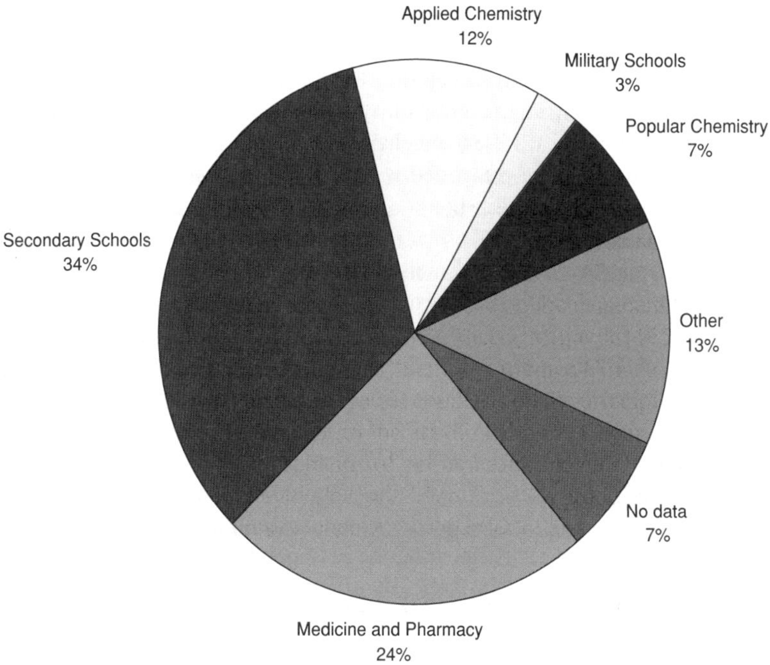

Graphic 2. Spanish Chemistry Textbooks, Audiences, 1830–1850. Source: Bertomeu (2000).

are changing some of these views by closer study of areas of 18th century chemistry, for instance, plant chemistry.[17] A survey of chemistry textbooks suggests similar conclusions: not only were medical students the main targeted public of chemistry textbooks, but among textbook authors physicians and pharmacists were an important category. But before speaking about authors, we should say some words about publishers and the textbook trade.

TEXTBOOKS AS MATERIAL OBJECTS: PRODUCTION AND CONTROL

Obviously, in the story of chemistry textbooks, publishers, printers, and booksellers are crucial actors. Publishers were not only a link between authors and readers, but also very influential participants in defining some of the salient features of chemistry textbooks: typographical character, size, format, and even contents and structure. Publishers were under the influence of multiple and strong forces: they had to take account of readers, and authors, interests, their books had to accord with the syllabus of educational institutions and they also had to compete with other publishers in book market. Moreover, particular laws regulated and controlled the textbook market.[18] At the end of the 18th century, chemistry textbooks were, as other books, under two types of control in Spain: first, a government license to publish, implying a censorship exercised by some scientific institutions or scientific advisors; and, second, the control of book circulation by the Inquisition.[19] Even when both types of control disappeared during the first third of the 19th century, textbooks were published under special rules of control. In assuming responsibility for education, the Spanish and French Governments established different mechanisms to control what was taught in the classrooms. Textbooks were controlled with special care and several methods were developed for that purpose, each method with a different influence on the textbook trade. The imposition of a single "official textbook" approved by the institutions or by the government was one of the first methods employed in France.[20] In Spain, several lists of "recommended textbooks" were published during the central decades of the 19th century.[21] Later on, textbooks were approved after being published, that is, they were accepted if any disapproving report was made by the government. Obviously, these last two methods of control placed publishers and authors in a more strategic position in textbooks trade.[22]

In addition to these governmental mechanisms, another system of control emerged from the scientific community: this was the "model book", a reference book that became an archetype copied and adapted by other textbooks. As a result, this "model book" played a major role in delineating the boundaries and the structure of the discipline. That role was played in France and Spain during the late 18th century by Antonie Fourcroy's textbooks; but the first "model textbook" was, indeed, the *Traité élémentaire de Chimie* by Louis Jacques Thenard (1777–1857), as in Germany and Sweden was the *Lärbok i chemie* by Jacob Berzelius.[23] These books went through several editions, with marked changes in their contents and structure. Authors of chemistry textbooks were obliged to adopt these changes in their books while adapting the model to their particular audiences and educational context. Adapting a changing book to many different contexts was a difficult task. Authors and publishers were forced to make decisions that led them to a creative appropriation of "model textbooks". They did not copy these "model books"—that was, in fact, impossible—but rather selected and adapted their contents to very different situations. That explains why textbooks were by the middle of 19th century so different from each other.

TEXTBOOKS WRITERS

All the issues mentioned here—formative requirements of targeted public, material and normative conditions determined by teaching institutions and government, mechanisms of trade book control and censorship, economic interests and technical possibilities of publishers, and so on—constrained the activity of the authors of chemistry textbooks. They were responsible for creating a suitable text that met the requirements and interests of readers and publishers. But, at the same time, their books were a result of personal decisions, linked to their own backgrounds and scientific and teaching activities. Consequently, historians of science need to explore the biographies of these usually unknown writers in order to understand the structure and contents of chemistry textbooks. We should study not only their scientific education, their teaching practices, and their relationship with current research programs, but also the professional, ideological, or economic interest that appealed them to write a textbook. It is not easy to answer to these obvious questions because these secondary actors have usually been forgotten in historical narratives.

Like the German textbook writers studied by William Clark, the majority of French and Spanish authors of chemistry textbooks are not mentioned in Poggendorf's or Gillespie's biographical dictionaries.[24] They are generally regarded as obscure individuals, whose activity was focused more on teaching that on original research. In fact, this image matches more later 19th century authors better than those who lived at the end of the 18th century. In the late 18th century, very well-known authors—such as Macquer, Baumé, Fourcroy, Lavoisier, or Chaptal—wrote chemistry textbooks. In Spain, although most chemistry textbooks were translations of French books, the few Spanish authors who wrote an original textbook, in fact held very important teaching and research positions. A good example is Pedro Gutiérrez Bueno, the most prominent textbooks writer during late 18th century, and teacher of chemistry, first, at the Royal Laboratory of Chemistry and, second, at the new School of Pharmacy in Madrid.[25]

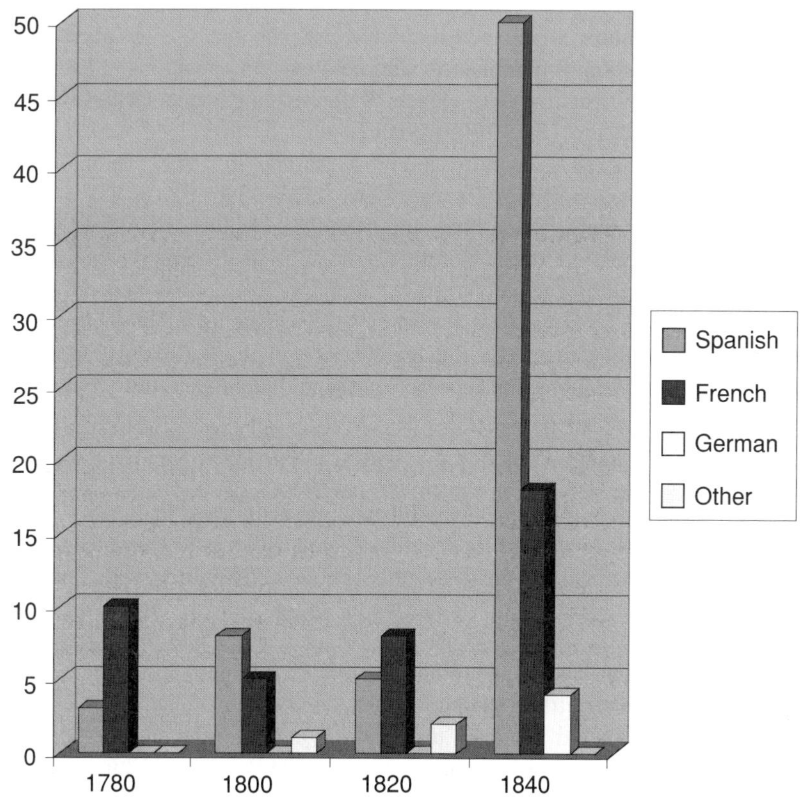

Graphic 3. Spanish Chemistry Textbooks, Original Languages, 1780–1860. Source: Bertomeu (2000).

The situation changed during the first third of the 19 century, with the new group of medical doctors and pharmacists who wrote chemistry textbooks for their students. Mateu Orfila (1787–1853) is a good example of this group: his book was perhaps the most frequently reissued textbook during the first half of 19th century.[26] It was translated into many European languages, among them German, Dutch, Italian, and Spanish. In Spain, Orfila's textbook was very influential not only due to the translations but also because in these years other authors used it as a model. In France, authors of chemistry textbooks such as Orfila, they attended the meetings of the *Société de Chimie Médicale* and published papers on medical chemistry in its journal and other medical journals. Their books were published by Nicolas Crochard, a bookseller and publisher related with the Paris Medical Faculty.[27]

In France, this group of physicians succeeded in controlling the textbook market during the first half of 19th century. Most of the Spanish chemistry textbooks of the same period were translations of French works, except for rare cases such as the *Corso analittico de chimica* by Giuseppe Mojon (1772–1837). The translators of these books were, indeed, physicians and surgeons or pharmacists such as Higinio Antonio Lorente or Francesc Carbonell i Bravo.[28] While in France the authors of most chemistry textbooks were medical doctors, in Spain that role was played during the 40s by a large group of pharmacists who held teaching positions in Faculties of Pharmacy or the new Faculties of Science: Rafael Sáez Palacios (1808–1883) and Carlos Ferrari y Scardini (1820–1890), Juan López-Chavarri (1813–1876), Manuel Rioz y Pedraja (1815–1887), Vicente Santiago Masarnau (m. 1879), Antonio Casares (1812–1888), and Ramón Muñoz Torres y Luna (1822–1890).[29]

Some of this group completed their studies in the Faculties of Sciences and held chemistry chairs. Their work was, therefore, crucial in the progressive consolidation of chemistry textbooks as a genre of chemical literature: their books were not aimed at pharmaceutical students but at new secondary school students following the new educational programs that emerged during the 40s. That explains also the great increase in the number of publications that appeared by the middle of the century (Graphic 3).

CHEMISTRY BOUNDARIES: BETWEEN PHYSICS AND PHARMACY

After analyzing readers, authors, and publishers, it seems clear that in the late 18th century and in the first third of the 19th century chemistry textbooks were read, written, and published in a medical and pharmaceutical context, in Spain as well as in France. What were the consequences of this medical context for chemistry textbook contents? Let us focus our discussion on two very important aspects: first, the boundaries of chemistry and its relationship with other disciplines, and second, the structure and classifications of the material included.

Textbooks offer a good approach to the question of the definition of discipline boundaries and its relationship with other sciences. At the end of 18th century, this problem was focused first on the relation between physics and chemistry, and second on the relation between chemistry and pharmacy. The emergence of experimental physics, particularly the theory of imponderable fluids, had created new links between chemistry and physics. As a result, an important group of textbooks published in France during the first years of the 19th century dealt with both physics and chemistry, chemistry being regarded as a part of physics. This approach was encouraged by the Société d'Arcueil research program, headed by Berthollet, and by the reform of education that created joint lectures on physics and chemistry in secondary schools during the revolutionary years. To authors such as Jean Baptiste Jumelin (1754–1807), chemistry was "la partie de la physique" that deals with the "affinité de composition". In the foreword to his *Traité élementaire de physique, de chimie et de physico-mathématique*, Jumelin asserted that he would have preferred to eliminate the word "chemistry" from the title, but he was afraid that readers would assume that "[the book] does not deal with matters that belonged to the part of physics that, until now, has been designated with this name [chemistry]".[30]

> "Quand l'attraction chimique ne fait simplement que rapprocher les molécules de matière et les tenir unies, on la nomme attraction ou affinité d'agrégation; quand elle change la nature des corps en les unissant ensemble, on la nomme attraction ou affinité de composition. On voit que ces deux effets différents sont dus à la même cause, et que, par conséquent, ils doivent être traités ensemble; et cependant, c'est de la connaissance des premiers dont on a fait la plus grande partie de la science qu'on nomme Physique; et de celle des seconds, la science qu'on nomme Chimie. On doit donc conclure que ces deux sciences prétendues doivent être confondues en une seule".[31]

Joseph Izarn (b. 1766), who become teacher at the *Lycée Bonaparte* in Paris, also defended the union of chemistry and physics in secondary teaching. He considered that the experience of the *Écoles Centrales*, where physics and chemistry were taught together, underlined the need for elementary textbooks to present both disciplines jointly; but he recognized that in more advanced treatises the two subjects had to be treated separately.

> "Tel est donc l'état actuel de nos connaissances physiques, que la plupart de leurs branches doivent avoir leurs Traités particuliers, tandis qu'elles doivent avoir le même Livre élémentaire".[32]

Textbooks such as Izarn or Jumelin's were scarcely published in Spain at all, mainly because secondary schools on the new French model had no jet been created.[33] In Spain other issues were much more pertinent: the boundaries between chemistry and pharmacy, and the medical applications of chemistry. However, a renewed interest in experimental physics emerged during the last third of the 18th century and chapters on "imponderable fluids" were included in chemistry textbooks, even those mainly aimed at medical and pharmaceutical students. The lectures that Pedro Gutiérrez Bueno gave in the Royal Laboratory of Madrid are a good example of this new chemistry courses.

Pedro Gutiérrez Bueno called himself "the first chemistry professor" in Madrid. He began his lectures in 1788 at the Royal Laboratory, an institution which was founded as part of a projected Academy of Science and also fulfilled a long-standing aspiration of the Madrid College of Pharmacists. Students of pharmacy attended the course as part of their professional training; but Bueno's lectures also attracted enlightened gentlemen, for whom the laboratory was a space for sociability and worthwhile recreation. To suit these diverse interests, Pedro Gutiérrez Bueno wrote a course on chemistry with a very long theoretical introduction and long chapters on subjects related to experimental physics. This was very different from old pharmaceutical chemistry textbooks, mostly intended to train practitioners and focusing on technical information about chemical preparation and procedures. Gutiérrez Bueno, himself a pharmacist, was thus obliged to claim for a renewed relationship between chemistry and pharmacy, as Antoine Fourcroy argued in the same years in France.[34] In fact, Gutiérrez Bueno and his students strongly contributed to the introduction of Fourcroy's ideas in Spain, through several translations and their own books. Like Fourcroy, Gutiérrez Bueno actively participated in the institutional reform of pharmaceutical training and practice, and taught chemistry in the new institutional context. The defense of new Pharmaceutical Schools against supporters of old apprenticeship system was intimately linked with the new chemistry textbooks that Gutiérrez Bueno wrote and the new relationship between chemistry and pharmacy that he sought to establish.[35]

Graphic 4. Pedro Gutiérrez Bueno (1745–1822). From his book "Prontuario de Química, Farmacia y Materia Médica", Madrid.

At the end of the 18th century, medical and pharmaceutical applications of chemistry were not widely accepted. Several papers and medical dissertations discussed whether chemistry was useful or worthless (even dangerous) in medicine. Some physicians argued against dangerous applications of chemistry in medicine by recalling the iatrochemical reveries of the 16th and 17th centuries. Even Fourcroy, who largely published about medical applications of chemistry, demanded caution in the use of chemistry in medicine, and he and others angrily criticized works such as Jean Baptiste Baumès's nosological system.[36] In addition to textbooks and translations by Pedro

Gutiérrez Bueno and his disciples, two substantial papers supporting Fourcroy views about the proper limits of chemistry were written by Spanish chemist: Juan Manuel of Aréjula (1755–1830) and Francesc Carbonell i Bravo.[37]

Aréjula and Carbonell like Fourcroy, criticized the use of chemistry as a core theory for physiology and nosology while making strong claims for its applications in other areas such as pharmacy, fumigation, or eudiometry. These ideas influenced Mateu Orfila, whose research contributed to the development of a new area of applications of chemistry in medicine: toxicology. Mateu Orfila's textbooks, one of the most influent chemistry textbooks in Spain as well as in France during the first third of 19th century, clearly reflected these discussions: he wrote a rather descriptive book, with long descriptions of the therapeutic and toxicological characteristics of chemical substances, but without addressing topics related to physiological chemistry. The new consolidated concept of "immediate principle" was used by Orfila as a useful criterion separating the domains of chemistry and physiology when dealing with animal chemistry, long before the term "organic chemistry" was widely employed in textbooks.[38]

STRUCTURE: FROM SIMPLE TO COMPLEX

Besides establishing the boundaries of the discipline, authors of textbooks should face another complex question: the organization of scientific knowledge in their books, i.e., the composition of a table of contents consonant with the leading didactic and epistemological trends and, at the same time, it had to fit with current chemical classifications. In other words, their table of contents had to adjust the order of things to the order of learning. The main expository principles used during these years were three: (1) progress "from what is known to what is unknown", (2) progress "from simple to complex", and (3) "group together chemical substances that have similar chemical properties". By the middle of the 18th century, authors such as Pierre Macquer set forth these criteria which were adopted by almost every French and Spanish authors such as Lavoisier, Fourcroy, or Pedro Gutiérrez Bueno:

> "Le plan que je me suis principalement proposé de suivre, est de ne supposer aucune connaissance chymique dans mon Lecteur; de le conduire des vérités les plus simples, et qui supposent le moins de connaissances, aux vérités les plus composées qui en demandent davantage. Cet ordre que je me suis prescrit, m'a imposé la loi de traiter d'abord des substances les plus simples que nous connaissions, & que nous regardons comme les éléments dont les autres sont composées, parce que la connaissance des propriétés de ces parties élémentaires conduit naturellement à découvrir celles de leurs différentes combinaisons; & qu'au contraire la connaissance des propriétés des corps composés, demande qu'on soit déjà instruit de celles de leurs principes. La même raison m'engage lorsque je traite des propriétés d'une substance, à ne parler d'aucune de celles qui sont relatives à quelqu'autre substance dont je n'ai point parlé"[39]

In the late 18th century, elementary analysis became a powerful tool to make the order from "unknown to known" compatible with the principle of the increasing complexity. However, since some elements were regarded as vehicles of particular properties—for example, oxygen as vehicle of acidity—a table of contents based on chemical composition was expected also to group substances with similar chemical properties together. Unfortunately, these three principles were not enough to establish a single standard arrangement of textbook contents. As Bernadette Bensaude Vincent has shown by comparing the textbooks of Lavoisier, Fourcroy, and Chaptal, very different tables of contents are equally compatible with these general positions.[40] Textbooks writers, therefore, required some additional criteria of organization. That is one of the reasons why some traditional groups of chemical substances, such as metals and salts, remained with no relevant changes during the 19th century. In fact, perhaps the most important issue that authors had to face was how to make compatible these traditional groups with the order of complexity from simple to complex based on the recently developed elemental analysis. Authors such as Louis Jacques Thenard solve this troublesome problem by taking for granted that chemical properties and composition were closely connected, that is, chemical properties were regarded as a consequence of chemical composition. Under the influence of Thenard's "model book", Spanish and French chemistry textbooks used oxygen as an organizing principle: non-metals were arranged according to their affinity for oxygen; metals were classified in six groups according to their reaction with oxygen and water; and even in plant chemistry, immediate principles were arranged in three groups taking into account their oxygen–hydrogen ratio.[41]

This "exaggeration of the role of oxygen"—as Ferdinand Hoefer called it—was a characteristic of French chemistry textbooks during early 19th century. But authors did not copy Thenard's classifications exactly: That was impossible. On the one hand, Thenard's textbook was not a stable object. Its contents and structure change from one edition to another, sometimes with minor amendments and additions but, in other cases, Thenard introduced radical changes as it happened in the last edition. On the other hand, authors sometimes disagreed with Thenard's points of view concerning important aspects. Orfila, for example, arguing against Thenard's views on combustion and adopting a position closer to British chemists' ideas on combustion, refused to group chemical substances in "combustible" and "burned" substances, a crucial division in Thenard's table of contents. Guided by his toxicological research, Orfila adopted an order of salts that was opposed to the order adopted by Thenard. Salts were grouped in classes sharing a common metal. This classification was especially interesting for toxicological activities related to forensic medicine, since metals were usually the object of inquiry.[42] Other physicians and authors of chemistry textbooks—such as Jean Lassaigne—also adopted this criterion, opposing, therefore, Thenard's "model book" arrangement. Lassaigne also criticized the order "from simple to complex", as did Berzelius during these years. Lassaigne suggested studying first the properties of air and water—the most usual substances, though not the simplest.

> "l'air et l'eau dont les propriétés et la composition doivent être connues pour l'explication d'une infinité de phénomènes qui se produisent ordinairement, et qui ne pourraient être compris si on ignorait leur véritable nature. Si la marche que nous adopterons dans nos descriptions n'est pas aussi méthodique que beaucoup d'autres, elle a, nous le pensons, l'avantage d'être plus à la portée des élèves. D'ailleurs, cette méthode, qui nous parait bonne sous bien de rapports, a été suivie par plusieurs chimistes distingués dans les leçons qu'ils professaient, et nous nous sommes fait un devoir de suivre la route qui nous avait été tracée par l'illustre professeur qui nous a précédé dans l'enseignement qui nous est aujourd'hui confié"[43]

Apart from Lassaigne and some other exceptions, most of the authors agreed with the general sequence "from simple to complex". However, even given these general criteria and the "exaggeration of oxygen" as organizing principle, authors had a mass of different possibilities for the arrangement of their books. During the beginning of 19th century, the deep influence of medical context explains some of the decisions that authors such as Orfila and Lassaigne were compelled to make.

CONCLUSIONS

Allen G. Debus has wrote that "if we are to learn more of the changes that have occurred in the history of chemistry, or the history of science for that matter, we must devote more time to textbooks employed and methods of teaching".[44] Even if he and other scholars have called for more research on the history of science teaching, substantial issues still remain unexplored, including, as we have seen, the complex history of scientific textbooks.

Our examples suggest that chemistry textbooks are the result of the interaction of several historical forces. In our analysis, we have taken into account the background, interests, and activities of their audiences, publishers, and authors. Chemistry textbooks, moreover, are located on an intersection between scientific knowledge and pedagogical views and have always come under strong social, economical, and political pressures. In Katherine Olesko's words, they are placed in a teaching space where economical, social, and political forces rush into the structure and function of scientific knowledge. That is why textbooks are so interesting sources for social and cultural history of science. They should be analyzed by paying attention to the approaches and conclusions of historians of books and reading as well as those of historians of education and science. We think that, in order to understand their fruitful complexity, we need not only more studies about particular textbooks but also fine-grained and microscale analysis about these sources, careful studies like those written about other genres of scientific literature. This means revising the image of science teaching as a secondary activity of scientists. This is the approach we are trying to develop in our current research on chemistry textbooks and whose early results we have summarized in this chapter.

NOTES

[1] The literature on scientific centers and peripheries is large. For a recent survey see, for example, K. Gavroglu (ed.), *The Sciences in the European Periphery during the Enlightenment* (Boston: Kluwer, 1999); A. Simoes, A. Carneiro, and M. Paula Diogo (eds.), *Travels of Learning. A Geography of Science in Europe* (Boston: Kluwer, 2003).

[2] O. Hannaway, *The Chemist and the Word. The Didactic Origins of Chemistry* (Baltimore and London: J. Hopkins University Press, 1975).

[3] See historiographical papers by J. R. R. Christie and J. V. Golinski "The Spreading of the Word: New Directions in the Historiography of Chemistry, 1600–1800", *History of Science* 20: 235–266 (1982); W. H. Brock, "Science Education", in R. C. Colby et al. (eds.), *Companion to the History of Modern Science* (London: Routledge, 1990), pp. 946–959; A. García Belmar and J. R. Bertomeu Sánchez, "Motivi, fonti e domande per una storia dei contenuti e delle pratiche dell'insegnamento della chimica", in *Atti del Convegno Nazionale di Storia e Fondamenti della Chimica* (Arezzo, 2000). Two very important studies are K. M. Olesko, *Physics as a Calling: Discipline and Practice in the Königsberg Seminar for Physics* (Ithaca: Cornell University Press, 1991); A. Warwick, *Masters of Theory: Cambridge and the Rise of Mathematical Physics* (Chicago: Chicago University Press, 2003). For a recent overview of historical studies on scientific pedagogy, see D. Kaiser (ed.), *Pedagogy and the Practice of Science, 1800–2000* (Cambridge: MIT Press, forthcoming).

[4] Several important studies on rhetorics of science have recently been published. For example, M. Pera and W. Shea, *Persuading Science: The Art of Scientific Rhetoric* (Canton, MA: Science History Publications, 1991); D. Locke, *Science as Writing* (New Haven: Yale University Press, 1992); P. Dear (ed.), *The Literary Structure of Scientific Argument* (Philadelphia: University of Pennsylvania Press, 1991).

[5] See Christie and Golinski *op. cit.* (3).

[6] For a general discussion on textbooks as historical sources, see A. Choppin, "L'Histoire des manuels scolaires: une approche globale", *Histoire de l'éducation* 9:1–25 (1980); A. Choppin, *Les manuels scolaires: histoire et actualité* (Paris: Hachette, 1992). Several studies on scientific textbooks have been recently published: Bettina Haupt *Deutschsprachige Chemielehrbucher (1775–1850)*, (Stuttgart, 1987); William Clark, "German Textbooks in the "Goethezeit". Part I–II", *History of Science* 35 (2–3): 219–239, 295–363 (1997). G. Lind, *Physik im Lehrbuch, 1700–1850* (Berlin: Springer, 1992); A. Lundgren and B. Bensaude-Vincent (eds.), *Communicating Chemistry. Textbooks and Their Audiences, 1789–1939*, (Canton: Science History Publications, 2000); B. Bensaude-Vicent, A. García Belmar, and J. R. Bertomeu Sanchez, *L'émergence d'une science des manuels. Les livres de chimie en France (1789–1852)* (Paris: Editions des Archives contemporaines, 2003). See also the studies included in J. R. Bertomeu Sánchez, A. García Belmar, A. Lundgren, and M. Patiniotis (eds.), Textbooks in the Scientific Periphery, *Science and Education, special issue* (forthcoming).

[7] For a recent review of studies on history of science and history of books, see Adrian Johns Science and the book in modern cultural historiography, *Studies on the history and philosophy of science* 29(2):167–194; A. Johns, *The Nature of the Book: Print and Knowledge in the Making* (Chicago: Univ. Chicago Press, 1998); M. Frasca-Spada and N. Jardine (eds.), *Books and the Sciences in History* (Cambridge: Cambridge University Press, 2000); J. Tophan, "Scientific Publishing and the Reading of Science in Nineteenth-Century Britain: A Historiographical Survey and Guide to Sources", *Studies in History and Philosophy of Science* 31:559–612 (2000).

[8] A. L. Lavoisier, *Tratado elemental de Chimica dispuesto en un órden nuevo segun los descubrimientos modernos. Escrito en francés por M. . . . y traducido al castellano para uso del Real Seminario de Minería de México* (México: por D. Mariano de Zúñiga y Ontiveros, 1797); *Id. Tratado elemental de química presentado baxo nuevo orden . . . Traducido al castellano por D. Juan Manuel Munárriz*. (Madrid: Imprenta Real 1798).

[9] J. R. Bertomeu Sánchez and A. García Belmar, "Spanish chemical textbooks (1800–1845). A bibliographical approach", in B. Bensaude-Vincent and A. Lundgren (eds.), *Comunicating Chemistry: Textbooks and their Audiences, 1789–1939* (Canton: History of Science Publications, 2000) and C. Fournier-Balpe, *Histoire de l'enseignement de la physique dans l'enseignement sécondaire en France au XIX siècle* (Paris: Thèse, 1994).

[10] An excellent study about the different appropriation of a scientific book by different readers is J. A. Secord, *Victorian Sensation. The Extraordinary Publication, Reception, and Secret Authorship of Vestiges of the Natural History of Creation* (Chicago: University Press, 2000).

[11] For example, P. A. Adet, *Leçons élémentaires de chimie à l'usage des lycées, ouvrage rédigé par ordre du gouvernement, par...* (Paris: Dentu, 1804); J. Izarn, *Leçons élémentaires de physique et chimie expérimentales, par... 1re partie...* (Paris: Levrault, 1805) Schoell et Cie and P. Jacotot, *Cours de physique expérimentale et de chimie, à l'usage des écoles centrales, et spécialement de l'Ecole centrale de la Côte d'Or, par...* (Paris: Richard, 1800). See Bensaude-Vincent, García Belmar and Bertomeu Sánchez, *op. cit.* (6)

[12] On this chemical laboratory, see R. Gago, "El plan del rector Blasco (1786) y la renovación e las dsiciplinas científicas en la Universidad de Valencia: la química y la enseñanza clínica", *Estudis* 6:157–170 (1977); A. E. Ten Ros "Un intento de renovación científica en la Universidad del siglo XVIII. La cátedra de química de la Universidad de Valencia", *Llull* 5:133–147 (1983); A. García Belmar and J. R. Bertomeu Sánchez, "El laboratorio químico de la Universidad de Valencia a través de sus gastos (1790–1808)", in J. M. López Piñero, H. Capel, and J. Pardo Tomás (eds.), *Ciencia e Ideología en la Ciudad* (Valencia: Generalitat Valenciana, 1992), pp. 123–133.

[13] A. Nieto-Galán, "Ciència a Catalunya a l'inici del segle XIX: teoria i aplicacions tècniques a l'escola de química de Barcelona sota la dirección de Francesc Carbonell i Bravo (1805–1822)", Ph. D., Barcelona, 1994.

[14] [Gutiérrez Bueno, Pedro] *Curso de química, teórico y práctica, para la enseñanza del Real Laboratorio de Química de esta Corte* (Madrid: Antonio Sancha, 1788); *Id. Curso de química, dividido en lecciones para la enseñanza del Real Colegio de San Carlos. Por D...* (Madrid: Villalpando, 1802); *Id. Práctica del Curso de Química dividido en lecciones para la enseñanza del Real Colegio de San Carlos* (Madrid: Villalpando, 1803); *Id. Prontuario de Química, farmacia y materia médica, dividido en tres secciones* (Madrid: Villalpando, 1815). He also translated the new chemical nomenclature into Spanish.

[15] Some recent publications on this subject are J. Golinski, *Science as Public Culture: Chemistry and Enlightenment in Britain, 1760–1820* (Cambridge: University Press, 1992); N. Brooks, "Public Lectures in Chemistry in Russia, 1750–1870", *Ambix 44*:1–10 (1997); A. García Belmar and J. R. Bertomeu Sánchez, "Pedro Gutiérrez Bueno (1745–1822), los libros de texto y los nuevos públicos de la química en el último tercio del siglo XVIII", *Dynamis* 21:351–374 (2001).

[16] A. García Belmar and J. R. Bertomeu Sánchez "French Chemistry Textbooks (1802–1852). New Books for New Publics and New Educational Institutions", in B. Bensaude-Vincent and A. Lundgren (eds.), *Comunicating Chemistry: Textbooks and Their Audiences, 1789–1939* (Canton: History of Science Publications, 2000), pp. 17–56.

[17] See F. L. Holmes, *Eighteenth-Century Chemistry as an Investigate Enterprise* (Berkeley: University of California, 1989); F. L. Holmes, "The Chemical Revolution and the Art of Healing", *Caduceus* 11(2):103–126 (1995).

[18] On this issue, see A. Choppin's books, quoted in note (6).

[19] J. R. Bertomeu Sánchez "La censura gubernativa en España durante el reinado de José I (1808–1813)", *Hispania* 54(188):917–954 (1994).

[20] For chemistry, this was the role played by P. A. Adet's *Leçons élémentaires de chimie à l'usage des lycées, ouvrage rédigé par ordre du gouvernement* (Paris: Dentu, 1804).

[21] *Boletín de Instrucción Pública*, 1847, X, 610, 643. Among others, the list included R. Kaeppelin, *Curso elemental de química teórico y práctico, por... traducido de la segunda*

edición y adicionado por los doctores en farmacia D. Rafael Saéz Palacios... y D. Carlos Ferrán y Scardini (Madrid: Boix, 1843) and A. Bouchardat, *Elementos de química con sus principales aplicaciones á la Medicina, á las artes y á la industria,..., por..., traducidos de la segunda edición y adicionados por los farmacéuticos D. Gregorio Lezana... y D. Juan Chavarri...* (Madrid: J.R. Calleja, 1845).

[22] See our papers on this issue that have been published in B. Bensaude-Vincent and Lundgren, *op. cit.* (6) and Bensaude-Vincent, García Belmar and Bertomeu Sánchez, *op. cit.* (6).

[23] See Bensaude-Vincent and Lungren, *op. cit.* (6). For German chemistry textbookssee Haupt, *op. cit.* (6).

[24] W. Clark, "German Textbooks in the "Goethezeit". Part I–II", *History of Science* 35(2–3):219–239, 295–363 (1997). Bensaude-Vincent, García Belmar, and Bertomeu Sánchez, *op. cit.* (6), pp. 140–143.

[25] On Gutiérrez Bueno, see García Belmar, *op. cit.* (15); J. R. Bertomeu Sánchez and A. García Belmar, "Pedro Gutiérrez Bueno y las relaciones entre la química y la farmacia durante el último tercio del siglo XVIII", *Hispania* LXI(2):539–562 (2001).

[26] M. J. B. Orfila, *Élémens de chimie médicale* [Paris: Crochard, 1817, 2 Vols. (8th ed., Paris: Labé Editeur, 1851)]. An abridged version was published in 1828: *Résumé de leçons de chimie de... appliquée à la médecine pratique et à la médecine légale, par M. Alexandre Pichon* (Paris: Ponthieu et cie, 1828). See J. R. Bertomeu Sánchez and A. García Belmar, "Mateu Orfila y las clasificaciones químicas: un estudio sobre los libros de texto en Francia a principios del siglo XIX", *Cronos* 2:3–35 (1999); J. R. Bertomeu Sanchez, A. García Belmar, and B. Bensaude-Vincent, "Looking for an Order of Things: Textbooks and Chemical Classifications in Nineteenth Century France", *Ambix* 49(2):227–251 (2002).

[27] J. R. Bertomeu Sánchez and A. García Belmar, "Mateu Orfila's Elémens de chimie médicale and the Debate About the Medical Applications of Chemistry in Early Nineteenth Century France", *Ambix* 47:1–25 (2000b).

[28] G. Mojon *Curso analítico de química escrito en italiano por... Traducido en castellano e ilustrado con los descubrimientos más modernos por el Dr. D. Francisco Carbonell y Bravo* (Barcelona: Antonio Brusi, 1818); J. A. Chaptal *Elementos de química... Traducidos al castellano por D. Hyginio Antonio Lorente* (Madrid: Viuda e hijos Marín, 1793–1794); J. A. Chaptal, *Química aplicada a las artes, por... Traducida del francés al castellano por el Dr. D. Francisco Carbonell y Bravo* (Barcelona: Brusi, 1816). Carbonell's son was also an active translator of French textbooks: J. P. L Girardin, *Lecciones de química elemental, tr. de las últimas ediciones francesas en armonía con los adelantos modernos, por D. Francisco Carbonell y Font* (Barcelona: Pujal, 1839–1841, 2 Vols.); J. P. L. Girardin, *Lecciones de química elemental, con figuras repartidas..., explicadas los domingos en la Escuela municipal de Ruan, por... traducidas de la segunda edición francesa, dada á luz en el año 1839 y adicionadas por D. Francisco Carbonell y Font* (Barcelona: José Matas y Bodallés, 1841, 2 Vols.).

[29] For example, R. Kaeppelin, *Curso elemental de química teórico y práctico, por... traducido de la segunda edición y adicionado por los doctores en farmacia D. Rafael Saéz Palacios ... y D. Carlos Ferrán y Scardini.* (Madrid: Boix, 1843, 547 p. + 16 h); J. J. Berzelius *Tratado de química, por J. J. Berzelius; nueva edición completamente refundida, según la cuarta edición alemana publicada en 1838 por B. Valerius ; Traducido por Rafael Sáez y Palacios y Carlos Ferrari y Scardini* (Madrid: Imp. y Libr. de Ignacio Boix, Editor, 1845–1851); A. Bouchardat *Elementos de química con sus principales aplicaciones á la Medicina, á las artes y á la industria, ..., por ..., traducidos de la segunda edición y adicionados por los farmacéuticos D. Gregorio Lezana ... y D. Juan Chavarri ...* (Madrid: J. R. Calleja, 1845). See Bertomeu Sánchez, García Belmar, *op. cit.* (9).

[30] J. B. Jumelin, *Traité élémentaire de physique, de chimie et de physico-mathématique, par...* (Paris: Duminil-Lesieur, 1806), p. 9.

[31] *Ibid.*

[32] Joseph Izarn. *Leçons élémentaires de physique et chimie expérimentales, par... 1re partie...*, (Paris: Levrault, Schoell et Cie, 1805), *Preface*. On Izarn's biography see Bensaude-Vincent, García Belmar and Bertomeu Sánchez., *op. cit.* (6).
[33] From this group of "physics and chemistry" books, only a textbook by J. M. Brisson was translated into Spain: *Elementos ó principios físico-químicos, destinados para servir de continuación á los principios de física escritos en francés por el C.... Traducidos al castellano por D. Julián Antonio Rodríguez, ... Tomo Quarto*. (Madrid: En la imprenta de la administración del Real Arbitrio de Beneficencia, 1804), 398 p.
[34] J. Simon, "The Chemical Revolution and Pharmacy: A Disciplinary Perspective", *Ambix* 45(1):1–13 (1998).
[35] J. R. B. Sánchez and A. G. Belmar, Los libros de texto de química destinados a los estudiantes de medicina y farmacia en España (1788–1845), *Dynamis* 20:457–489 (2000); Bertomeu Sánchez and García Belmar, *op. cit.* (25).
[36] See F. L. Holmes, *op. cit.* (17); Bertomeu Sánchez and García Belmar, *op. cit.* (27).
[37] Juan Manuel de Aréjula *Discurso sobre la necesidad de la Química en la teoría y práctica de la Medicina. Leido en el Real Colegio de Medicina y Cirugía de Cádiz, el primero de octubre de 1795. Por Don ...*, (Cádiz: Manuel Bosque [1795]) 47 p.; Francesc Carbonell i Bravo *De Chemiae ad Medicinam applicationis usu et abusu*, [Monspelii: Apud G. Izar et A. Ricard, an IX (1804)]. Carbonell also translated the influential text by A. Fourcroy Discurso sobre la unión de la química y la farmacia, traducido del francés por el Dr. D. Francisco Carbonell y Bravo, Madrid, Repullés. See J. R. Bertomeu Sánchez, *op. cit.* (29).
[38] On this issue, see Bertomeu Sánchez, García Belmar, *op. cit.* (27).
[39] J. Pierre Macquer *Elemens de chymietheorique, par... Nouvelle edition* (Paris: Jean Thomas Herissant, 1753), pp. xvi–xvii.
[40] B. Bensaude-Vincent "A View of the Chemical Revolution Through Contemporary Textbooks: Lavoisier, Fourcroy and Chaptal", *British Journal for the History of Science* 23(4):435–460 (1990). For a study about a Spanish textbook, see J. R. Bertomeu Sánchez and A. García Belmar, "El *Curso de química general aplicada a las artes* (1804–1805) de José María San Cristóbal y Josep Garriga i Buach", in J. L. Barona *et al.* (eds.), *La Ilustración y las ciencias*, (Valencia: PUV, 2003), pp. 179–237.
[41] On plant chemistry, immediate principles and classification, see R. Löw, *Pflanzenchemie zwischen Lavoisier und Liebig* (Straubing, München: Donau Verlag, 1977). See also U. Klein, *Experiments, Models, Paper Tools. Cultures of Organic Chemistry in the Nineteenth Century* (Stanford: Stanford University Press, 2003).
[42] See Bertomeu Sánchez and García Belmar, *op. cit.* (26).
[43] J. Louis, Lassaigne *Abrégé élémentaire de chimie considérée comme science accessoire à l'étude de la médecine, de la pharmacie et de l'histoire naturelle, par...* (Paris: Bechet jeune, 1829), p. 62.
[44] A. G. Debus, "The significance of chemical history", *Ambix* 32:1–14 (1985).

CRISTINA SENDRA-MOCHOLÍ

BOTANY IN UNIVERSITY STUDIES IN THE LATE 18TH CENTURY. THE CASE OF VALENCIA UNIVERSITY*

INTRODUCTION

In the 16th and 17th centuries, botany had not yet been established as a subject and although plants were studied in natural history, providing descriptive knowledge of nature, most scientific research on plants was conducted in more or less direct connection with their medical uses. Consequently, the only university chairs in existence in Europe that concerned the study of plants, were those known as chairs of simples, i.e., of simple medicines or materia medica.

The increasing importance of the study of plants during the 18th and early 19th centuries in western Europe, led to the first chairs of botany connected to certain institutions, such as universities, botanical gardens, schools of surgery, etc. being established. In this period consequently, botany became a scientific discipline and the rules of conduct that were to govern its activity began to take shape.[1]

Two chairs of botany were established at the University of Valencia—the Chair of Chemistry and Botany and the temporary Chair of Botany—and the former chairs of simples and herbs was done away with in 1787. The aim of this paper is to analyze certain aspects of the process that led to the creation of these chairs and highlighted the experience of the University of Valencia in this period when modern botany was established. We have based our research mainly on two manuscripts: the first concerns a syllabus drawn up in 1772, which was never actually put into practice; the second is a report about the founding of a botanical garden written in 1779 by Tomás Manuel Villanova Muñoz y Poyanos, a professor at the University of Valencia. As we shall see, the analysis of these texts permits us to answer, from the point of view of the academic world in Valencia at the time, different questions such as what the words botany or botanist were understood to mean, or what professors considered should be taught in this subject at university, as well as others that we shall point out throughout the paper.

THE UNIVERSITY OF VALENCIA AND STUDIES OF MEDICINE. THE 1772 SYLLABUS AND BOTANY

At the end of the 18th century, the University of Valencia was made up of an arts faculty which taught philosophy prior to admittance into any of the main faculties, and three main faculties: Theology, Civil and Canon Lawn, and Medicine.[2] The study of plants

M. Feingold and V. Navarro-Brotóns (eds.), Universities and Science in the Early Modern Period, 259–272.

belonged entirely to the faculty of medicine. Since the foundation of the University in 1502, the so-called *segona cadira de medicina* (second chair of medicine) had concentrated on teaching anatomy in autumn and winter, and herbs and other simple medicines in spring and summer. Around the year 1560, this chair was divided into two; one for anatomy, the other for simples. In 1567, when Juan Plaza (1525–1603) was appointed, the latter became known as the chair of simples and herbs. As we will see further on, this chair remained in existence in the school of medicine until 1787.[3]

In the mid-18th century, by virtue of the 1733 Constitutions, medical studies were organized under eight chairs. The chair of simples and herbs, was accompanied by those of Anatomy, Hippocratic Aphorisms, Surgery, Practice, and three of "courses" or theoretical subjects. The first 2 years, students were taught Anatomy and Simples, and Hippocratic Aphorisms in the third year. Along with these subjects, the theoretical subjects were taught in the first 3 years. Practicals took place during the third and fourth year and surgery was only studied in the fourth year.[4] Many years later, during the reign of Charles III, a far-reaching university reform was implemented in Spain resulting in many new syllabi being drawn up by the main Spanish universities. The syllabus of the University of Seville was adopted in 1769, those of Valladolid, Salamanca, and Alcalá in 1771, that of Santiago in 1772, Oviedo in 1774, Granada in 1776, and lastly Valencia in 1786.[5]

When the Valencia syllabus, known as the Blasco syllabus since its creation and implementation was largely due to Vicente Blasco, the rector of the university at that time, was approved in 1786 and published in 1787, it brought the teaching of science into line with the modern requirements typical of that period. In addition to botany classes, it also established chairs of chemistry, mathematics, mechanics, and experimental physics and astronomy.

Botany had featured in several syllabi drawn up during Carolingian times (the reign of Charles III). Consequently, the syllabus submitted by Gregorio Mayáns with the aim of unifying the syllabi of all Spanish universities in 1767, included botany and chemistry, as did the syllabi of the Universities of Valladolid and Granada. However, the Blasco syllabus, which as we have already said included two new chairs of botany in the school of medicine, was the only one to be completely implemented.[6]

In order to understand the process that led to the creation of these chairs of botany, it is useful to analyze the syllabus drafted in the University of Valencia in 1772, immediately preceding the Blasco syllabus. This previous syllabus also stemmed from the Carolingian reformation program, although it was not officially approved, it did give rise to protracted discussions as to whether botany should be included in studies of medicine, rather than under the former chair of simples and herbs.

The 1772 syllabus also revealed the opinions of professors as regards what university teaching should be. It was the professors themselves who drew up this syllabus. José Gascó, who held the chair of simples and herbs between 1749 and its abolition when the new syllabus was implemented in 1787, was one of the persons commissioned to draft the part concerning the studies of medicine together with Manuel Mañes, Vicente Adalid and Agustín Vicens, the occupants of the chairs of surgery, practices, and Hippocratic Aphorisms, respectively. Furthermore, it made provision for allowing individual votes from anyone not in agreement with any of the points on the syllabus. As we shall see, in the case of medical studies, only one vote was cast, that of the professor of medicine José Albertós.

José Albertós y Sanz was the course or theory professor of the University of Valencia from 1752 until his death in 1776. He was also the judge of the sub-delegation of the Royal Tribunal of the *Protomedicato* of Valencia, a body designed to regulate the professional practice of medicine. The only printed work by Albertós that we are aware of deals with illness caused by eating salted fish.[7] In his report appended to the 1772 syllabus, Albertós explained the need to incorporate botany into the study of medicine and offered a detailed plan for the chair of simples and herbs.

The source documents used for the present analysis of the process that led to the creation of chairs of botany at the University of Valencia, are the text belonging to the part of the syllabus that concerns the chair of simples and herbs and also the individual vote cast by professor José Albertós.[8] Use also was made of a small manuscript also written in 1779, by another professor of medicine at the University of Valencia, Tomás Manuel Villanova Muñoz y Poyanos, who penned several books and manuscripts on botany and was to occupy the chair of chemistry and botany following the implementation of Rector Blasco's syllabus.

TOMÁS MANUEL VILLANOVA MUÑOZ Y POYANOS AND BOTANY

Born in Bigastro (Alicante) into a farming family, Tomás Villanova studied medicine at the University of Valencia from 1757 to 1764. In 1766, he was named public academic or extraordinary professor (a sort of assistant professor of a subject that complemented the subject established as a chair) in order to teach the *Institutiones* of Andrés Piquer.[9]

In the late 1760s and early 1770s, Villanova undertook a 2-year study journey, visiting at least France, Italy, Germany, and Hungary, taking particular interest in natural history, experimental physics, and mathematics.[10]

The recent discovery of a manuscript copied by Villanova has confirmed that he visited Pisa in 1771. It is a sort of treatise on plant materia medica which lists the most commonly used autochthonous medicinal plants, according to Tournefort's classification system. In it, Villanova describes himself as a pupil of the department of medicine and botany at the Angelo Tilli University in Pisa.[11]

Upon his return to Valencia, he sat the competitive exams for different chairs of medicine. He also taught medicine, surgery, physics, and mathematics in several private academies. He acted as a stand-in for professors in different chairs of medicine until 1780, when he was appointed to the chair of Hippocratic Aphorisms. Later, with the implementation of the new syllabus in 1787, he occupied the chair of Chemistry and Botany.[12]

In addition to medicine, his written and printed oeuvre also deals with disciplines including physics, astronomy, chemistry, arithmetic, algebra, geometry, and botany.[13] His printed works on botany include his *Problema physicum [...]*, published in 1774 and certain tables with corrections to the Linnaean classification of some of the plants mentioned by the Italian Luigi Tessari in his materia medica.[14] Most of his writings remained in manuscript, however, including the foreword and general syllabus of *Flora Valentina*, two Latin-Spanish lexicons of botanical terms, both unfinished, and a series of teaching notes.[15]

The fact that Tomás Villanova had belonged since his student days to the academic world of Valencia, traveled for academic purposes for 2 years and visited some of the main universities of Europe, had cultivated botany amongst other subjects, and was to occupy the recently established chair of chemistry and botany when the Blasco syllabus was implemented, made him an author of particular interest for our research.

One of the texts he wrote on botany chosen as a source document for our research was his *Dictamen sobre un Jardín Botánico* (Guidelines for botanical gardens), written in 1779, between the syllabus of 1772 and the Blasco syllabus. He was commissioned by the municipality to write this report on the creation and upkeep of a university botanical garden.[16] Villanova deals with all the requisites for setting up a botanical garden on some land near the Alameda, one of the main boulevards in Valencia. It also contains the author's ideas about what botany is, and what a botanical garden and the work of a botanist should be. Attached to this report is a color layout of the future botanical garden indicating its different elements: the building containing the classroom, library, natural history museum, and the professor's and gardener's quarters; the greenhouse or enclosure; different zones for the garden to contain delicate flowers, medicinal plants, wild plants and water plants and plants from woodlands, irrigated land and meadows; and finally the wood area or plant nursery.[17]

BOTANY IN THE ACADEMIC WORLD OF VALENCIA

To begin with, let us see these texts reveal about what was understood by botany in the academic world of Valencia. The writers of the syllabus defined botany as "the study of plants in the way it is taught by those who profess this faculty", thus recognizing the existence of a group of authors dedicated to botanical studies.[18]

The report by Albertós is even clearer. According to this author, botany, like any other scientific field, consists of certain principles or bases which make it a stable practice. In botany, these principles are the terms used to describe plants: corolla, petal, nectary, stamen, anther, pistil, style, stigma, etc. These terms of theoretical botany are essential, for Albertós, to what he calls "practical" botany, the identification of plants. Knowing that "this is wheat, this is a cabbage, this is a radish" is not enough, one must also know the theoretical principles of botany and understand the books that discuss plant genera and species, for otherwise, "one has knowledge of nothing".[19]

Tomás Villanova also maintained in his *Dictamen* that botany, like natural history in general, consists of "knowing and distinguishing between the genera and species of natural productions". The countless differences between plants make its study necessary.[20]

MEDICINE, BOTANY, AND THE TEACHING OF BOTANY: THE CHAIR OF SIMPLES AND HERBS

Let us move on to a second question: Should botany be taught at university?

In 18th-century Spain, most chairs of botany were established in institutions other than universities such as the Royal Botanical Gardens in Madrid (1755) and those of Cartagena (1787), the Colegio de Cirugía (School of Surgery) in Cádiz (1790) and in Barcelona (1793), the Royal Association of Medicine and other Sciences in Seville (1782), and the *Real Sociedad Económica de Amigos del País de Zaragoza* (1797), inter alia.[21]

As we saw earlier, in the case of universities, certain syllabi also envisaged including chairs of botany, generally attached to studies of medicine. However, only was this rearrangement at the University of Valencia actually implemented. But what did professors of medicine think? Should Botany be included in medical studies? The authors of the syllabus considered that "this arduous, vague, and cumbersome subject, because of the way botanical authors deal with it, is neither useful nor helpful to first or second year students" and although botany "is used in many parts of Europe, it is of greater benefit to vanity and curiosity than public health".[22] In support of this argument, they quoted the work *Methodus Studii Medici* by the Dutch physician Hermann Boerhaave (1668–1738), published and enlarged upon by his disciple Albercht von Haller (1708–1777) in 1751.[23] According to those who drew up the syllabus, Boerhaave and Haller had affirmed that "botany is not absolutely necessary in the training of a medical practitioner" and that the "profession of a botanist is different from that of a medical practitioner".[24]

José Albertós, the dissenting voice on the 1772 syllabus, defended the need for medical students to study botany. He deemed it just as necessary for the knowledge of materia medica as anatomy is for physiology: the knowledge and description of the parts that make up the human body are necessary in order to understand their use and functions, just as familiarity with plants and their parts is necessary to understand their uses and effects. Furthermore, he argued that since plants constitute the majority of the remedies used to treat illness "it is absolutely essential to be familiar with them, because otherwise it is impossible to differentiate between useful and harmful ones".[25] He too quoted the authority of Boerhaave, this time to defend the study of botany citing Boerhaave's *Historia Plantarum*, published anonymously, which contained the classes given by Boerhaave whilst director of the Botanical Gardens in Leiden.[26]

Albertós added that failure to study botany could lead to the situation condemned almost a century earlier by the Neapolitan doctor Leonardo di Capua (1617–1695), "physicians know not what they are prescribing, pharmacists know not what remedy they are dispatching, and the rustic herbalists, who can hardly read, pick the medicinal simples out blindly".[27] However, Albertós does not reject the classical authors in his defense of botany; after quoting Lionardo di Capua, whose work was impregnated with a bitter criticism of Galenic medicine, he resorts to Galen himself and volumes six, seven, and eight of his work *De simplicium medicamentarum facultatibus* which list and describe the virtues of medicinal simples. As we can see, tradition and renovation are both invoked by Albertós in his defense of botany.[28] Finally, Villanova applies the usefulness of botany not only to medicine "but to all the other arts and customs of human life", and recommends "promoting, facilitating, and extending its study".[29]

Let us now answer a third question: What place should botany occupy in the Chair of Simples and Herbs? As we have already mentioned, the authors of the syllabus deemed there to be no space for the teaching of botany, "which is ostentatious and curious rather than useful or beneficial in the treatment of disease".[30] The professor of Simples and Herbs had to explain both simple and compound medicaments, their properties, the dosage needed to heal and, finally, how to make out prescriptions. Since this "could never be managed by means of any of the methods put forward by botanical authors", the textbook they recommended was the materia medica included in the work *Medicina vetus et nova* with the title *Pharmacia galenico-chymica ad tyrones* by Andrés Piquer, complemented by the work of the French physician Étienne Françoise Geoffroy (1672–1731).[31] Herborizations or field studies however had to continue, albeit only in the shape of outings to the fertile lands and seaside areas nearest the city. These outings to collect herbs had been a mainstay of the classes given in the Chair of Simples since the mid-16th century, and had later become a compulsory part of medical studies, as is indicated in the 1611 and 1733 *Constituciones*. As we will see later, the authors of the syllabus reiterated, as they had been doing since the mid-16th century, the need to create a botanical garden where professors could teach the "names, nature, and effects of plants". Hence, the direct examination of plants, including observation in their natural habitat, had continued to form part of the classes taught by the Chair of Simples.[32]

José Albertós was in favor of the teaching of botany and advised against the use of Andrés Piquer's medical treatise because it "mixes simples and compounds without explaining their nature [...] it is all about healing methods, not about the knowledge of simples".[33] Once again he recommended, firstly, Galen's fourth and fifth books on simples. According to Albertós, Galen is still an essential source for the knowledge of simples. Along with the Galenic texts, Albertós recommended the materia medica of Heinrich Johann Nepomuk von Crantz (1722–1799), a professor at Vienna University and a member of the generation of students trained by Gerhard van Swieten (1700–1772) following Boerhaave.[34] He also recommended the materia medica of the Italian Luigi Tessari.[35]

Finally, for teaching students the basic principles of botany, Albertós recommended *Philosophia botanica* (1751) by Linnaeus.[36] As regards classification systems, Albertós conceived of them as a way of recognizing plants, whether by means of the corolla, like Tournefort, or by the stamen and pistils, like Linnaeus, without attaching much importance to which system was used. He rejected all causes or explanations that could stem from the application of the systems. Consequently, he recommended excluding from teaching the first two chapters of Linnaeus' *Philosophia botanica* "which deal with the author's library and art systems", recommending instead the use of Crantz's materia medica, precisely because it omits "impertinent and systematic issues, which only serve to obscure understanding and entertain".[37]

Villanova no longer speaks of a chair of simples and herbs, but of a chair of botany involving not only the study of plants but also "by extension and as a complementary subject, teaching about animals and minerals". This idea can be understood if the chair of botany is recognized as the successor of the chair of simples. For the same reason, as we will see below, Villanova declared that a botanical garden must contain a museum with the "productions of the animal and mineral kingdom".[38]

BOTANICAL GARDENS

In the mid-16th century, when Juan Plaza was appointed to the chair of simples and herbs, his duties included for the first time, in addition to conducting field studies, the creation of a garden of medicinal plants:

> *[...] e perquè tinga compte ab un ort en lo qual se planten les erbes que a d'aquell parexeran necessàries, donant-li loch opportú hon se fassa dit ort e ertolà que tinga càrrech de ultimar aquell e porte compte de les botigues de apothecaris.*[39]

We have no proof of the creation of this garden, it would have been the first university garden of its type on the peninsula and amongst the first in Europe. In any case, the fact that the duties of the chair included its creation, underlines once again the importance of observing plants directly in the teaching of that period.[40]

In the 17th century, two *horts de les herbes medicinals* were established in the University of Valencia. The first was founded at the suggestion of the senate of professors of medicine and particularly the professor of simples, Gaspar Pons in 1632. The second one, run by the professor of simples Gaudencio Senach, was created in 1684,[41] but it cannot have lasted many years since the byelaws of 1733 said that the city should set aside "a garden for medicinal herbs".[42]

The writers of the 1772 syllabus once again underlined the need to set up a botanical garden, like those in all the famous universities of Europe. The professor should go there at least once a month to teach "the name, nature, and effects of the plants". Furthermore, the authors used the term "botanical garden" as opposed to "medicinal herb garden", used in the previous byelaws of 1733.[43]

Villanova also used the term botanical garden and pointed out that it should not consist merely of a collection of medicinal plants "because even those whose medicinal faculties have been unknown to date are not lacking in them, in fact they are being discovered every day [...] and the same applies to their use in the arts". Every plant species had to be studied, as it might somehow contribute to human life. Villanova consequently aimed to have all known vegetable species and their main varieties in his garden. Other specimens from the animal and mineral kingdoms that were also useful to man, should also be exhibited in a "museum or collection room", together with a herbarium containing plants that could not be grown in the garden.[44]

The main purpose of a botanical garden was, according to Villanova, to facilitate the study of plants, "which is not obtained from books alone, although they must also be seen to be necessary, but mainly from the observation of the (natural) produce itself".[45]

The garden envisaged by Villanova was undoubtedly mainly intended for teaching purposes. The distribution of plants was to enable them to be "walked between during lessons".[46] The basic aim of its layout and design was to facilitate the growth and care of the plants. In this respect, to enable a large number of plant species to be cultivated, Villanova planned to "create several areas according to the main differences of the plants as regards their original habitat". A total of six areas lined up one next to the other, were to be used to cultivate plants from the hills, countryside, irrigated lands, meadows, the sea, and finally water plants. Medicinal plants were to have a separate area in "the first and main part of the garden". An enclosure was to house the plants from the hottest climates.[47] Similarly, the plants were to be grown in neat rows running across the garden, ideally in a mathematically ordered number. This simple design, quite different from a garden for recreation, must have been that of a botanical garden designed mainly for teaching and, although Villanova made no direct mention of it in his guidelines, for experiments, "I am not unaware that in certain botanical gardens, walkways and passages are built inside the cultivation areas with different figures and creative works, but I also know that such strange things, being superfluous and distracting to the eye, are precisely being eliminated from the most famous gardens of Europe [. . .]".[48]

Medicinal plants were still the main feature of this botanical garden, a legacy of the classical gardens of medicinal plants. They were to be grown separately from the other plants and constituted the main part of the garden. In addition, they had an intricate irrigation system and the full name of the plants was to appear next to each one (other plants were identified by a number referring to the general catalog of the garden) and finally it was hoped that their sale could contribute to the economic upkeep of the garden itself.[49]

BOTANISTS

Finally, we examined these texts to answer one last question: What was a botanist?

As we have seen, the authors of the syllabus acknowledged the existence of a group of writers dedicated to cultivating botany with a series of methods and principles of their own. Albertós used the term *botanophile*, as did Linnaeus, to designate the amateur botanist, that is to say, one who makes observations of plants but without following the norms or principles of scientific botany, since only the people who are familiar with these "can be proper botanists rather than *botanophiles* or rustic herbalists". Albertós' definition of a botanist, then, is one who uses the right terms to describe vegetable species.[50]

In his *Dictamen*, Villanova provided a complete description of what a botanist should be: botanists are those who dedicate themselves especially to the knowledge of all kinds of plants, observe them in different circumstances, examine their parts, and speak about them to different people. Their activities include making new observations, confirming old ones, carrying out experiments, perfecting and finding new uses for plants as food, medicine, yarns, dyes, liquors, fodder, timber, etc. In this way, Villanova justified the botanist's task, for only botanists are capable of "perfecting the uses of plants and finding new ones".[51]

CONCLUSIONS. BOTANY AND THE CHAIRS OF BOTANY IN THE 1787 SYLLABUS

The first chairs of botany were established in Europe during the 18th century. Such chairs were often established in recently created institutions other than universities, such as schools of surgery, royal botanical gardens, and different scientific associations. University chairs of botany however, usually stemmed from faculties of medicine and were therefore a legacy of the old chairs of simples or materia medica, the only university chairs related to the scientific study of plants during the 16th, 17th, and part of the 18th century. Such was the case of the University of Valencia.

As mentioned earlier, the professor of the theory of medicine, José Albertós—on the basis of authors such as Hermann Boerhaave, Albercht von Haller, Leonardo di Capua, and Galen himself—suggested that botany be incorporated into the classes taught by the chair of simples. Consequently, together with the materia medica of Heinrich Johann Nepomuk von Crantz and Luigi Tessari, Linnaeus' *Philosophia botanica* would be the textbook used by pupils to learn the principles of botany. Students were not to know about plants in an empirical fashion—"this is chicory, this is wheat, this is cabbage, this is a radish [...]"—although they would have acquired such knowledge from nature, but were to know the elements or foundations of scientific botany. These elements were botanical terms: corolla, petal, calyx, nectary, pistil, stamen, etc. Knowing them makes it possible to understand the descriptions of different plants provided by "botanical writers" and identify any plant by means of the methods put forward by such authors.

These ideas promoted by Albertós prevailed in the Blasco syllabus and as a result, botany was incorporated into studies of medicine by means of the classes given by the temporary chair of botany to be attended by first year medical students. The textbook used in this department was *Curso elemental de botánica [...]* (An elementary botany course) by Casimiro Gómez Ortega and Antonio Palau y Verdera, the first and second occupants of the chair at the Royal Botanical Gardens of Madrid, in 1785.[52] Like Albertós, the main concern of the authors of this textbook was for students to learn about the different external parts of plants and the correct terms for describing them. The basic aim was that at the end of the course, students themselves would be able to identify any plant, in order to "use them to consult any text related to the nature, uses, and virtues of each plant".[53]

Medicinal plants did however continue to be the core subject. As a result, first-year medical students were to attend classes in the chemistry and botany department in addition to those of the temporary botany department. Chemistry classes finished in March and all medical students, not only first-year students, had to take botany classes in April and May in the botanical garden on the "knowledge and properties of plants for medicinal purposes as stated in Murray [...]".[54] The textbook referred is *Apparatus medicaminum [...]* by Linnaeus' pupil, the professor of medicine and botany at Göttingen University, Johan Andreas Murray.[55] Studies at his department of botany covered the properties of medicinal plants, their uses, and the different ways of preparing them. Although students of medicine began by studying the system of identifying any plant, they then concentrated on studying medicinal plants through the classes taught in this department of chemistry and botany.

Similarly, as Villanova pointed out, the university botanical garden should aim to contain all existing plants and not merely those with known medicinal properties; because any plant could be of use to human life, they should all be studied. Medicinal plants did however continue to be the mainstay of the garden. They were to be grown in the main part of the garden and were the object of most care. Their full name was to appear next to them and their sale was to contribute to the economic upkeep of the garden itself. This idea could also be seen in the syllabus of rector Blasco since the professor of chemistry and botany was only to teach about medicinal plants in the botanical garden. Consequently, 18th-century university botanical gardens even though more comprehensive must also be considered to be the legacy of the classical gardens of medicinal herbs.

In short, in the last third of the 18th century, the importance of the knowledge of terms used by "botanical authors" to describe each known plant species (since those who have empirical knowledge of different plants "know nothing") in conjunction with the idea that any plant could be of use in human life, extended the scientific study of plants to encompass all plants because, in the words of Villanova, "those whose medicinal properties have yet to be discovered, do nevertheless have them, and are indeed discovered every day [...] and the same occurs as regards their use in the arts [...]". This was the situation studied here; these texts led to the creation of chairs of botany which, in the case of the University of Valencia, were designed to teach botanical basics and terms, in the temporary department of botany, and the properties, applications, and preparation of different medicinal plants, in the department of chemistry and botany.

NOTES

* This paper was made thanks to a grant from Caja Madrid Foundation.

[1] See J. M. López Piñero and J. Pardo Tomás, *La influencia de Francisco Hernández (1515–1587) en la constitución de la botánica y materia médica modernas* (Valencia: Instituto de Estudios Documentales e Históricos sobre la Ciencia. Universitat de Vsalència—C.S.I.C., 1996), pp. 25–26; J. M. López Piñero and M. L. López Terrada, *La influencia española en la introducción en Europa de las plantas americanas (1493–1623)* (Valencia: Instituto de Estudios Documentales e Históricos sobre la Ciencia. Universitat de Vsalència—C.S.I.C., 1997), pp. 11–14; and J. M. López Piñero and M. L. López Terrada, "La botánica en el reinado de Felipe II", in C. Añón and J. L. Sancho (eds.), *Jardín y Naturaleza en el reinado de Felipe II* (Madrid: Sociedad Estatal para la Conmemoración de los Centenarios de Felipe II y Carlos V, 1998), p. 278. According to these authors, in the 16th and 17th centuries, botany did not exist as a scientific discipline, the conversion of the study of plants into a profession was not contemplated in any form, nor were any rules governing the scientific activity related to botany defined. As a result, figures as distinguished as Fuchs, Clusius, Hernández, the Bauhins, Tournefort, and in the 18th century, Linnaeus, the Jussieus, Mutis, Ruiz, and Pavón, amongst others, continued to practice as physicians and apothecaries. A similar viewpoint appears in Karen Meier Reeds' works, "Renaissance Humanism and Botany", *Annals of Science 33*: 591–542 (1976); and *Botany in Medieval and Renaissance Universities* (New York and London: Garland, 1991) on botany, humanism, and the university in Renaissance Europe.

2 For an overview of the University of Valencia in the 18th century, see S. Albiñana, *Universidad e Ilustración. Valencia en la época de Carlos III* (Valencia: I.V.E.I., Universitat de València, 1988).
3 For the chair of simples and herbs at the University of Valencia, see J. M. López Piñero, *Ciencia y técnica en la sociedad española de los siglos XVI y XVII* (Barcelona: Editorial Labor, 1979), pp. 98–99; J. M. López Piñero, "Los saberes médicos y su enseñanza", in *Historia de la medicina valenciana*, J. M. López Piñero, dir., 3 Vols. (Valencia: Vicent García Editores, 1988–1992), Vol. I, pp. 111–115 and Vol. II, pp. 11–12; J. M. López Piñero and V. Navarro, *La història de la ciència al País Valencià* (Valencia: Ed. Alfons el Magnànim, 1995), pp. 67–68, 207–208, 301; J. M. López Piñero, "Las plantas del mundo en la historia. Ciencia, botánica y vida humana", in J. M. López Piñero and M. Costa Taléns, dirs. (eds.), *Las plantas del mundo en la historia. Ilustraciones botánicas de cinco siglos* (Valencia: Fundación Bancaja, 1996), pp. 25–27.
4 *Constituciones de la insigne Universidad Literaria de la Ciudad de Valencia, hechas por el Claustro Mayor de aquella en el año 1733* (Valencia: Imprenta de Antonio Bordázar de Artazu, 1733), p. 86.
5 S. Albiñana, "Antecedentes del plan de estudios del rector Blasco", in A. Ten (ed.), *Plan de Estudios aprobado por S.M. y mandado observar en la Universidad de Valencia* (Valencia: Ayuntamiento de Valencia, 1984), p. 26.
6 A. Ten, "El plan de estudios del rector Blasco y la renovación científica en la universidad española de fines del siglo XVIII", in A. Ten (ed.), *Plan de Estudios aprobado por S.M. y mandado observar en la Universidad de Valencia* (Valencia: Ayuntamiento de Valencia, 1984); J. M. López Piñero and V. Navarro, "Estudio histórico", in J. M. López Piñero et al. (eds.), *La actividad científica valenciana de la Ilustración*, 2 Vols. (Valencia: Diputación de Valencia, 1998), Vol. I, pp. 64–67.
7 The work is entitled *Exacta historia de la enfermedad ocasionada por el pescado calcinado llamado pagel, donde se explica todo lo perteneciente a la cal, padecida y formada por el Dr. D [. . .]* (Valencia, 1776). Mention of José Albertós is made in the bibliographic repertoires by J. P. Fuster, *Biblioteca valenciana de los escritores que florecieron hasta nuestros días. Con adiciones y enmiendas a la de D. José Ximeno*, 2 Vols. (Valencia: Imprenta y Librería de José Ximeno, 1827–1830), Vol. II, p. 87; and A. Hernández Morejón, *Historia Bibliográfica de la Medicina Española*, 7 Vols. (Madrid: Imprenta de la Viuda de Jordán e Hijos, 1842–1852), Vol. VII, p. 353. Also in S. Albiñana, *Universidad e Ilustración [. . .]* (footnote 2), p. 288.
8 These texts are part of the *Plan de Estudios de la Universidad de Valencia que presenta al Real Supremo Consejo en cumplimiento de la Real Orden de 27 de enero de 1772 [. . .]* housed at the AMV (Municipal Archives of Valencia: Libro de instrumentos ordinario del año 1772), D-132, fols. 443r.–580r. A variety of aspects of this syllabus have been studied by S. Albiñana, *Universidad e Ilustración [. . .]* (footnote 2), pp. 190–205; J. M. López Piñero and V. Navarro, *La història de la ciència [. . .]* (footnote 3), pp. 307–312; J. M. López Piñero and V. Navarro, "Estudio histórico" (footnote 6), pp. 62–63.
9 In this period, the teaching of medicine was marked by the influence of the later work of A. Piquer and Arrufat (1711–1772), one of the most influential physicians of that time. He wrote a series of rigorous, up-to-date textbooks that were published from 1735 onward. Villanova must have glossed his *Institutiones Medicae ad usum Scholae Valentina*, Matriti, Ioachimus Ibarra, 1762. For the influence of Piquer at the University of Valencia, see J. M. López Piñero and V. Navarro, "Estudio histórico" (footnote 6), pp. 85–88.
10 All this is told by Villanova himself in the *Libro de méritos de los opositores a cátedra*, 117, fols. 320v.–321r., housed in the AUV (Archives of the University of Valencia).
11 Manuscript by T. M. Villanova, *Distributio plantarum medicinalium. Auctore Angelo Tilli M.D. et Botanices in Pisana Universitate Professore. Pisis Anno MDCCLXXI.* 115 fols., Biblioteca Nacional de Madrid, Ms. 2242. My grateful thanks to J. M. López Piñero and F. Jérez Moliner for enabling me to locate this manuscript and others belonging to Villanova.

[12] Noteworthy historical studies about Villanova include those by J. P. Fuster, *Biblioteca valenciana [...]* (footnote 7), Vol. II, pp. 247–251; A. Hernández Morejón, *Historia Bibliogáfica [...]* (footnote 7), Vol. VII, pp. 340–344; M. Colmeiro, *La botánica y los botánicos de la península Hispano-lusitana* (Madrid: M. Rivadeneyra, 1858), pp. 10, 84, 169; F. Barberá Martí, *Sesión apologética dedicada al doctor Don Tomás Villanova Muñoz y Poyanos. Discurso leído en la apertura del Instituto Médico Valenciano el día 20 de octubre de 1888, por el Dr. D. [...]. Socio de mérito de esta corporación* (Valencia: Ferrer de Orga, 1888) and S. Albiñana, "Las cátedras de medicina en la Valencia de la Ilustración", *Estudis* 14:197–200 (1988). For his scientific activity in the realm of botany, see C. Sendra Mocholí, "La enseñanza de la botánica en la Valencia del último tercio del siglo XVIII", *Cronos* 1:122–124 (1998).

[13] R. Gago et al., "El plan de estudios del rector Blasco (1786) y la renovación de las disciplinas científicas en la Universidad de Valencia: la química y la enseñanza clínica", *Estudis* 6:63 (1977). These authors grouped his manuscripts by subjects as follows: medicine (five), chemistry (eight), botany (four), hydrology (seven), mathematics (four), astronomy (ten), physics (five), and Arab chronology (one). S. Albiñana, "Las cátedras de medicina [...]" (footnote 12), p. 199, counted seventeen printed works and almost fifty manuscripts.

[14] The first work is the *Problema physicum de mirabili quodam repulsionis affectu ex succi tithymali in aquam instilatione [...]* (Valencia: José Estevan Dolz, 1774); the second, by the Italian L. Tessari, was published in Valencia as *Materia medica contracta [...] In usum Scholae Valentinae* (Valentinae: In Officina Iosephi et Thomae de Orga, 1791), together with Villanova's tables of corrections *De Materia medica contracta Ludovico Tesari nuperrime in hac civitate pro scholae usu recusa monitum ad tirones [...].*

[15] Both *Flora Valentina inchoata, sive Plantarum in Valentino Regno [...]* and one of the glossaries and most of the teaching notes concerning the study of plants included under the title of *Adersaria Botanica* are housed in the AMNCN (Archives of the National Natural Science Museum) in Madrid, Ms. caja 180. The other glossary is in the Archives of the Royal Botanical Gardens, Madrid (ARJBM), entitled *Vocabularium Botanicum Latino-hispanicum ex variis auctoribus collectum. [...] 1780,* Ms. leg. I, 1, 6: 3.

[16] The University of Valencia was part of the municipality until 1827. See S. Albiñana, *Universidad e Ilustración [...]* (footnote 2), pp. 24–35.

[17] *Dictamen sobre un Jardín Botánico*, Valencia, June 12th 1779, AMNCN, Ms. caja 180. To which the *"Plano del huerto de la M.I. ciudad de Valencia sito al lado izquierdo de su mayor Alameda"* is attached. AMNCN, M8-CD3/411.

[18] *Plan de Estudios [...] 1772* (footnote 8), fol. 500r.

[19] *Ibid.*, fols. 519v.–521v.

[20] T. Villanova, *Dictamen sobre un Jardín [...]* (footnote 17), Ms., fol. 1r.

[21] See F. J. Puerto Sarmiento, *La ilusión quebrada. Botánica, sanidad y política científica en la España Ilustrada* (Madrid: Serbal, C.S.I.C., 1988), pp. 222–242.

[22] *Plan de estudios [...] 1772* (footnote 8), fol. 500v.–501r.

[23] *Hermanni Boerhaave [...] Methodus Studii Medici. Emaculata & Accessionibus locupletata ab Alberto ab Haller [...]*, 2 Vols. (Amstelaedami: Sumptibus Jacobi Wetstein, 1751).

[24] *Plan de Estudios [...] 1772* (footnote 8), fol. 500v.

[25] *Ibid.*, fol. 519r.

[26] The full title of the work is *Historia plantarum, quae in Horto Academico Lugdunis-Batavorum crescunt cum earum characteribus, & medicinalibus virtutibus.* Desumptis ex ore clarissimi Hermanni Boerhaave..., 2 Vols. (Romae: Apud Frasciscum Gonzagam, 1727). Different editions of this work are known to have been published in Leiden (1727), Rome (1727, 1731 y 1738), London (1731, 1738) and Venice (1766). See G. A. Lidenboom, "Bibliographia Boerhaaviana. List of publications written or provided by H. Boerhaave or based upon his works and teaching. Systematically arranged and compiled by [...]", in G. A. Lindeboon (ed.), *Analecta Boerhaaviana* (Leiden: E.J. Brill, 1959), pp. 78–79; W. T. Stearn, "Boerhaave as a botanist", in A. G. Lindeboom (ed.), *Analecta Boerhaaviana. Boerhaave and his time. Papers*

read at the international symposium in commemoration of the tercentenary of Boerhaave's birth, Leiden, 15–16 November, 1968 (Leiden: E.J. Brill, 1970), p. 115.

[27] *Plan de Estudios [. . .] 1772* (footnote 8), fol. 520r. Lionardi di Capua was a founder member of the researchers' academy in Naples and later of the Arcades Academy in Rome. See N. F. J. Eloy, *Dictionnaire Encyclopédique des sciences médicales*, 4 Vols. (Paris: P. Asselin, G. Masson, 1973), Vol. I, pp. 534–536. This quotation is probably from his book *Del parere del signor Lionardo di Capoa divisato in otto raggionamenti ne'quali narrondosi l'origine et progresso della medicina e l'incertezza della medicina si fa manifesta* (Naples, 1689). This book sets forth the mistakes and doubts in which the doctrines of classical authors incurred and also those of many modern writers. In addition to this, he recommended that botany be studied because plants are the source of most curative remedies for our illnesses.

[28] *Plan de Estudios [. . .] 1772* (footnote 8), fol. 520r.

[29] T. Villanova, *Dictamen sobre un Jardín [. . .]* (footnote 17), Ms., fol. 1r.

[30] *Plan de Estudios [. . .] 1772* (footnote 8), fol. 501r.

[31] It must have been the fourth edition of this work by Piquer: *Medicina vetus et nova. Postremis curis retracta, & aucta. Ad usum Scholae Valentinae. Editio quarta* (Matriti: Apud Joachimum Ibarram Typographum, 1768). Étienne Françoise Geoffroy taught chemistry at the *Jardin du Roi* and medicine at the *Collège de France*, besides being a member of the *Académie des Sciences*, Paris and the *Royal Society*, London. His *Tractatus de materia medica* (1741) was not published until ten years after his death. See W. T. Smeaton, "Geoffroy, Étienne François", in C. C. Gillispie (ed.), *Dictionary of Scientific Biography*, 18 Vols. (New York: Charles Scribener's sons, 1972), Vol. V, pp. 352–354.

[32] *Plan de Estudios [. . .] 1772* (footnote 8), fol. 499v.–500r.

[33] *Ibid.*, fol. 513r.

[34] See A. Dechambre, "Crantz, Henri-Joachim-Népomucène", in A. Dechambre, dir., *Dictionnaire Encyclopédique des sciences médicales*, Vol. XXII (Paris, p. Asselin and G. Masson, 1879), p. 722; Kleinwaechter, "Crantz, Heinrich Johann Nepomuk von", in *Biographisches Lexikon der hervorragenden Ärzte aller Zeiten und Völker*, 5 Vols. (Berlin: Urban & Schwarzenberg, 1930), Vol. II, pp. 137–138; E. Lesky, "Primera escuela vienesa", in P. Laín Entralgo, dir., *Historia Universal de la Medicina*, 7 Vols. (Barcelona: Salvat Eds., 1973), Vol. V, pp. 87–88. Albertós' reference is to *Materia medica et chirurgica, juxta systema naturae digesta*, 3 Vols. (Vienna, 1762). Other editions: Vienna (1766) and Lovania (1772).

[35] As we have already seen, many years later Tomás Villanova took care of publishing this book, which in compliance with the new syllabus of 1787, was to be studied in the fourth year of medicine. See J. L. Peset "Los estudios de medicina", in A. Ten (ed.), *PLan de Estudios aprobado por S.M: y mandado observar en la Universidad de Valencia* (Valencia: Ayuntamiento de Valencia, 1984), p. 70.

[36] *Philisophia botanica in qua explicantur fundamenta botanica [. . .]* (Stockholmiae: Apud Godofr. Kiesewetter, 1751). Another edition prior to 1772 published in Vienna (1763). A book that contained 365 aphorisms, one for each day of the year, divided into twelve chapters and accompanied by explanations, comments, references, and examples in smaller print. For an in-depth analysis of this work, F. Stafleu, *Linnaeus ans the Linnaeans. The spreading of their ideas in Systematic Botany, 1735–89* (Utrecht: Oostheek's Uitgeversmaatschappij, 1971), pp. 25–78.

[37] *Plan de Estudios [. . .] 1772* (footnote 8), fols. 516v.–517r., 523v.

[38] T. Villanova, *Dictamen sobre un Jardín [. . .]* (footnote 17), Ms., fols. 1v.–2r.

[39] F. M. Grajales, *El doctor Juan Plaza. Estudio biográfico* (Valencia: Imprenta de Manuel Alufre, 1893), p. 28, reproduces this provision, agreed upon by the city magistrates on May 16th 1567.

[40] S. García Martínez, "Gaudenci Senach i la càtedra valenciana de Botànica Mèdica 1682–1694", *Afers* 5/6:372 (1987), points out that authors such as M. Velasco and Santos, *Reseña histórica de la Universidad de Valencia* (Valencia, 1868), V. P. Cervera, *Noticia histórica del*

catedrático valenciano de materia médica Dr Juna Plaza (Valencia, 1895) and F. M. Grajales, *Hijos ilustres de Valencia. El doctro Melchor de Villena. Noticia biográfica* (Valencia, 1916) declared that this garden did not even last as long as Plaza's own time. P. Lechón and Moya, *Sesión apologética dedicada al Dr Melchor de Villena* (Valencia, 1884) wrote nonetheless that this same garden grew considerably years later.

[41] S. García Martínez, "Gaudenci Senach [. . .]" (footnote 40), pp. 373–374, 383–384.

[42] *Constituciones [. . .] 1733* (footnote 4), p. 86.

[43] *Plan de Estudios [. . .] 1772* (footnote 8), fols. 499v.–500r.

[44] T. Villanova, *Dictamen sobre un Jardín [. . .]* (footnote 17), Ms., fol. 1v.–2r.

[45] *Ibid.*

[46] *Ibid.*, fol. 6r.

[47] *Ibid.*, fols. 3v.–7r.

[48] *Ibid.*, fol. 6r.

[49] *Ibid.*, fol. 7.

[50] *Plan de Estudios [. . .] 1772* (footnote 8), fol. 523v. Linnaeus used the term *botanophili* in his *Philosophia botanica* to describe amateur botanists. See F. Stafleu, *Linnaeus [. . .]* (footnote 36), p. 35.

[51] T. Villanova, *Dictamen sobre un Jardín [. . .]* (footnote 17), Ms., fols. 1v.–2r.

[52] C. Gómez Ortega and A. Palau y Verdera, *Curso elemental de botánica, teórico y práctico, dispuesto para la enseñanza del Real Jardín Botánico de Madrid* (Madrid: Imprenta Real, 1785).

[53] *Ibid.*, p. xl.

[54] *Plan de Estudios aprobado por S. M. y mandado observar en la Universidad de Valencia* (Edición facsimilar a cargo de Antonio Ten. Valencia: Ayuntamiento de Valencia, 1984), p. 8.

[55] This textbook was published by the University of Valencia as *Apparatus medicaminum tam simplicium quam praeparatorum et compositorum in praxeos adjumentum consideratus Scholae valentinae recusus*, 2 Vols. (Valentiae, In Officina Josephi Estevan et Cervera, 1790–1791).

AGUSTÍ NIETO-GALAN AND ANTONI ROCA-ROSELL

SCIENTIFIC EDUCATION AND THE CRISIS OF THE UNIVERSITY IN 18TH CENTURY BARCELONA

1714: BARCELONA "WITHOUT" A UNIVERSITY

In 1938, on the occasion of the centenary of the restoration of the University of Barcelona, the Catalan historian Ferran Soldevila, published an admirable book entitled *Barcelona sense Universitat i la restauració de la Universitat de Barcelona.*[1] In fact, the University of Barcelona was closed by the Bourbon dynasty after the War of the Spanish Succession from 1714 until 1837. Catalan historiography, including Soldevila, has often perceived this event as a tragedy, as a punitive measure taken by the new dynasty against the supporters of the old Habsburg regime.[2] From the political point of view, there is no doubt that the War was very damaging for the Catalans, but the view of the closure of the University as a symbol of cultural humiliation is probably exaggerated.

What we shall try to show in this paper is how the absence of the University was offset, at least in the case of the natural sciences, by the emergence of alternative institutions which gave rise to an original scientific culture in Barcelona and in Catalonia as a whole. As in other European countries,[3] the crisis of the Catalan universities facilitated a general educational reform within the framework of the projects of enlightened despotism. Under the patronage of the new Bourbon dynasty, in particular, during the reign of Charles III, new local academies and "colleges" provided new scientific curricula. In Barcelona, it seems that the absence of a University did not hinder the introduction of the experimental sciences. The alternative educational framework that we shall try to describe can even be understood as favorable to scientific "modernization".

In Catalonia, the technical challenges of early industrialization called for the training of craftsmen and industrial workers, which could be provided mainly through the apprenticeship system but also in alternative routes opened up by skilled experts on various arts.[4] Nevertheless, it was often believed that this kind of training could be better provided by regular institutions, such as the University or the new academies. Thus, on several occasions, the city government of Barcelona called for the re-opening of the city's University, pointing out that the new industries demanded higher scientific education.[5] In opposition to the Spanish government's plans of modernization at the expense of the guilds of craftsmen, the city government included the University among its demands for an efficient training of craftsmen. In 1776 it stated:[6]

M. Feingold and V. Navarro-Brotóns (eds.), Universities and Science in the Early Modern Period, 273–288.

> It seems that it would not be unreasonable to formally advocate the resumption of Higher studies[7] or the planning of a Literary[8] University which would be as desirable in the city as in the rest of the Province [of Catalonia]. [The University in Barcelona] would obtain additional advantages for popular industry unobtainable at its present location at Cervera.

In 1768, the social reformer Francesc Romà i Rossell highlighted the role of the academies in modernizing agriculture and industry, drawing attention to the difficulties of the Academy of Sciences, which had been founded in Barcelona a few years before.[9] Romà i Rossell, and the Catalan Board of Commerce (*Junta de Comerç*) attributed the shortcomings of Catalan craftsmen to their lack of training and suggested that they should be educated in academies of physics.[10] These public demands for alternative training met with a different response in the context of a city without a University, and within the general framework of the 18th century Bourbon reforms.

SCIENCE "WITHOUT" A UNIVERSITY: THE REFORMS OF THE NEW BOURBON DYNASTY

The scientific institutions of the Spanish Enlightenment have attracted the interest of numerous scholars in recent years.[11] The Bourbon plans for modernization affected the public administration, the economy and culture, in a society which had been isolated since the decadence and fall of the 17th century Spanish Empire. In the period from 1700 to 1789, technical and medical schools, botanical gardens, *Sociedades Económicas de Amigos del País*, and new academies provided more adaptable channels for the introduction of new sciences such as physics, geology, botany, and chemistry.[12] These plans were reinforced by a new policy of appointing foreign experts and sending young students abroad (*pensionados*).[13] Despite the opposition of the old, scholastic, traditional universities (Salamanca, Alcalá, Santiago, etc.), the new network of societies of useful learning contributed to the creation of alternative channels of reform.

A new University in the city of Cervera

Experts agree that, at least, in the Spain of the 18th century, University curricula were far removed from experimental sciences and mathematics.[14] These curricula trained for the professions of Law and Theology, in the service of the monarchy and the church, and their syllabi were wedded to the old scholasticism, which, in the utilitarian context of the Enlightenment had come to be perceived as useless. Even the Faculties of Medicine, despite the considerable level of scientific training, required an external evaluation of the professionals they provided by the so-called *Protomedicato,* a board of supervisors.

In 1714, after the War of the Spanish Succession, all the Catalan universities (Barcelona, Girona, Lleida, Tarragona, Tortosa, Solsona, and Vic) had been closed, and in 1717 only one new University was created in the small town of Cervera, near

Lleida.[15] Although the creation of the University of Cervera reflected the political repression of the Catalans for their loyalty to the Habsburg dynasty, it also represented a good opportunity for the unification of all the Catalan universities. It seems that King Philip V himself had envisaged the creation of a new University at Cervera in accordance with the general plans of reforms against the old scholasticism.

In terms of scientific content, a number of authors have shown how some professors of Cervera, such as Tomàs Cerdà or Mateu Aymerich, both Jesuits, introduced some elements of modern physics into their chairs of philosophy.[16] Before 1767, it seems that the *Compañía* tried to transform Cervera into a Jesuit University. Even after their expulsion in that year, some signs of modernity persisted e.g., in the rejection, in 1770, of the plan for philosophy proposed by the University of Salamanca. Nevertheless, the fact that the chair of Mathematics was never occupied, and that the Faculty of Medicine was not properly established underlines the difficulties of creating a new University in a city without an academic tradition. To our knowledge, after the initial impulse of the new Bourbon monarchy, even the *Consejo de Castilla* lost interest in Cervera as a real alternative. In terms of the general plans of University reforms, the results of the University of Cervera differed little from the general pattern of inefficiency, with minor exceptions, such as the University of Valencia.[17] It should be pointed out that we have inherited a stereotyped image of the decadence of Cervera, largely through the experience of Antoni Martí i Franquès (1750–1832), one of the "great luminaries" of the Catalan scientific Enlightenment, and a prominent experimental natural philosopher, who regarded his years at Cervera as a waste of time and energy, and decided to pursue his interest in pneumatics in his own private laboratory.[18]

As Soldevila (1938) himself emphasized, the old University of Barcelona had not in fact disappeared completely after the *Nova Planta*. Given the importance of the *Hospital de la Santa Creu*, the Barcelona Faculty of Medicine, under the control of the *Protomedicato*, retained some of its activity. Grammar, rhetoric, and mathematics were taught by the Jesuits at the *Colegio de Cordelles*, and philosophy at the *Colegio de Sant Pau del Camp*. Even in 1783, the students of the *Seminario de Barcelona*, as well as those from *Cordelles*, obtained at Cervera official recognition of their curricula.[19] But the routes of scientific teaching and learning went much further.

The military academies

Philip V's centralized plans for the control of the territory of the new monarchy contributed to the creation of new military academies, such as the *Academia Militar de Matemáticas de Barcelona* (1715), and the *Academia de Guardamarinas of Cádiz* (1728).[20] The Bourbons needed to modernize the Army and the Navy to retain their dominance of the American colonies and to promote the manufacturing industry following the policy of Colbert. Both projects required new professionals trained in the rigid ambience of military academies.[21]

In contrast to the University scholasticism, the military academies became leading institutions in disciplines such as geography, astronomy, navigation, drawing, mechanics, and architecture,[22] and they facilitated the introduction of the new experimental and mathematical sciences. Useful applications of physics, mathematics, or chemistry for military purposes seemed far removed from the abstract discussions between the defenders of scholasticism and the *novatores*, on the "suitable" natural philosophy to be taught at the universities.[23] In the end, the enlightened despotism was based on concrete material reforms of the organization of the state, and on military education which opened up a promising market for the new professionals.[24]

In the particular context of Barcelona, following the defeat in 1714, the *Academia Militar de Matemáticas*, which had links with a late 17th century academy, was set up, incorporating the old chair of mathematics of Madrid for artillery and engineering.[25] The *Academia*, which was to set the educational standards for Spanish officers, was directed by Jorge Próspero Verboom, who imported from Brussels a teaching plan based on mathematics, fortification and drawing. In 1751, an Academy of Artillery was also founded in Barcelona, and, in 1760, the old *Academia de Matemáticas* became the *Academia de Ingenieros*. The high investment in military education, as well as in the excellent libraries and collections of scientific instruments of the academies is evidence of the important role of these institutions in the introduction of the new sciences as well as in the consolidation of a utilitarian modern discourse.[26] As H. Capel has emphasized, the 18th century debate on the utility of modern science and on "suitable" science and mathematics was mainly held in the context of the Royal Navy, the military Engineers and the Artillery.[27] As far as we know, in the Catalan society of the 18th century, the actual influence of these military academies was small. We know that some noblemen studied at the Barcelona academies and that some officers played an active part in new scientific projects; but these may have been isolated cases.

Medicine and surgery at the new academies

Although the new University of Cervera offered an official degree in Medicine, it did not attract much interest, and the education of physicians and surgeons followed alternative routes of training at universities outside Catalonia such as Huesca, Toulouse or Montpellier. Moreover, in 1760, the *Real Colegio de Cirugía de Barcelona* was created in order to enhance the training and licensing, and hence the professional status, of the surgeons. Although this was primarily intended to train Army surgeons, it included civilian objectives.[28] In the context of the professional disputes between surgeons and physicians, the new educational plans of the *Colegio*, and the prominent role of leading figures such as Pere Virgili and Antoni Gimbernat facilitated the introduction of chemistry and experimental physics. After the creation of the new *Colegio de Cirugía* in Barcelona, a group of medical doctors founded, in 1770, the *Academia Médico-Práctica*. This complemented the official medical curricula, and stimulated debates on therapies, nosologies, and new studies on epidemics and public health.[29] A student of Medicine, after taking an official degree at Cervera under the approval of the *Protomedicato*, typically followed an eclectic cursus including either the *Colegio de Cirugía* or the *Academia Médico-Práctica*.

New experimental sciences were also integrated in the teaching syllabus of the Colleges of Surgery. This was, for example, the case of Antoni Cibat (1770–1812), who held the chair of experimental physics at the Barcelona *Colegio de Cirugía*.[30] In 1804, Cibat wrote a textbook, *Elementos de Física Experimental*,[31] which included Newtonian physics, pneumatics, electricity, optics, magnetism, measuring instruments. The book was inspired by Musschenbroek's *Elementa Physicae* text, and was adapted to the level of the students of the *Colegio*.[32]

Some years earlier, in the period 1797–1798, in the *Academia Médico-Práctica*, Francesc Carbonell, the future director of the School of Chemistry of Barcelona, had given weekly lectures on pneumatic chemistry with its potential applications to Medicine, following Jean-Antoine Chaptal's *Éléments de Chimie*.[33] Members of the *Academia Médico-Práctica*, other physicians, and public in general, attended and funded Carbonell's lectures.[34] Moreover, in 1802, Vicenç Mitjavila (1759–1805)[35] delivered at the *Academia Médico-Práctica* a "Semestre Médico Clínico o primeras lecciones de medicina clínica", placing especial emphasis on the principles of chemistry and pneumatics, which were relevant to the art of healing.[36] Chemistry was closely linked to various 18th century problems of public health, medical topographies and nosologies,[37] and the Barcelona *Academia Médico-Práctica* produced meteorological tables,[38] and dispensed lectures on chemical properties of airs and waters.[39]

SCIENCE "WITHOUT" A UNIVERSITY: SOME PARTICULARITIES OF THE LOCAL SCIENTIFIC CULTURE

Despite the apparent consistency of the Bourbon plans of modernization, as presented in the rhetoric of Royal protection and the standard regulations (*Ordenanzas*), some variations were evident in local contexts, such as Barcelona. The remains of the old University, the strong presence of the military, the institutional eclecticism of the medical academies, or the industrial dynamism of the Catalan economy were all probable factors in the special characteristics which will be described below.

Informal gatherings: tertulia, and the church

The intellectual atmosphere of the late 17th century probably favored the public discourses in defense of a new natural philosophy and of the introduction of the new experimental sciences.[40] In this regard, numerous informal scientific gatherings and meetings—private *tertulias* among apothecaries, book handlers, clock makers—were held in Barcelona.[41] Prominent figures from the Army, the church, and the aristocracy also gathered in private venues. During the years of the War of the Spanish Succession a group of apothecaries, physicians and surgeons met at the shop of the naturalist Jaume Salvador (1649–1740).[42] In 1764, the apothecary Francesc Sala gathered, at his shop, the 16 founders of the *Conferencia Físico-Matemática Experimental*, which became the immediate precursor of the future Academy of Sciences of Barcelona.[43]

From the perspective of the new experimental sciences, the absence of a University in Barcelona and the poor results of Cervera were partly offset by the teaching activity of the religious congregations.[44] Whereas the reforms of the monarchy faced considerable opposition from the Tribunal of the Inquisition and the *Compañía de Jesús*,[45] Bacon and Newton were praised by friar B. J. Feijoo, in the first half of the 18th century, and, in Barcelona, the *Congregaciones Marianas* provided a favorable environment for intellectual and scientific discussions.[46] In the framework of the Christian Enlightenment,[47] for instance, Josep Rubió i Nadal, a priest from a small town in Tarragona, published a short treatise on the variation of declination of the magnetic compass.[48]

The Jesuits also used the University of Cervera to promote themselves as teachers of higher education and to enlarge their syllabuses.[49] After teaching philosophy at Cervera, Tomàs Cerdà, a prominent member of the *Compañía*, introduced experimental physics and mathematics at the *Colegio de Cordelles*, and contributed to the emergence of a new style of scientific education which some years later was to have a great impact on the founders of the Barcelona Academy of Sciences.

The Real Academia de Ciencias Naturales y Artes

In 1764, an organization that at first was known as a *Conference* for experimental physics and mathematics was founded in Barcelona by a group of physicians, apothecaries, priests and lawyers; it shortly became the Royal Academy of Sciences of Barcelona (the *Real Academia de Ciencias Naturales y Artes*, RACAB).[50] Its aims soon went beyond the old natural philosophy of the universities.[51] This academy became a sort of complementary institution for the teaching of modern science and, from the very beginning, it devoted considerable effort toward the introduction of the new experimental physics and the promotion of a new scientific culture in Catalonia.[52]

In 1770, after receiving official Royal patronage, the RACAB was organized into nine different sections (*Direcciones*), which reflected its interest in the experimental sciences: 1. Algebra and geometry, 2. Statics, hydrostatics, and meteorology, 3. Electricity and magnetism, 4. Optics, 5. Pneumatics and acoustics, 6. Natural history, 7. Botany, 8. Chemistry, and 9. Agriculture. In the *Dirección* of chemistry, for example, phlogiston and oxygen theories were rapidly assimilated;[53] in 1768 the Academy created a chair of mathematics, which was a continuation of its equivalent at the *Colegio de Cordelles*.[54] In fact, as a natural replacement of the University, the Academy of Sciences never gave up in its efforts to introduce new chairs of chemistry, physics, botany, agriculture, etc. The Baconian tradition of experimental natural philosophy had already inspired its foundation under the banner of Musschenbroek's experimental physics.[55] From the very beginning, its scientific activities were always accompanied by instruments and machines to perform experiments.[56]

After rejecting Cervera's intellectual atmosphere, Martí i Franquès became one of the most renowned members of the RACAB in the late 18th century. His contributions to pneumatic chemistry were disseminated in many international journals [*Journal de Physique* (1801), *Philosophical magazine* (1801), *Gilbert's Annalen* (1805), *Annales de Chimie* (1807), etc.]. In 1790, he read a Memoir entitled "Sobre la cantidad de aire vital que se halla en el aire atmosférico y sobre varios métodos de conocerla", in which he proposed a new experimental method which used a new reagent ("liver of sulfur") and a new eudiometer to calculate the "exact" quantity of "vital air" (oxygen) in hundreds of samples of atmospheric air taken in many areas of Barcelona. He stated that vital air represented 21% of the whole and "*mofeta*" (nitrogen) 79%, with a constant result in the majority of the samples.[57] It is for good reason that, in his taxonomy of the scientific academies in the 18th century, James McClellan presents the RACAB as a very favorable environment for the rapid reception of Lavoisier's chemistry.[58]

The technical schools of the Junta de Comerç

After a substantial increase in population and agricultural productivity in Catalonia in the 18th century,[59] merchants invested their revenues in textile mills, which provided an attractive milieu for the development of the cotton industry, and calico-printing in particular. This process reinforced the economic basis of the Barcelona *Junta de Comerç*, within the general framework of the Spanish *Junta General de Comercio y Moneda* for the promotion of the arts and manufactures of the Kingdom.[60] From the late 18th century onward, the Barcelona *Junta* set up an important network of technical schools (navigation, design, mechanics, physics, chemistry, agriculture and botany, mathematics, etc.) to educate craftsmen and early industrial workers.[61] Under its political control, tradesmen and early entrepreneurs enjoyed a considerable financial independence. The *Sociedad Económica de Amigos del País*, a standard model of the new utilitarian institutions of the central Bourbon plans of reform, played a minor role in Catalonia. In Barcelona, in particular, this did not exist in the 18th century,[62] and the official centralized organization of the *Junta General de Comercio y Moneda* left some leeway for political and economic autonomy to the *Junta Particular* in Barcelona. Unlike the traditional University, the Barcelona network of schools had explicit utilitarian objectives and most of their syllabuses were based on contemporary foreign scientific and technological developments.[63] Although other local *Juntas* also promoted scientific and technological education,[64] the Barcelona *Junta* created a solid system of education which lasted for decades, and was to become the core of the new *Escuela Industrial*, founded in 1850.[65]

Under the direction of Sinibald de Mas, the Navigation School of Barcelona (*Escola de Nàutica*), founded in 1770, developed an original 2 year teaching program which introduced the "sciences of the art of sailing" to a growing number of students, offering courses in geometry, cosmography, the use of instruments, trigonometry, and astronomy.[66] Although the school provided practical training, Navigation has always called for a considerable training in scientific and mathematical developments, especially for the problem of measuring longitude at sea. Under the supervision of the Navy, this was included in the syllabus of the school, which offered a number of practical solutions, recommending the method of lunar distances.[67]

When, in 1791, as part of the general policy of the Spanish Royal Navy, Francisco Javier Winthuysen, the *Comisionado para la inspección de las Escuelas Náuticas establecidas en el Reyno*, visited the Barcelona School of Navigation,[68] Sinibald de Mas defended its independence of the military regulations, its civil utility along the

Catalan coast, and its particular way of organizing the scientific curriculum. Although some of the academic reforms proposed by Winthuysen were finally accepted, de Mas retained considerable autonomy for the *Escola de Nàutica*.[69]

Late 18th century plans of the Barcelona *Junta* considered the appointment of an outstanding foreign chemist, Joseph-Louis Proust,[70] to the chair of the Barcelona School of Chemistry, but to ensure the continuity of the chair, the *Junta General* suggested sending local scientists to France for training. Francesc Carbonell, who had been educated in Montpellier under the wing of the French chemist Jean-Antoine Chaptal, was appointed director of the new School of Chemistry, which opened its doors in 1805.[71] Here, the relationship between the Barcelona *Junta* and the *Junta General de Comercio* seemed to be more fluid than the relations of the *Escola de Nàutica* with the Navy. The difficulties with the Navy provide a good example of the frequent tensions between the military and the mercantile industrial culture of the city. Perhaps because of their autonomy, the schools of the *Junta de Comerç* were able to continue their activity in Barcelona during the early decades of the 19th century, and, overcoming serious political upheavals, they established an original system of education in which scientific education played a crucial role.

Already in the 19th century, lectures at the School of Chemistry, for instance, were aimed at tradesmen, manufacturers, and members of the bourgeoisie. Wine producers, spinners and weavers, calico-printers, and dyers contributed to the Junta policy, highlighting the need for technical education and the new utilitarian sciences. Carbonell's intended social relationship between the school and the city was formally established through the *Public Exercises of Chemistry*.[72] To an audience from many walks of life, students gave talks on chemical elements, affinity, metallic oxides, distillation apparatus, calico-printing, dyeing, and animal and vegetable matters. Their formal lectures were followed by experimental demonstrations and a discussion with questions raised by the public.[73] These "exercises" alerted the audiences to the potential of chemistry in a range of applications and aroused considerable interest in economic, technological, and intellectual circles. The lectures gave impetus to the pursuit of social recognition of chemistry as a new profession. Knowledge of chemistry reached an even wider public via the translation and publication of textbooks and journals and the appearance of a new technical periodical, *Memorias de Agricultura y Artes*,[74] the work of the School of Agriculture and Botany, Mechanics, and Chemistry between 1815 and 1821.

The links of the new chemistry with society were also strengthened by the students who attended lectures and conducted laboratory experiments at the School of Chemistry from 1805 to 1822. There were almost 400 pupils, divided by profession as follows: Surgeons (27%), pharmacists (20%), craftsmen (18%), tradesmen (13%), physicians (9%), others (13%).[75]

It seems that medical and industrial interests were the driving force behind the social recognition of Chemistry in Barcelona in the early decades of the 19th century. For technical advice "*in situ* ", and in order to resolve large scale problems, Carbonell often visited his students' workshops.[76] Foreign technical innovations and new theories of chemistry were imported and introduced into industries and manufacturing processes.[77]

NEW PROGRAMS FOR SCIENTIFIC EDUCATION

In this context of institutional eclecticism, projects for "alternative universities" were common. For example, in 1767, after the expulsion of the Jesuits, Francesc Subiràs took charge by Royal order of the *Colegio de Cordelles*.[78] Subiràs had been one of the founders of the Academy of Sciences in 1764, but he had been obliged to leave Barcelona for Madrid to lecture at the *Real Colegio de Nobles*. Although he sought to maintain Cerdà's former chair of mathematics, he had more ambitious plans. In 1772, perhaps anticipating some of the *Junta*'s projects, he planned to create 16 new chairs in the *Colegio de Cordelles* which would embrace subjects such as architecture, physics, astronomy, geography, navigation, natural history, chemistry, metallurgy, agriculture, drawing, design, and arts.[79] Subiràs considered that the *Colegio* should act as a driving force for social and educational reforms. Although it was a center for the aristocracy, he would open its doors to craftsmen and tradesmen, and, in fact, to young people, in general. Moreover, in his plan, the *Colegio* would not be under Church control. As regards scientific training, Subiràs proposals were conceived as a response to the need for general scientific and utilitarian education.

Subiràs stated that:

> "As the precepts of our holy Religion are invariable and the rules of good breeding are always the same, the education of a Christian and a citizen must be uniform in all the dominions and for all the subjects of our Catholic King. Nevertheless, as regards common and particular utility, this education must be imparted adapted to the circumstances of each Country and the condition of its subjects; so, this part of the education should be suited in Catalonia to the nature and quality of its people. The situation of the Principality [80] and its limited extension oblige the Catalans to be occupied in farming, in manufacturing, in the commerce of goods and fruits; a large number [of Catalan people are occupied] in the sciences and many more in the Army, serving His Majesty on land and sea. In consequence, the education of the Catalan nation should be oriented to the Sciences, Arts, Agriculture, Trade and also to the Army".[81]

Subiràs considered that the conditions of Catalonia led to a new economy based on industry and, therefore, he regarded scientific education as a basic need for all people, including the nobility. His reform of the *Colegio de Cordelles* never materialized but the subsequent emergence of the schools of the Junta, from 1770 onward, can be seen as a realization of Subiràs's original conceptions.

In the early years of the 19th century, in the midst of the crisis of the Spanish Ancien Regime, modern science came to be regarded as a powerful tool in the projects of a modern state, where new values of individual freedom and public education needed to be established. In 1813, for example, Francesc Carbonell, the director of the Barcelona School of Chemistry of the *Junta*, published a national plan for the inclusion of natural sciences in the general educational projects of the new liberal political manifesto declared at the *Cortes de Cádiz*, in 1812.[82] Science would be taught at primary level in schools; this would then be linked to all medical activities, and, subsequently, connected to agriculture and the arts.[83] This triple aim was again to be achieved without the intervention of the University, through a centralized coordination of scientific and technical exchanges of teachers and students in industries, academies and colleges.[84] Following the RACAB model, Carbonell planned to establish an Academy of Sciences in every Spanish province, and proposed a detailed program for the promotion of natural sciences which included: scientific travel within Spain for teachers; correspondent fellows; the diffusion of scientific journals; prizes to encourage innovation; public examinations and demonstrations; closer links with industries, etc.[85] Carbonell may have been seeking to apply the Barcelona model of the *Junta de Comerç* schools on a national level.

CONCLUSION

Although Soldevila's historical study of the period 1714–1837, during which Barcelona was without a University, could be substantially improved upon, many of his arguments are still valid today. The academic humiliation of the Catalans was not as evident as he suggested, however, and at least in Barcelona a number of alternative institutions filled the gap. Mathematics, experimental physics, or modern chemistry were taught for military, industrial, or agricultural purposes; the result was a renewal of natural philosophy within a complex network of institutions, both public and private initiatives, with different styles of organization and values. Divided between its medieval objectives of elite education and the new demands of urban capitalism, the 18th century University system throughout Europe found itself in a serious general crisis, from which Spain did not escape. In Catalonia, this crisis may in part account for the closure of the old universities and the creation of Cervera—an experiment whose failure has left a negative image of an anti-scientific, old fashioned institution.[86]

Nevertheless, alternative centers, dispensing modern scientific and technical education, were created by the central state and also by local institutions. The Army, representing the Bourbon state, was one of the promoters. After the War of the Spanish Succession, when Catalonia was occupied by the Bourbon Army, Catalan tradesmen and new industrialists were not slow to exploit the new situation. The Army became a very important customer: the troops needed food and clothing, and provided an impetus for local economic development.[87] If the *Colegio de Cirugía* is regarded as a milestone in the new professionalization of medicine and surgery, future historical research ought to evaluate the role of the Army academies as a whole in the Catalan society of the 18th century.

Modern science was also promoted through private *tertulias* which provided a favorable atmosphere for the subsequent emergence of new institutions, such as the RACAB. Although a number of authors have suggested that the upper classes did not take much interest in scientific education in 18th century Spain,[88] the new scientific and technological centers, in Barcelona, were oriented toward new middle, urban classes (skilled artisans, tradesmen, lower nobles) with greater interest in training. In 1851, the final year of the schools of the *Junta*, more than 2000 students had enrolled.

This high number of students demonstrates that scientific and technical training, including training in the arts, had deep roots in Catalan society. In that year, the schools of the *Junta* with a specifically scientific orientation were incorporated into a new industrial school, which was soon to become a school of industrial engineering.[89]

Some years ago, Jack Morrell advocated a rigorous treatment of national histories of science, integrating the development of scientific disciplines and their organizational aspects, making use of models and discussing in each case the significance of localism.[90] Morrell based his paper on the development of science in Scotland in the 18th and 19th centuries in the context of British science, in which the role of Scottish contributions is clearly recognizable. Our paper has no such theoretical aims; it attempts instead to supply some models for the development of science and technology in 18th century Catalonia.

We must regard the existence of a Catalan system of scientific education in the 18th and 19th centuries as a product of various circumstances and strategies developed by diverse actors. Despite the loss of its political autonomy, Catalonia followed a distinct economic development based on modern agriculture and industry. The leading social classes were made up of tradesmen, farmers, and industrialists, including the lesser nobility, the growing bourgeoisie, and guilds of craftsmen. These social sectors had difficulty in gaining access to Spanish political power, which was mainly in the hands of the Castilian nobility; nevertheless, they managed to make the most of this situation.

The *Junta de Comerç* was the main instrument used by these social sectors for the promotion of economic development. Its schools offered a wide range of subjects for technical and scientific training. Although the schools of the *Junta* were related to the general scientific policy of the Spanish state, they developed apart from other similar centers created at that time. The schools of the *Junta* benefited from the existence of other private scientific institutions, such as the *Real Academia de Ciencias Naturales y Artes* of Barcelona, whose members were in charge of some scientific schools; these private initiatives complemented State policy. We have pointed out that the state installed in Catalonia some academies for scientific training, in keeping with the scientific policy of the Bourbons, developed mainly through the Army. At the same time, the State abandoned the University of Cervera, reducing the resources for its consolidation.

Catalonia was not rich in natural resources; its new modern economy depended on the transformation of agriculture, the development of commerce and the establishment of textile industries, and knowledge was seen as providing the remedy for the lack of natural resources. Examples of this concept have been mentioned above. The city government of Barcelona, and some individual social reformers such as Francesc Romà i Rossell and Francesc Subiràs, regarded the new sciences as a distinctive activity of the Catalan nation and frequently called for initiatives to promote science. Although the development of these ideas led to dissemination of practical scientific knowledge in Catalan society, it would prove very difficult to consolidate significant research institutions. In fact, academic science had considerable difficulty in growing during the 19th and 20th centuries. The historical explanation of this is still open to discussion.[91]

It may be concluded that the different routes of a regional, utilitarian science responded to the new demands of a society which, after losing most of its political power and even without a University, came to be one of the driving forces of the process of Spanish industrialization.

NOTES

[1] F. Soldevila, *Barcelona sense Universitat i la restauració de la Universitat de Barcelona (1714–1837)* (Barcelona, 1938).
[2] For a historiographic review of the University of Cervera, see J. Prats, *La Universitat de Cervera i el Reformisme Borbònic* (Lleida,1993).
[3] For a general perspective, see: N. Hammerstein, "Epilogue: The Enlightenment", in H. De Ridder-Symoens (ed.), *Universities in Early Modern Europe (1500–1800)*, Vol. II (Cambridge, 1996), pp. 621–640.
[4] James J. K. Thomson, *La Indústria d'indianes a la Barcelona del segle XVIII*, (Barcelona, 1990); Ángel Calvo Calvo, "Transferencia internacional de tecnología y condicionamientos nacionales: la industria sedera catalana durante la transición al régimen liberal", *Quaderns d'Història de l'Enginyeria*, 3:131–168 (1999).
[5] Ajuntament de Barcelona, *Informes y representaciones* (1776), cited by Ernest Lluch, *El Pensament econòmic a Catalunya, 1760–1840: els orígens ideològics del proteccionisme i la presa de consciència de la burgesia catalana*, (Barcelona, 1973), pp. 125–126. The dates of these documents are 1749, 1767, 1776, and 1777, as they have been collected by Montserrat Ventura-Munné, "El Col·legi de Cordelles sense els jesuïtes, un projecte fracassat", in C. Martínez-Shaw (ed.), *Historia Moderna, historia en construcción,* Vol. I (Lleida, 1999), p. 539 (footnote #27).
[6] "Parece no será violenta digresión la que en consecuencia se dirija a tratar formalmente de un restablecimiento de estudios generales o de la planificación de una Universidad Literaria tan necesaria en esta ciudad, como utilísima al resto de la Provincia con otras ventajas a favor de la Industria Popular, que no las que consigue ni permite su actual mansión en la de Cervera", mentioned by Lluch (1973), 199.
[7] "Studium Generale": a medieval university.
[8] "Literaria" had a general meaning, i.e., belonging to knowledge or science.
[9] Francesc Romà i Rossell, *Las señales de la felicidad de España y medios de hacerlas eficaces*, preliminar study by E. Lluch (Barcelona, 1989), pp. 96–102. Facsimile edition of the 1768 printed edition.
[10] Junta de Comerç de Barcelona, *Discurso sobre la agricultura, comercio e industria del Principado de Cataluña,* Edition of the 1780 manuscript by E. Lluch (Barcelona, 1997), p. 81.
[11] M. Sellés, J. L. Peset, and A. Lafuente (eds.) *Carlos III y la Ciencia de la Ilustración*, (Madrid, 1988); G. Anes et al., *Actas del Congreso Internacional sobre "Carlos III y la Ilustración"* (Madrid, 1989).
[12] See Sellés et al. (1988). For the case of chemistry, Ramón Gago, "The new chemistry in Spain", *Osiris*, 2nd series, 4 (1988), pp. 162–192.
[13] A. Rumeu de Armas, *Ciencia y Tecnología en la España ilustrada. La Escuela de Caminos y Canales* (Madrid, 1980), pp. 110–112.
[14] M. Peset, J. L. Peset, *La Universidad española (siglos XVIII y XIX)* (Madrid, 1974); S. Albiñana, *Universidad e ilustración. Valencia en la época de Carlos III* (València, 1988).
[15] Cervera is a Catalan city near Lleida, 100 km from Barcelona.
[16] J. Agustí Cullell, *Ciència i tècnica a Catalunya en el segle XVIII o la introducció de la màquina de vapor* (Barcelona, 1983); I. Casanovas, *La cultura catalana del segle XVIII* (Barcelona, 1932).
[17] Prats (1993).

[18] For the biographical details of Antoni Martí i Franquès (1750–1832), see A. Quintana, "Estudi biogràfic i documental", *Memòries de l'Acadèmia de Ciències Naturals i Arts de Barcelona,* Vol. 24 (1935), pp. 63–304.

[19] Soldevila (1938), pp. 57–59.

[20] H. Capel, *Geografía y matemáticas en la España del siglo XVIII* (Vilassar de Mar, 1982); H. Capel, J. E. Sánchez, and O. Moncada, *De Palas a Minerva. La formación científica y la estructura institucional de los ingenieros militares en el siglo XVIII* (Barcelona, 1988).

[21] A. Lafuente, J. L. Peset, "Las Academias militares y la inversión en ciencia en la España ilustrada (1750–1760)", *Dynamis* 2:193–209 (1982); A. Lafuente and J. L. Peset, "Militarización de las actividades científicas en la España Ilustrada (1726–1754)", in J. L. Peset (ed.), *La ciencia moderna y el Nuevo Mundo* (Madrid, 1985), 127–147; Capel (1982); Capel et al. (1988).

[22] Lafuente, Peset (1985), p. 130.

[23] For the Spanish *novatores*, see, for example, À. Martínez Vidal and J. Pardo Tomás, "*In tenebris adhuc versantes*. La respuesta de los novatores españoles a la invectiva de Pierre Régis", *Dynamis* 15:301–340 (1995).

[24] For the problem of "utility" in the Enlightenment, see: R. Olson, *Science Deified and Science Defied. The Historical Significance of Science in Western Culture.* (Vol. 2: *From the Early Modern Age through the Early Romantic Era ca. 1640 to 1820*). (Berkeley, 1990); J. A. Maravall, "El principio de la utilidad como límite de la investigación científica en el pensamiento ilustrado", in J. A. Maravall *Estudios de la Historia del pensamiento español del siglo XVIII* (Madrid, 1991), pp. 476–488.

[25] Capel et al. (1988); H. Capel, "Las Academias de Ingenieros", in Sellés et al. (ed.) (1988), pp. 187–204.

[26] J. Riera, "L'Acadèmia de Matemàtiques a la Barcelona Il·lustrada (1715–1800)", in *II Congrés Internacional d'Història de la Medicina Catalana*, Vol. II (Barcelona, 1977), pp. 73–128.

[27] Capel (1988), pp. 190–193.

[28] On the Spanish Colleges of Surgery, see M. Astrain, *Barberos, cirujanos y gente de mar. La sanidad naval y la profesión quirúrgica en la España ilustrada* (Madrid, 1996).

[29] For the Academia Médico-Práctica, see: S. Montserrat and M. Carreras, *Historia de le Real Academia de Medicina de Barcelona* (Barcelona, 1954); Francisco Guillén Grima, Fernando de San Eustaquio Tudanca, "La salud pública y la administración municipal. El dictamen de la Academia Médico Práctica de Barcelona", in *Actas del VIII Congreso Nacional de Historia de la Medicina*, Vol. 3 (Murcia, 1986) pp. 1.239–1.254; N. Gorina, "La Academia médico-práctica en la epidemiología barcelonesa del setecientos (1770–1800)", *Medicina e Historia* 22:5–28 (1988). See also the work of A. Zarzoso, *Societat i Salut Pública a la Catalunya del segle XVIII* (Lleida, 2004); A. Zarzoso, "Mediating Medicine through Private Letters: the Eighteenth-Century Catalan Medical World", in W. de Blécourt and C. Usborne (eds.), *Cultural Approaches to the History of Medicine, Mediating Medicine in Early Modern and Modern Europe* (Basingstoke, 2003), 108–126.

[30] A. Cibat, "Memoria sobre los efectos que causa el oxígeno del aire atmosférico en la vida y constitución del hombre" (1805); A. Cibat "Memorias físicas sobre el influjo del gas hidrógeno en la constitución del hombre" (1805), mentioned by M. Usandizaga, *Historia del Real Colegio de Cirugía de Barcelona* (Barcelona, 1964), pp. 215–236.

[31] A. Cibat, *Elementos de Física Experimental* (Barcelona, 1814). The edition was prepared for the teaching activities of the Escuela de Física de la Junta de Comerç, which was founded in 1814.

[32] For A. Cibat's work, see: Agustí (1983), pp. 50–55.

[33] J.-A. Chaptal, *Élémens de chimie* (Montpeller, 1790).

[34] *Diario de Barcelona*, 7-V-1821, 127, 5.

[35] J. Danon, "Notas Biográficas IV: Vicente Mitjavila", *Medicina e Historia* 47:3–4 (1975).

[36] "Las nociones sobre los rudimentos o principios químico-neumáticos que debe tener el médico práctico explicados con extensión y claridad", *Diario de Barcelona*, 5-VII-1803, 86, 853–854.
[37] G. Risse, "Medicine in the age of Enlightenment", in A. Wear (ed.), *Medicine in Society. Historical Essays* (Cambridge, 1992), pp. 149–195.
[38] *Estatutos de la Real Academia Médico-Práctica de Barcelona* (Barcelona, 1786). (new edition, 1797), pp. 9–10. Follets Bonsoms 1110. Biblioteca de Catalunya, Barcelona.
[39] *Diario de Barcelona*, 7-V-1821, 127, 5.
[40] V. Navarro, *Tradició i canvi científic al País Valencià modern (1660–1720): Les Ciències Físico-Matemàtiques* (Valencia, 1985), pp. 49–50.
[41] J. Iglésies, "La Real Academia de Ciencias Naturales y Artes en el siglo XVIII", *Memorias de la Real Academia de Ciencias y Artes de Barcelona*, 34:38 (1964).
[42] J. M. Camarasa, *Botànica i botànics desl Països Catalans* (Barcelona, 1989), pp. 86–91.
[43] A. Nieto-Galan and A. Roca-Rosell (eds.) *La Reial Acadèmia de Ciències i Arts de Barcelona als segles XVIII i XIX. Història, Ciència i Societat* (Barcelona, 2000).
[44] Agustí (1983); D. Goodman, "Science and the Clergy in the Spanish Enlightenment", *History of Science*, 21:111–135 (1983); M. G. Tomsich, *El jansenismo en España* (Madrid, 1972).
[45] R. Herr, *España y la Revolución del siglo XVIII* (Madrid, 1988), pp. 9–30: Chapter II: "Regalismo y Jansenismo en España", (1st English version, Princeton 1960).
[46] L. Gassiot, "Tomás Cerdà i els inicis de l'Acadèmia de Ciències de Barcelona", in A. Nieto-Galan and A. Roca-Rosell (eds.), *La Reial Acadèmia de Ciències i Arts de Barcelona als segles XVIII i XIX* (Barcelona, 2000), pp. 125–133.
[47] On the Catalan Christian Enlightenment, see Joaquim M. Puigvert (ed.) *Bisbes, Il·lustració i jansenisme a la Catalunya del segle XVIII* (Vic, 2000).
[48] A. Roca-Rosell, R. Puig, and J. M. Grau (eds.) *L'obra de Josep Rubió i Nadal, científic il·lustrat, rector de Vilanova de Prades (1792–1807) "Hipótesis ... con la que se descubre la causa de la declinación y variación de la Aguja de Marear. 1807*, (Vilanova de Prades, 1999).
[49] Prats (1993).
[50] Iglésies (1964), 190; Nieto-Galan; Roca-Rosell (2000).
[51] R. Gago, "El plan de estudios del rector Blasco (1786) y la renovación de las disciplinas científicas en la Universidad de Valencia: La química y la Enseñanza clínica", *Estudis* 6:157–167 (1977).
[52] F. Barca, "La Reial Acadèmia de Ciències i Arts de Barcelona com a cos docent", in A. Nieto-Galan and A. Roca-Rosell (eds.), *La Reial Acadèmia de Ciències i Arts de Barcelona als segles XVIII i XIX* (Barcelona, 2000), pp. 165–195.
[53] Iglésies (1964), 97, 244.
[54] Josep Balari i Jovany, *Historia de la Real Academia de Ciencias y Artes* (Barcelona, 1895), 163–164; Francesc Barca Salom, "La Càtedra de Matemàtiques de la Reial Acadèmia de Ciències i Arts de Barcelona. Més de cent anys de docència de les matemàtiques", in Víctor Navarro et al. (eds.), *Actes de les II Trobades d'Història de la Ciència i de la Tècnica* (Barcelona, 1993), pp. 85–91.
[55] Iglésies (1964), 82.
[56] C. Puig-Pla, "Els primers socis-artistes de la Reial Aacadèmia de Ciències i Arts de Barcelona (1764–1824)", in A. Nieto-Galan and A. Roca-Rosell (eds.), *La Reial Acadèmia de Ciències i Arts de Barcelona als segles XVIII i XIX* (Barcelona, 2000), pp. 287–310.
[57] Antoni Martí i Franquès, "Sobre la cantidad de aire vital que se halla en el aire atmosférico y sobre varios métodos de conocerla", printed in 1935, *Memòries de l'Acadèmia de Ciències Naturals i Arts de Barcelona* 24:21–36 (1790).
[58] "Spain did possess a significant provincial scientific society in the Real Academia de Ciencias Naturales y Artes of Barcelona.... Like its French neighbors the Barcelona Academy was in these regards a provincial institution ... it followed closely developments in Lavoisier Chemistry. The horizon of its interests was more limited to Catalan, and after 1790, it took up

questions of industrial growth and development", James E. McClellan, *Science Reorganized. Scientific Societies in the Eighteenth Century* (New York, 1985), p. 139.
[59] P. Vilar, *Catalunya dins l'Espanya moderna* (Barcelona, 1987), 4 Vols. (1st French edition, 1962); James J. K. Thomson, *Els orígens de la industrialització a Catalunya. El cotó a Barcelona 1728–1832* (Barcelona, 1994). (Original English version, Cambridge University Press, 1992).
[60] Á. Ruiz Pablo, *Historia de la Real Junta particular de Comercio de Barcelona 1760–1847* (Barcelona, 1919); J. Iglésies, *L'obra cultural de la Junta de Comerç (1760–1847)* (Barcelona, 1969); J. Monés, *L'obra educativa de la Junta de Comerç 1769–1851* (Barcelona, 1987).
[61] It was thought that the union of science and new industry was necessary: "Las artes se hacen cada día más científicas en Europa, como los que las profesan, y esto impone en los nuestros la necesidad útil de instruirse y agregar a su laboriosidad todo el aprovechamiento posible de las muchas enseñanzas gratuitas que en Cataluña dispensa su Junta de Comercio…", Antoni Bonaventura Gassó, *España con industria fuerte y rica* (Barcelona, 1816).
[62] E. Lluch, "El caso de la no fundación de la Sociedad Económica de Amigos del País de Barcelona", *Revista de Occidente* 115:51–70 (1971).
[63] These were, for example, the cases of the chairs of Chemistry and Mechanics. For the School of Chemistry, see A. Nieto-Galan, "Ciència a Catalunya a l'inici del segle XIX. Teoria i aplicacions tècniques a l'Escola de Química de Barcelona sota la direcció de Francesc Carbonell i Bravo (1805–1822)", Ph. D. Thesis (Barcelona, 1994); for the Chair of Mechanics, see Agustí (1983).
[64] For an overview of the technical education in 18th century Spain, see A. Escolano Benito, *Educación y economía en la España ilustrada* (Madrid, 1988).
[65] G. Lusa, "La creación de la Escuela Industrial Barcelonesa (1851)", *Quaderns d'Història de l'Enginyeria* 1:1–51 (1996).
[66] R. Fernández and H. Sierco, "Ensenyament professional i desenvolupament econòmic. L'Escola Nàutica de Barcelona", *Recerques* 15:7–31 (1984). Some other schools of navigation were founded in Spain during the 18th century. See M. Sellés and A. Lafuente, "La formación de pilotos en la España del siglo XVIII", in J. L. Peset et al. (eds.), *La ciencia moderna y el nuevo mundo* (Madrid, 1985), pp. 149–191.
[67] See for example, Francesc Barca Salom, "La longitud, una coordenada conflictiva", in Juan José Achútegui et al. (eds.), *I Simposio de Historia de las Técnicas* (Santander, 1996), pp. 266–277.
[68] "Instrucción para la disciplina, estudios y exámenes sobre que deben arreglarse las escuelas particulares de Náutica…", Archive of the Junta de Comerç (Biblioteca de Catalunya, Barcelona), file 27, box 38.
[69] The Barcelona *Junta de Comerç* argued that its School of Navigation was not a permanent center. It was only a "provisional" structure to train pilots. Arxiu de la Junta de Comerç, file 27, box 38. On Winthuysen intervention, see, also, Barca (1996).
[70] Hired by the Spanish Bourbon monarchy to improve various applications of chemical knowledge, Joseph-Louis Proust was one of the most outstanding figures. See Gago (1988).
[71] Ramón Gago, "Cultivo y enseñanza de la química en la España de principios del siglo XIX", in J. M. Sánchez Ron (ed.), *Ciencia y sociedad en España* (Madrid, 1988), pp. 129–142.
[72] See, for example, F. Carbonell, *Ejercicios públicos de Química que sostendrán en la casa Lonja los alumnos de la Escuela gratuita de esta Ciencia establecida en la ciudad de Barcelona por la Real Junta de Comercio del Principado de Cataluña* (Barcelona, 1818).
[73] *Diario de Barcelona*, 249 (1818), 1.971–1.974.
[74] The journal *Memorias de Agricultura y Artes que se publican de orden de la Real Junta de Gobierno del Comercio de Cataluña* appeared from 1815 to 1821.
[75] Archive of the Junta de Comerç, box 254. (Biblioteca de Catalunya, Barcelona). Cited in A. Nieto-Galan, "Searching an identity for chemistry in Spain: Medicine, Industry, University,

the Liberal State and the new 'Professionals' ", in David Knight and Helge Kragh (eds.), *The Making of the Chemist* (Cambridge, 1998), pp. 177–190.
[76] Archivo General de Simancas. Consejo Supremo de Hacienda. Junta de Comercio y Moneda, file 272.
[77] *Diario de Barcelona*, 309 (1816), 1.545.
[78] Iglésies (1964), pp. 129–130.
[79] F. Subiràs, "Plan general de la educación que se puede dar en el Imperial y Real Seminario de Nobles de Cordelles de la Ciudad de Barcelona", mss, Audiència, Reg. 567, ff. 102r–150v (original version is in file 72, dated 11th March 1772) Arxiu de la Corona d'Aragó (Barcelona, 1772); for a transcription, see: J. G. Panadés, "La Pedagogía Catalana del Antiguo Régimen". Ph.D. Thesis (Barcelona, 1977), 441–486. There is a partial transcription in Manuel García Doncel "Els quatre enfocaments inicials de l'Acadèmia", in Agustí Nieto-Galan and Antoni Roca-Rosell (eds.), *La Reial Acadèmia de Ciències i Arts de Barcelona als segles XVIII i XIX* (Barcelona, 2000), pp. 81–124.
[80] Catalonia was a Principality.
[81] "Como los preceptos de nuestra santa Religión son invariables, y las reglas de la buena crianza siempre las mismas, la educación en lo Cristiano y político debe ser uniforme en todos los Dominios y a todos los vasallos de nuestro Católico Monarca. Pero, en lo perteneciente a la utilidad común y particular, debe proporcionarse a las circunstancias del País donde se da y a la condición de los sujetos que la reciben; por consiguiente, esta parte de educación en Cataluña debe contraerse a la naturaleza y calidad de sus naturales.// La situación del Principado y su corta extensión obligan a los Catalanes, a que unos se ocupen al cultivo de las tierras, otros a la fábrica de Manufacturas, otros al tráfico de géneros y frutos; una grande parte a las ciencias, y mucho mayor a las Armas, sirviendo a Su Magestad en los ejércitos de mar y tierra.// Por tanto, a las ciencias, a las Artes, a la Agricultura, al Comercio y aun a la Milicia debe dirigirse la educación de la nación Catalana.", Subiràs (1772), ff. 1v–2r.
[82] Francesc Carbonell, *Ensayo de un plan general de enseñanza de las Ciencias naturales en España por el Dr Don Francisco Carbonell Bravo* (Palma de Mallorca, 1813).
[83] Carbonell (1813), p. 7.
[84] Carbonell (1813), pp. 18–19.
[85] Carbonell (1813), p. 16.
[86] Some decades later, the government tried again to reform the universities. See M. Peset and J. L. Peset, *Gregorio Mayans y la reforma universitaria* (Valencia, 1975); Albiñana (1988).
[87] Junta de Comerç de Barcelona, *Discurso sobre la agricultura, comercio e industria del principado de Cataluña* (1780). (Barcelona, 1997). See also E. Lluch, *La Catalunya vençuda del segle XVIII. Foscors i clarors de la Il.lustració* (Barcelona, 1996).
[88] A. Lafuente, "La enseñanza de las ciencias durante la primera mitad del siglo XVIII", *Estudios dedicados a Juan Peset Aleixandre* (Valencia, 1982) II, 447–493.
[89] G. Lusa and A. Roca-Rosell, "Doscientos años de técnica en Barcelona. La técnica científica académica", *Quaderns d'Història de l'Enginyeria* 3:101–130 (1999).
[90] J. Morrell, "Reflections on the History of Scottish Science", *History of Science* 12:81–94 (1974), reprinted in J. Morrell, *Science, Culture and Politics in Britain, 1750–1870* (Aldershot, 1997).
[91] It would be difficult to summarize this discussion here. There is an approach to the question from a biographical perspective in the two volumes work: J. M. Camarasa and A. Roca-Rosell (eds.), *Ciència i tècnica als Països Catalans. Una aproximació biogràfica als darrers 150 anys* (Barcelona, 1995).

LUIS CARLOS ARBOLEDA APARICIO AND DIANA SOTO ARANGO

THE THEORIES OF COPERNICUS AND NEWTON IN THE VICEROYSHIP OF NUEVA GRANADA AND THE AUDIENCIA DE CARACAS DURING THE 18TH CENTURY

The enlightened current of thought that permeated the Viceroyship of Nueva Granada, in the second half of the XVIII century, coinciding with the reign of Carlos XVIII, gave rise to a new "useful" philosophy where the teachings of Newtonian natural philosophy were implicit and, as a result, theories and conceptions opposed to scholastic philosophy, in which reality was explained through observation and experiments. Traditional philosophy was ousted by the teaching of modern physics that broke with tradition, thus creating a different philosophical dimension in the new generation.

Before its introduction into the curriculum of the universities of the Viceroyship of Nueva Granada, heliocentric ideas circulated, for example, in activities related to scientific missions, geodesic expeditions, establishment of boundaries and frontiers, and building of fortifications.[1] Among the most renowned expeditions during the Bourbon reign was that geodesic one of La Condamine to the Viceroyship of Peru, that included such Spanish scientists as Antonio Ulloa and Jorge Juan. This expedition arrived at the city of Quito around 1736, creating a propitious environment for the discussion on modern science in this city.

In the 1740s, the Universidad Gregoriana of the city of Quito became an important center for the development of a scientific spirit. In Santafé, there was a similar situation, as the Universidad Javeriana was the first of the educational institutions to open its doors to the new scientific and philosophic ideas. In the University of Caracas, the introduction of the theories of Copernicus and Newton took place only later, toward 1778, with the scholar Marrero. In Santafé de Bogotá, from 1773 onward, there was an on-going debate between José Celestino Mutis and the Dominican congregation about the Copernican theories, which continued for almost half a century. This controversy was linked to political and social interests, in particular to a conflict between two groups each seeking to control education.

The attempts to introduce and develop enlightened philosophy in the universities of Quito and Caracas developed differently in Santafé. Scholars like Fathers Juan de Hospital, Miguel Antonio Rodriguez, José Mejía Lequerica in Quito, Marrero in the Caracas University and Mutis, Valenzuela, Vallecilla, Vásquez, Padilla, among other professors of the Santaferenian University, taught the theories of Newton and Copernicus in their lectures on philosophy or mathematics. These professors were successful in awakening scientific interest in the fossilized university life of Santafé,

M. Feingold and V. Navarro-Brotóns (eds.), Universities and Science in the Early Modern Period, 289–311.

Quito, and Caracas; their teachings exemplify the institutionalization of enlightenment ideas and scientific knowledge in university classrooms—a process that unfortunately were not completed in the colonial era.

In this chapter, we will analyze some developments in the progress and the defense of enlightenment thought in Santafé, Quito, and Caracas. We will take into account the diversity of circumstances in those cities in the period of the introduction, development and crystallization of enlightened philosophy, with its encyclopedic component. Therefore, the analysis will focus on an extensive period during which the systems of Copernicus and Newton were debated in institutions of higher learning in Quito, Santafé, and Caracas, with significant long-term effects on the cultural environment of the viceroyship.

THE BOURBON REFORMS IN RELATION TO THE GEO-POLITICAL DIVERSITY OF THE REGION

During the Bourbon era, the Viceroyship of Nueva Granada initiated a series of reforms within a policy of enlightened despotism. In the administrative-political sphere, these reforms were instituted in 1739. After several changes of names and administrations, in 1786 the Real Audiencia of Caracas was created.[2] It should be taken into account that the Captain General of Venezuela exercised authority in his territory equivalent to that of the Viceroy in Santafé. Therefore, each territory was completely independent from the other, and dealt directly with the respective ministers of Spain.[3]

The main sectors of the colonial economy were mining, land, and commerce. The public income of the Viceroyship of Nueva Granada was three million pesos, a sum that was just enough for its own maintenance.[4] In Venezuela, "out of the two million two hundred and eighty one thousand seven hundred and three pesos of income", there was something left for the metropolis.[5] Political power was the exclusive monopoly of nobles, although toward the end of the century, consolidation of the Creole and Mestizo groups was established. The first of these groups in particular, "began to be conscious of historical initiative, culturally influenced by the enlightened movement and the scientific climate of inquiry generated by the Botanical Expedition".[6] In the middle of the 18th century, higher education in the Viceroyship of Nueva Granada was dominated by the clergy. This was due to the inability of the Viceregal state and the local civil elite to establish, in an enduring manner, the secular forms of teaching that launched the new philosophy. Besides, the medieval tradition of education, characteristic of the alliance between church and state for the domination of the colonies, still persisted.

It is true that after the expulsion of the Jesuits, the Creole elite considered that their time had come to direct and control education. However, they were defeated by the power of the religious orders, who, paradoxically, were supported in this by the Crown. In Quito, when the public university was opened in 1778, a lay chancellor, Doctor Nicolás Carrión Vaca, as well as three secular Creoles, were appointed to the general assembly to carry out unpaid teaching duties in the university. But in Madrid it was soon decided to remove the Chancellor and the three professors, and to support the bishop in the appointment of new staff.[7]

A more significant demonstration of the power of some religious communities is seen in the case of public prosecutor Moreno Escandón, who in spite of his determination failed to institute a public university and a new curriculum in Santafé. The reforming winds of colonial political regimes, supported by the Creole elite, blew in directions different from those in Nueva Granada. At the beginning of the 19th century, the alliances between state, church, and the military were consolidated, while in New Spain for example, viceregal power had already agreed on socio-cultural projects together with local economic groups.[8]

In the Viceroyship of Nueva Granada, the alliance between the state and the Creole elite outside the university was only evident in the exceptional case of José Celestino Mutis, who received state support for his botanical expedition from 1783. In contrast, in the Viceroyalties of Peru and New Spain, such an alliance outside the universities was agreed on—for example, the foundation of the School of Mines, the Botanical Garden and the School of Surgery and Pharmacy.[9]

An obvious but usually forgotten fact is clear from the above situations: the cultural dynamics of the different viceroyalties cannot be analyzed in isolation from the economic priorities of each. On the other hand, the desire of each religious order to dominate education, led them into disputes over the right to confer academic degrees.[10] In the Viceroyship of Nueva Granada, before the expulsion of the Jesuits, the debate was above all between the Jesuits and the Dominicans. After 1767, the latter claimed the right to determine educational policies and therefore sought to get hold of the establishments and prerogatives of the Jesuits.[11] However, around this time, the Crown exercised the right of "Royal Patronage" and established the *Regium Executor* . This was a policy that gave the control of higher public education to the State, under the King's dominion rather than the Pope's. This was distinct from mass education, which was both free and compulsory; nor did it affect poor sectors of the population,[12] in the sense that this term has acquired today. On the contrary, during the years in question, higher education was characterized by its exclusivity; it was reserved for members of the Spanish and Creole elite, who were destined to occupy positions in local administration. In effect, this policy was continued with the rigorous "information" for admission to the University, and the cost of doctoral degrees was increased.

Thus, the Dominicans' hope of controlling former Jesuits property conflicted with the Crown's decision to retain it. For this reason, ten *Juntas de Temporalidades* were set up throughout the kingdom of Spain, including the colonies. One of these corresponded to the Nuevo Reino de Granada, controlled by the Viceroy of Santafé himself, and another was set up in the city of Caracas. From the point of view of the religious communities, the right to confer academic degrees guaranteed not only control over education, but, more importantly, political control. A review of relevant historical cases shows that the Creole elite who graduated from such institutions had a better prospect, by virtue of the education, of being appointed to the few positions available in local administration.

Although it is true that the Crown tried to reform and control education, in practice the attempts of the civil sector to take control of higher studies in Santafé and Quito were thwarted by the political-economic power of the ecclesiastics, in particular by the Dominican order. After extensive debates in Santafé and others which were less prolonged, such as that of the Bishop of Quito, the Dominicans obtained the support of the Crown, and thus were able to control to a great extent what was taught in universities. They were also able to obtain prerogatives and endowments for their respective schools. In the same way in the University of Caracas, which although it had a more secular tradition was nonetheless under the patronage of the Dominicans, who retained their two chairs which dated from 1742.[13] Thus, the desire of the Creole elite, especially in Quito and Santafé, to control the universities, were thwarted by the Crown, which had probably already recognized the new realities of its American colonies: foreshadowing a nationalist project that had its roots in the new trends in philosophy and science.

PIONEERS IN THE TEACHING OF COPERNICUS AND NEWTON IN THE VICEROYSHIP OF NUEVA GRANADA

In relation to the introduction of the Copernican and Newtonian theories in the Viceroyship of Nueva Granada, we believe that these new theses, before reaching the university cloisters, had already circulated in scientific missions, geodesic expeditions, and fortifications.[14] They may also have been propagated at an institutional level in the Military Academy of Mathematics established in Cartagena de Indias from 1731, and in the Academy of Geometry and Fortification created in Caracas in 1760.[15]

As far we know, the Jesuits were the first to teach the theories of Descartes, Copernicus and Newton in the universities of the Viceroyship of Nueva Granada in a systematic way.[16] Around the mid-18th century the Society of Jesus was extremely powerful, with important representation in educational institutions throughout the world. It was thus able to deploy its educational strategies internationally, perhaps more effectively than any other religious order. As a truly international force in culture and education, the Order was able to adapt its strategies in both fields to the demands of the new alliances, so as to maintain its power. In the era of international capitalism, the new science was more pragmatic than speculative, more physics than metaphysics. Yet, the adaptation to these new developments did not take place without resistance. If the needs of power inclined teaching toward change and modernity, tradition and old ways of thinking only allowed a kind of pseudo modernity. Even so, because of its power in culture and teaching, its internationalization and interactions with the elite and courts, the Society of Jesus was capable of self innovation and creativity at the highest level, in accordance with the scientific standards of the enlightenment. This innovative capacity was transmitted to the nodes of the international Jesuit network, through its various types of missions and, particularly, through its official channels of communication.

The cultural adaptation of Newtonianism was not an automatic development; it was a process which took place over several decades catalyzed mainly by the pressures of socio-political changes in Europe and particularly in France. The new theories that were taught during that same period in the Universidad Javeriana of Santafé and the Gregoriana in Quito, were adopted by missionaries educated abroad or in tune with the transformations that were occurring in European institutions.[17] This, undoubtedly, facilitated the interchange and spread of new ideas. It is not surprising, then, that in a culturally effervescent center, such as Quito, the Jesuits were the intellectuals most closely involved with the Franco-Spanish geodesic mission, led by La Condamine and Jorge Juan. It was probably as a result of his contacts with La Condamine[18], for example, that Father Magnín was stimulated to write *Millet in harmony with Descartes or Descartes reformed* (1774), in which he defends the Copernican system on the basis of Newtonian laws.[19] Magnín's treatise is perhaps the first evidence of the French adaptation in Nueva Granada referred to above. Undoubtedly, the atmosphere of internal reform in the cloisters of the community of Saint Ignatius in Quito—catalyzed by the external factor of the geodesic mission—contributed, in spite of local difficulties, to gradually opening the way for the likes of Magnín to maintain new theoretical positions.

As in Quito, it is significant that in Santafé Copernican theories were being taught in the Universidad Javeriana in 1775, by means of the philosophy course entitled *Physica specialis et curiosa*.[20] It is evident that the Descartes-Newton debate was not alien to the university cloisters of Santafé and Quito. In the surviving records of these courses we can see, for example, that one controversial scientific theme was the shape of the earth.[21] This subject could not escape the attention of the Jesuits, especially the Quito community, because of their direct connection with members of the expedition. Thus, pressured by the social and intellectual impact of the activities of the geodesic mission in the local enlightened environment, the Jesuit professors at the Universidad Gregoriana, were obviously encouraged after 1740 to teach in accordance to the European trends. In so far as this experience was viable in Quito, it was transmitted in the following decade to Santafé. The delay may be explained by the fact that in Santafé there was no powerful catalyst such as the geodesic mission to steer a change of mentality.

Another case relating to the spread of the new system can be seen in the teachings of the Jesuit Francisco Javier Aguilar at the Universidad Gregoriana. In his philosophy course, taught between 1753 and 1756, Aguilar explained general and particular physics.[22] In relation to particular physics, he stated that knowledge of "the world, the heavens and meteors" was "a matter of fact, an agreeable treatise which a philosopher ignores at his peril". He taught natural phenomena by means of two disputations concerning the world and the heavens. In his course, he presented the five systems of the world that he considered fundamental: that of Plato, of the Egyptians, Ptolemy's, Copernicus's and that of Tycho Brahe. He focused, however, only on the last three. Aguilar explained that toward the year 1477 Copernicus visualized a world system different than Ptolemy's, and emphasized that his theses had been forbidden by the Roman Inquisition since 1616, although they were allowed to be proposed as hypotheses. He allegiance, however, remained with Brahe's system.[23]

As mentioned above, in the Universidad Javeriana of Santafé the philosophy professor gave a course entitled *Physica specialis et curiosa* in 1775. This Jesuit organized his teaching in a similar way to that adopted by Father Aguilar, but while Aguilar presented the Tychonic system as the most adequate, the Javeriana professor recognizes in his second disputation on the constitution of the main parts of the world, that the Copernican system is nowadays publicly debated in Italy, France, and in some regions of Germany. The author follows the "logical" order of explaining first the most ancient systems: those of Pythagoras and Plato. Next, he refers to what he calls "some new systems": those of Tycho Brahe and Copernicus.He refutes Tycho's system and emphasizes that the Copernican system is the simplest, compared with former models, in explaining the planetary system. Certainly, it is simpler than the Ptolemaic system of cycles and epicycles with the Earth at the center. Tycho Brahe's system was also simpler than Ptolemy's; for this reason, the aim of Copernicans such as the author of *Physica specialis* and Mutis in 1776, was to refute it.

As has already been mentioned, in the Jesuit universities of Quito and Santafé, the competing theories of the cosmos were discussed eclectically in this period (1740–1760). Father Juan Bautista Aguirre, who replaced Father Aguilar as professor of philosophy, continued to defend his predecessor's position in the course he taught between 1756 and 1758. He explained the Copernican system but opted for the Tychonic because it did not contradict "the Sacred Scriptures". The merit of Father Aguirre's lectures was in their discussion of recent experiments and discoveries.[24] In 1757, in collaboration with his pupil José Mará Linati, he defended 257 philosophical theses. The sixteenth, devoted to particular physics, states that "The Ptolemaic world system is contrary to what is observed in astronomy and therefore should not be accepted. The Copernican system is opposed to Scripture and therefore should be refuted. Therefore, Tycho's system should be preferred to those of Ptolemy and Copernicus".[25]

It should not be forgotten that although Copernicus's *De revolutionibus* was placed on the Index in 1616, around 1758 it was removed from the revised Index. However, cultural events of a normative order, even events as important as this, are unlikely to be reflected immediately in the social dynamics of institutions far away from the centers of influence. Independent of the local consequences of this act, apparently certain philosophy professors at Quito and Santafé chose to continue defending Tycho's theories in this stage of transition of ideas and adaptation, like Father Aguirre. At the crossroads of cognitive strategies characteristic of this period, it remained possible for the different protagonists to promote in apparent harmony different educational programs. Thus, while Fathers Aguilar and Aguirre supported Tycho's system, the Jesuit from the Universidad Javeriana opted for the Copernican system.

MUTIS AND HOSPITAL DECLARE THEMSELVES COPERNICANS

As we have just pointed out, the formal liberalization of the teachings and dissemination of *De revolutionibus* did not guarantee that the traditional sectors would accept or defend heliocentric theories rather than those in which they had been educated and which were more in agreement with traditional and eclectic cosmologies such as those of Ptolemy and Tycho. Thus, it is significant that Father Juan de Hospital taught the Copernican system and Newtonian mechanics in his philosophy course in Quito during 1759–1762 and refuted the older systems.[26] One of the results of his teaching was the public "conclusions" discussed, under his guidance, by Manuel Carvajal on 14th December 1761. Carvajal defended 12 theses; the 11th questioned the world systems of Ptolemy and Tycho and recommended the Copernican system "for being more in agreement with astronomical observations and the laws of physics".[27]

Father Hospital tried to establish a scientific circle, in order to disseminate the basic principles and methods of modern physics. Up to now, we do not have sufficient information to establish the direct relation between Father Hospital and the Pichinchense Academy, created in 1766. Nevertheless, it has been stated that the physics course given between 1760 and 1761 had profound effects on the cultural activity of Creole and Quitenian scholars, such as Eugenio Espejo, Carvajal, and Joaquín Rodríguez. The latter was the father of Miguel Antonio Rodriguez, who introduced the teaching of anatomy into the physics course, and attached great importance to mathematical studies in the University of Quito.[28]

For his part, before the expulsion of the Jesuits in 1767, José Celestino Mutis had discussed the Copernican system in the Colegio de San Bartolomé without provoking any public repercussions. Moreover, Mutis emphasized the tradition of the Universidad Javeriana in the teaching of new science and natural philosophy: "If the love for truth has detained me more than necessary in expressing my inclination for the Copernican system, it is right to conclude by celebrating these happy times in which we see the renaissance of natural philosophy in this kingdom".[29] The dissemination of the heliocentric theory and the Newtonian world system by Mutis between the years 1760 and 1770 was more in accord with the ideas that already prevailed in metropolitan centers, and had great local impact. Nevertheless, even before the arrival of reformist Spanish ideas to Nueva Granada during the 1740s, as a result of enlightened Bourbon policies, we have evidence of continued local appropriation of the new science, within the Aristotelian and Cartesian eclectic natural philosophy.

When Mutis began institutionalizing the discourse of the new philosophy, he obviously did not start from zero. During the previous 20 years, the cultural elite of Quito and Santafé and its centers of influence had recognized the existence of new techniques and new scientific knowledge. A new theoretical-experimental vision was displacing old philosophical conceptions concerning the intricacies of the natural world. This vision was eclectic and different from that of the Cartesians; but such differences were not always thought of in terms of opposition or conflict. These new knowledge and techniques were recognized, on many occasions, as operationally valuable, providing explanation and clarification of natural phenomena , and able to function without ontological scruples within the Aristotelian and peripatetic natural philosophy.

The contribution of Mutis, 20 years after the beginning of this local tradition, was to teach that the new physics could not be submerged or adapted to the old metaphysics. A different matter is that Mutis may have assumed in his scientific activities, and spread through his teachings, a new kind of eclectic attitude in respect to the modern philosophy; particularly in reference to the sensitive issues of the relationship between physics and natural religion, and between mathematics and experiment. Such attitudes could be explained both by the changes in his intellectual education and by the ambivalent circumstances of the local environment in which he had developed his scientific work. Before 1767, Mutis expounded his so called *Reflexiones sobre el sistema Tycónico,* in the Colegio San Bartolomé, in which he defended the Copernican system using two propositions: (1) That it is the earth that moves with the other planets, while the sun and the fixed stars remain static with the exception of the sun's rotation on its axis. (2) That the Copernican system is not contrary to Scripture. In general terms, we can place the *Reflexiones* toward the end of the period that we have called the adaptation of the new philosophy to the traditional worldviews. It constitutes a clear attempt to make the Copernican thesis viable, surpassing the traditional options represented principally by Aristotelian and Cartesian natural philosophy. No doubt, Mutis bore in mind the difficulties imposed by the conservative institutions in the hands of the Dominicans on all local projects of educational and cultural reform. In spite of his personality and academic prestige as professor of the new philosophy, Mutis was conscious of the institutional force of this power and therefore avoided direct confrontation. This dispute was settled years later in 1773, through a less eclectic discourse, more in tune with the theses of Copernicus and Newton. During this period, Mutis had time to develop his thoughts and to produce a more systematic conceptual appropriation of these theses, based on the direct study of the works of Newton in the areas of physics and mathematics.[30]

In 1773, Mutis publicly declared himself a follower of Copernicus. On this occasion, he dedicated his public "conclusion" to the Vice-reine in the Colegio del Rosario. Mutis said: "Having instructed myself in the fine knowledge and the clear light that I was never able to discover in the shadows of the old philosophy, I confess publicly that I am a Copernican". Mutis then defended 16 theses in favor of the heliocentric system. It is interesting to note that he devotes at least 11 theses to proving that these theories had not been banned, and that, on the contrary, their teaching was a part of the new reform proposed by Carlos III.[31]

Equally, we should emphasize a debate that took place in the city of Caracas between a university professor, the Count of San Javier, and a "certain Valverde, philosopher of noble condition and ecclesiastical status, on active service to the king and graduated from the Thomistic University".[32] This dispute began on 1st of August 1770 with a discussion between Valverde and the count on the usefulness or uselessness of Aristotle's philosophy. Valverde replied in writing on 7th August of the same year, setting out his few philosophical position in this respect. In the development of his central theme, Valverde shows that he is familiar with the Copernican and Newtonian theses and with the new science. There is no evidence, however, that Valverde's position was based on firm knowledge of these ideas. Valverde employed theological arguments and criticized the old philosophy from the perspective of the new science, but the document lacks firm foundations. He attempts to use what looks like "new" points of view to introduce his theological position. For instance, when he calls on the principle of transitivity of the equivalence, if A = B and B = C, then A = C, it is in order to attack the Arians, who refused to accept the notion of The Trinity.[33] Likewise, Valverde makes clear his unconditional adherence of Thomistic philosophy.

The case of Mutis is different from that of Valverde, at least in his 1773 dissertation. This is presented in scientific terms, designed to answer possible inquisitors who might object to his theories, as indeed was the case. It is not then an incidental strategy of Cartesian distraction, like the dispute between Valverde and the Count of San Javier.

In this process of dissemination of modern scientific ideas about physics in the colonial period, we also find several illustrious figures whose work need interpretation. They may have been hiding their true way of thinking because they feared the Inquisition, which was seen by intellectuals as a very powerful institution in Spain, judging by its actions and number of victims. One example is given by the Olavide trial, which embodied "the problems of freedom, culture and above all showed the enormous power that the Inquisition maintained in Spain".[34]

THE DOMINICANS CONCEDE IN QUITO AD ATTACK SANTAFÉ

It is necessary to make clear that the Dominicans' attitude in regard to the public conclusions on the Copernican system was completely different in Quito and in Santafé. The differences were marked by the balance of political power and, in particular, the control of education. We can locate a first stage, both in Quito and in Santafé, between 1740 and 1767, in which the Copernican system was freely discussed in public disputations by the Jesuits and by Mutis. At this stage, the Dominicans kept silence. It is possible that the main reason for the Dominican reticence was the Jesuits' power over the Inquisition. From 1746 to 1767 the Inquisitor general was López de Prado, a friend of the Jesuits. On the other hand, self-interests and the power for the control of education were by then well defined, and the situation was of mutual respect between the religious orders. In effect, Jesuits and Dominicans enjoyed the same privileges and rights to confer degrees in their respective universities from 1702 onward.[35]

In order to understand the background of the debate initiated by the Dominicans against Mutis' defense of Copernicanism, some relevant points should be recalled. In contrast to the former decade, by the early 1770s things had changed with the expulsion of the Jesuits. The Dominicans petitioned the *Junta de Temporalidades* claiming the educational privileges and property of the Jesuit Order.[36] In Quito, the *Junta de Temporalidades* was responsible for the foundation of a public university with the property of the Universidad Gregoriana and the seminary-school of San Luis, which formerly belonged to the Jesuits. The Dominicans managed to satisfy their old aspiration of controlling higher education in Quito. On 23rd August 1776, the Junta decided "to transfer the Universidad Santo Tomás with its income and possessions, to the building of the seminary-school San Luis, declaring it as the only official university in the Audiencia de Quito". On the other hand, the Dominicans secured in the Royal Decree of 1786 that the patron saint of the new public university would be Saint Thomas of Aquinas.[37]

In Santafé, on the other hand, the Dominicans opposed the Moreno and Escandón educational project, as they thought that the new university he proposed would remove their privileges in the Viceroyship.[38] The project derived from the royal decree of university reform and was aimed at placing control of education under the dominion of the king and not the pope, would give the civil sector the right to inspect education and, with this, control the teachings and appointments in educational institutions. In presenting his project, Moreno wrote that the religious orders have taken hold of the center of science, and control the appointment of chancellors, deans of study, and grade examiners and act as arbiters to confer degrees, leaving the lay staff to live always as inferiors without hope of getting rid of such a burden. The main reason for the debate is made clear: the open struggle between the civil sector and the religious sector, in this case the Dominicans, who were not prepared to allow their monopoly to be removed. Their resistance had been anticipated by Moreno in his proposal of educational reform: the Order of Preachers is the only order that, because of their concern not to lose their present right to confer degrees, may say that they disagree, but only after they have accepted these powerful privileges.[39]

Thus, the opposition by the Dominicans to Mutis' defense of the Copernican system in the 1770s involved more than philosophical arguments. Inevitably, what was at stake was the creation of a public university, and its control by the laity. Thus, informing the debate was a direct confrontation between two organized groups over the control of higher education.[40] In fact, what the Dominicans really hoped for was to gain time to use their influence in Madrid, in order to block the curriculum proposed by Moreno and Escandón, which seriously jeopardized their interests. Thus, the debate on the Copernican system was immediately linked to the ongoing debate about the reform of the curriculum. On this occasion, the debate over Copernicanism passed from the Santafé Inquisition, to the Santo Tribunal of Cartagena and, finally, reached the Supreme Inquisition of Castile, on 6th of March 1775.[41] Apparently, Mutis succeeded in convincing his judges that the heliocentric system was not opposed to dogma.[42]

THE CREOLES OF QUITO, CARACAS, AND SANTAFÉ TAKE UP NEWTON AND CPERNICUS AGAIN

While it is true that the enlightened court of Carlos III elevated secularization university policy to a juridical and institutional level, in practice the situation was different. The party of enlightenment was not always able to restrain the onslaught unleased against them by specific religious communities and conservative sectors. Other factors in the complex Spanish situation contributed to this tension, but the religious and inquisitorial power by itself was very strong. Besides, the political environment had changed since the French Revolution. The Spanish Crown had developed measures of control and repression against university professors, students and books circulating in the Viceroyship.

Santafé de Bogotá, the capital city of the Viceroyship, had by this time four institutions of higher education: the Universidad de Santo Tomás, the Colegio Mayor de Nuestra Señora del Rosario, the Colegio Mayor de San Bartolomé and San Nicolás de Bari.[43] The only university that could confer degrees on the laity, after the expulsion of the Jesuits, was the Universidad de Santo Tomás. Quito, the capital of the Audiencia, only had one university which the lay population could attend after the expulsion,[44] the Universidad de Santo Tomás, which had been a public university since 1775.[45] The University of San Fulgencio which had both royal and papal approval, had conferred degrees only up to 1775.[46] In Caracas, we find three convents dedicated to teaching: San Francisco, San Jacinto, and Las Mercedes.[47] All three opened their classrooms to the laity.[48] The only university that functioned during the colonial period and conferred degrees in Caracas was the Universidad de Santa Rosa de Lima, with the seminary-school that had given it its academic origin. In 1784, the school separated from the University and took on a more secular character, although every 2 years the chancellorship was shared with an ecclesiastic.[49]

The differences relating to educational institutions and privileges, were mainly between Quito and Santafé. In Quito, the Dominicans had been lords and masters since the expulsion of the Jesuits. At the same time, in Santafé they had begun a debate with the civil sector for the control of education. In Caracas, the Dominicans exercised a relative influence in the university, as this institution had a more secular character, particularly from 1784 onward. The debates they initiated were focused to gain more sinecures there; their opponents were the Franciscans as well as the civil sector.

In the culture of the Audiencia of Quito, the generation educated under the influence of Father Hospital became influential toward 1779. In that year, Espejo published *El nuevo Luciano de Quito*, in which he criticized the dominant philosophy.[50] Although Espejo did not have a teaching post in the university, he was at the time the director of its library and he helped to guide the studies of several Creoles of the new generation. One of these was Miguel Antonio Rodriguez,[51] the son of Espejo's classmate of the same name, with whom he had attended philosophy lectures delivered by Father Hospital. This young professor, who belongs to the generation that undertook to reconcile scientific ideas with religion, brought Copernicus and Newton to the fore once again. However, only in 1797 was Copernicus defended in public, when his student Pedro Quiñonez y Flórez, expounded his *Theses Philosophicas sive Philosophia universal*.[52]

It is surprising that the Dominicans did not launch a polemic against Professor Rodriguez. Their restraint may perhaps show, behind the ideological stance of the Dominicans in Quito and Santafé, their fundamental problem: that of controlling their respective universities. A similar situation occurred in Caracas; we believe that the Dominicans there must have been reluctant to teach enlightened ideas[53] in their philosophy classes at the university, even though, as we shall see, they did not engage in Controversy. In Quito, the reform presented by Father Pérez Calama in 1791, which were only partially applied, guaranteed that the teaching of theology of *prima y visperas* was controlled by a Dominican. We have already pointed out, this was the case in the Universidad de Caracas as well, where the Dominicans had been in control of the teaching of philosophy and holy scripture since 1742.

In the theology curriculum drawn up by Pérez Calama, Saint Thomas is studied to shape a system in which the young learn well what the saint clearly teaches. With this method, the pupils will be great and solid theologians. In physics, the Copernican system was still not accepted. Based possibly on the 1786 Constitution and on the Pérez Cálama plan of studies, and faced with the creation of new statutes for the university in 1803, the Dominicans demanded that the Chancellor and the cloister of this university should impose the doctrine and traditional structure on the already secularized university. It seems that by this time, the influence of the Dominicans had declined in the university in Quito, because the cloister's response was eclectic and recommended the studies of mathematics and physics as a priority.[54]

By 1791, several events took place in Santafé that illustrate the educational spirit of the time. First, Francisco Antonio Zea published, under the title of *Avisos de Hebephilo*, a critique of scholastic teaching, and the students of San Bartolomé requested the appointment, at their expense, of a teacher of mathematics and natural philosophy.[55] Outside the university, the publication of the newspaper *Papel periódico de la ciudad de Santafé de Bogotá* began under the direction of Manuel del Socorro Rodríguez, who at the same time was the director of the Public Library and coordinator of the *Tertulia Eutropélica*. We can name among other tertulias the *Arcano de la Filantropía*, written by Antonio Nariño, and the *Tertulia del Buen Gusto* by Doña Manuela Santamaría. In the same year, the priest, José Domingo Duquesne wrote a document entitled *Historia de un Congreso Filosófico tenido en Parnaso por lo tocante al Imperio de Aristóteles*; it is written in a confrontational style like those already published earlier in Quito by Espejo. It is clear, then that enlightened ideas spread among the Creole elite through the literary circles and the Botanical Expedition, but was not accepted yet in the university institutions.

The most important fact in the university environment, however, and the one with most repercussions, was the debate between professor Manuel Santiago Vallecilla and professor Santiago Gregorio de Burgos, Director of the Colegio del Rosario. Vallecilla defended "useful" philosophy, taught under the Moreno and Escandón plan, and was not prepared to accept orders from Burgos, because, according to Vallecilla, the reasoning of the Director of the Colegio del Rosario was of no use in controlling the functions of teachers and in evaluating the doctrines to be read.[56] In 1796, Professor Francisco Vásquez Gallo adopted a similar position when he did not wish to read or defend the *Suma doctrina* of Saint Thomas and Master Goudin,[57] and defended the heliocentric system. As in the case of Vallecilla, on this occasion Viceroy Ezpeleta supported the director against the professors who defended enlightened ideas. On 15th June 1776, the Board of Studies condemned the excesses of Professor Vásquez and those who would dare in the future present, even as a hypothesis, such sacrilegious system as Galileo's. But although the Board of Studies reprimanded the Vásquez, the professor continued defending the heliocentric system and attacking the peripatetic Goudin. The Director of the Colegio del Rosario suggested that it is convenient not only to dismiss him from the school but from the city as well, so that he does not corrupt the youth by his influence and by such fatal disobedience. Vásquez's tenacity in maintaining his position reflects the consciousness of the irreducible cultural value of certain theories which has by then taken root among the scholarly elite of Santafé.[58]

The reprisals in Santafé did not put an end to the controversy over Copernicanism. The debate restarted on 20th June 1801 when, at the request of Viceroy Mendinueta, Mutis wrote a report on the Copernican thesis that was defended in public disputations by the unreformed Augustines of the University of San Nicolás of Bari.[59] This university had been well known since Father Vasquez's reform in 1773, for the teaching of new theories.[60] This reform of studies in the university was implemented through the following injunction:

> the peripatetic philosophy and theology which is full of useless issues that only serve to waste time in a fruitless manner, should be abolished, and a useful and beneficial philosophy should be taught which would be able to provide students with the means of the fruitful study of other subjects and to discover truth, the aim we are all striving for.[61]

It seems that the new ideas and method that attacked scholastic positions were implemented in San Nicolás de Bari by June 1776, when ecclesiastic higher studies were organized in the unreformed Augustinian convents of Cartagena and Santafé. In the latter, Father Diego Francisco Padilla was elected moderator of the teaching of philosophy, and remained in charge till 1778. On his return from a trip to Europe in 1788, Padilla was appointed Dean of Studies and he reorganized the teaching of philosophy by introducing the systematic teaching of new ideas.[62] This modernizing process was consolidated with the establishment around 1800 of a course in mathematics in which Newton and Copernicus were taught.[63] In this context, Mutis' 1801 report to the Viceroy legitimized the teaching of Copernicus and Newton in Santafé.

Around 1790, the situation was equally dreary in Caracas, where innovators confronted traditionalists who were supported by the Crown. This was a period of political repression in the American colonies, when the metropolis made clear its intention to exercise total control over studies, teachers, books and students. The struggle with the priest Baltasar de los Reyes Marrero[64] is not an isolated event either in Caracas or in Spain. An apparently insignificant—the ejection of a pupil from the classroom, because he had not done his mathematics homework according to the modern approach—was used by the Consejo Real as pretext for imposing exemplary policy. Behind the teaching of the new enlightened theories, the regime detected the ability of the rising social elite to subvert the colonial order, which was intolerable. Hence the rigid measures of control over teachers, such as the demand to monitor the "class notes" which students were required to memorize before the lesson; the official visits to classrooms every second month, and to the students' homes to "seize" any book likely to cause "harm" to religion and to the Crown. These repressive policies were particularly severe in Caracas which, because of its geographical position and its commercial freedom since 1788,[65] was the point of entry for circulation of books on the new philosophy.[66]

Although it is true that the decisions about teachers in Santafé and Caracas were promulgated under the same royal seal, in Caracas the members of the Real Audiencia and the chancellor supported Marrero.[67] In Santafé neither Vallecilla nor Vásquez

Gallo had the support of the Chancellor or the *Junta de Temporalidades*. They were supported by, among others, the Vice-chancellor and, later on, by Camilo Torres. In Santafé, Viceroy Ezpeleta gave unconditional support to those chancellors who represented tradition and scholasticism. The Consejo Real took a similar stand in its trial against Marrero, supporting Cayetano and the Chancellery, in spite of the Real Audiencia's support of Marrero and its Chancellor Juan Agustin de la Torre (1750–1804).[68] It should be taken into account that the Cabildo, with the support of the faculty, had been pressing for the teaching of mathematics for several years. As in Santafé, the Creole elite discussed and put into practice the useful knowledge of the new philosophy, both inside and outside the classrooms, seeking to take advantage of any opportunity that might be useful to their social and cultural project.

By the time of this incident, in 1790, the Academy of Public and Spanish Law[69] had been created in Caracas; its first president was Miguel José Sanz.[70] In the same year, the Chancellor of the University of Caracas, Doctor Juan Agustin de la Torre, published "Economic Discourse, Love of letters in relation to Agriculture and Commerce". In this treatise, the Chancellor asserted, like Zea in Santafé, the urgent need for a "patriotic science".[71] Science is classified as an instrument related to life, which possesses the secret of strengthening and dignifying it. Like Mutis earlier, de la Torre insisted on the need to create a teaching post, and "to develop the books, instruments and machines that are indispensable to this teaching, as without these aids, it is difficult to demonstrate to students the effects of their application".[72] Marrero taught that experimental philosophy was to be preferred to a merely rational philosophy, because the first is based on reason and experiment, while the second is only based on reason.

Besides, this enlightened professor pronounced several public statements in defense of the Copernican and Newtonian systems. Authors like Kepler, Huygens, Volta, Lavoisier, Musschenbroek, Buffon, Sigaud de la Fond, Bails, Jacquier, Nollet, Brisson, among others, find that their theses are defended in the Universidad de Caracas, not only during the period of Marrero's teachings (1788–1790) but they are also debated and defended by Pimentel and other academics that continued with the teaching of philosophy.[73]

In general, it can be said that as in Quito, with the teachings of Hospital, in Caracas with Marrero, in Santafé with Mutis, Felix Restrepo and the application of the Moreno and Escandón plan, the new theories did not fall on deaf ears. A new generation was introduced to and adopted the new experimental philosophy. At the University in Caracas, among other students of Marrero, we can cite Rafael Escalona who later continued his master's line of thought in his own teachings.

NOTES

[1] In this work, we are not addressing Copernicus cosmological theories; when mentioning the thesis or the Copernican system, we are referring to heliocentrism.

[2] The Real Cédula of 20th August 1739 created in a definitive form the Viceroyship of Nueva Granada, with the integration of the territories of Nueva Granada, Venezuela and Quito.

Venezuela left the Viceroyship in 1742, which was changed into a *Gobernación*. In 1777 Venezuela became *Capitanía General*, and in 1786 the Real Audiencia de Caracas was created. Toward 1786 the Provincia de Venezuela had around 333,359 inhabitants. The *Padrón del Arzobispado de Santafé*, that comprised several municipalities, had between 1780 and 1781 a total population of 399,446 inhabitants. Toward 1808 the population of the Presidencia de Quito was 600,000 inhabitants, and in total, there was in the Viceroyship a population of two million. The Audiencia de Caracas was not counted in this figure; see F. B. Figueroa, *Historia Económica y Social de Venezuela,* Vol. III (Caracas, 1978), p. 1203; J. M. Restrepo, *Historia de la Revolución en Colombia* (Medellín, 1974), pp. 48–49, and J. O. López, *Historia básica de Colombia* (Bogotá, 1984), p. 127.

[3] The Vice-Roy remained in office for 4 years, exercised superior military-political control and control of finances of Nueva Granada. The President of Quito remained in power for 5 years and was subject to the Viceroy's authority. The Captain of Venezuela had office for 7 years. See Restrepo, *op. cit.,* pp. 26–27, A. E. López, *La Real Audiencia de Caracas, su origen y organización (1786–1805)* (Mérida, 1976), and C. L. Curiel, *El discurso de la fidelidad. Construcción social del espacio como símbolo del poder regio (Venezuela siglo XVIII)* (Caracas, 1970).

[4] The economic financing of the Viceroyship was based on the rents of mining, mainly gold; but its highest income come from the taxing of sugar-cane aguardiente, though it should be pointed out that smuggling was predominant; see A. Delgado, *La colonia. Temas de historia de Colombia* (Bogotá, 1974), pp. 140–141.

[5] Cf. Restrepo, *Historia de la Revolución en Colombia* (note 4, above), p. 28.

[6] Cf. B. T. Zambrano, "El pensamiento historiador colombiano sobre la época colonial", *Anuario Colombiano de Historia Social y de la Cultura* 10:35 (1982).

[7] Cf. "Expediente sobre la Universidad Pública de Quito". Session of 19th April 1800, Archivo General de Indias (A.G.I.), Sala Segunda, f. 13 (our numbering). The Royal Order of 4th April 1786 established the appointment of alternative chancellors among seculars and ecclesiastics of Universidad de Quito. Before this the Real Cédula of 7th October 1784 had already established the same norm for the Universidad de Caracas. The Bishop offered to finance the teaching of subjects in Quito, and in this way, the economic ecclesiastical power subjugated the civil sector.

[8] In the Universidad de Caracas, by Royal Order of 4th May 1815, a visit to this institution was organized in order to reform the constitutions and to draw up a new pensum. This visit and the plan was the responsibility of Don José Manuel Oropeza, Lieutenant Governor of the Audiencia de Caracas. On 20th December 1815 the lieutenant presented the new pensum. See AGI, Audiencia de Caracas, Sección V, legajo 109, Doc. II (our numbering). In respect to the introduction of modern science in Nueva España, see J. J. Saldaña, "The Failed Search for Useful Knowledge: Enlighted Scientific and Technological Policies in New Spain", in J. J. Saldaña (ed.), *Crosscultural Diffusion of Science: Latin America* (México, 1987), pp. 35–58.

[9] The Royal Botanical Garden of Nueva España was founded in 1788 and in this same year chemistry teaching was created. In 1792, in the Real Seminario de Mineria, a Chemistry teaching post was also established. See P. Aceves, "Política botánica metropolitana en los Virreinatos del Perú y Nueva España", in J. F. Pérez and I. G. Tascón (eds.), *Ciencia, Técnica y Estado en la España Ilustrada* (Zaragoza, 1990), pp. 249–255.

[10] For example, in Quito, the debate on conferring degrees in the schools and universities of the Jesuits and Dominicans was regulated by the Royal Decree of 1696, in which it was established that both communities could confer degrees. In Santafé this situation was equally regulated by Royal Decree. Felipe V sanctioned it on 25th November 1704. See AGI, Audiencia de Quito, Legajo 196, see also A. Ariza, *El Colegio Universidad de Santo Tomás de Aquino en Santa Fe de Bogotá* (Bogotá, 1980), p. 119.

[11] In Santafé, for instance, Father Fray Jacinto Antonio Buenaventura says that by reason of the narrowness and poverty of the school to buy science books they ask to be given the properties of the Jesuits, in order to maintain the teachings in the school, with all their belongings and

rights and privileges that the former Jesuit university had, besides all other prerogatives that the Fiscal Moreno requests for the public university. See "Expediente sobre la universidad pública", Santafé, 30th June 1777, AGI, sección quinta Audiencia de Santafé, Legajo 759, Doc. 13, f. 15 (our numbering).

[12] The Fiscal Moreno says that if many because of poverty did not have the necessary money to obtain a Doctorate and to pay for the customary celebrations, they should be happy with the degree of Bachelor or Licensee, that is sufficient to become an employee and this would be more appreciated and less common that a Doctorate is at present. See "Informe del Fiscal Moreno", 25th October 1771, Santafé 1771. AGI, sección quinta Audiencia de Santafé, Legajo 759, f. 11.

[13] In 1538, the Dominican community founded the first university of the American Spanish Colonies, in the city of Santo Domingo, Isla Espanola, today the Dominican Republic. In the Viceroyship of Nueva Granada the first university was founded in the city of Santafé de Bogotá, in 1580; in Quito in 1686, and in Caracas from 1592 onward, convent studies were established. In the Universidad de Caracas the Dominicans were responsible for the teaching of the two subjects by authorization given by the Real Orden of 1742. In the new public university of Quito under the plan of studies of the Bishop Pérez Calama they kept control of some teaching posts.

[14] Scientific curiosity in Europe contributed to the foundation of several academies to gather and compile information from travelers in order to stimulate interest for natural science. One of the first visitors under Bourbon rule was Amédie de Frezier, in the Viceroyship of Peru from 1712 to 1714. In 1735, under the auspices of the French Academy and the Spanish Crown the Geodesic expedition led by Charles de la Condamine begun, which included scientists such as Pedro Bouguer, Luis Goudin, Seniergues, Joseph Jussieu, and the Spaniard Antonio de Ulloa and Jorge Juan. Later on, from 1777 to 1778, Hipólito Ruiz and José Antonio Pavón undertook an expedition to the Viceroyship of Peru, and from 1799 to 1804, Alejandro Humboldt and Amado Bonpland traveled to Spanish America. See J. J. Saldaña, "Nacionalismo y Ciencia Ilustrada en América", P. Aceves, "Política botánica metropolitana en los Virreinatos del Perú y Nueva España" and M. A. Puig-Samper, "La ciencia metropolitana y la conciencia nacional en las colonias", in *Ciencia, Técnica y Estado en la España Ilustrada* (note 11, above). See also P. L. Astuto, "La Ilustración en Quito y Nueva Granada", in *Eugenio Espejo (1747–1795): Reformador ecuatoriano de la Ilustración* (México, 1969); and C. Minguet, *Alejandro de Humboldt historiador y geógrafo de la América española (1799–1804)* (México, 1985), pp. 266–270.

[15] Cf. E. M. Dorta, *Materiales para la historia de la cultura en Venezuela, 1523–1828* (Caracas, 1967), p. 260. The Academy of Caracas functioned from 1760 to 1768. G. H. de Alba, *Documentos para la historia de la educación en Colombia,* Vol. IV, 1777–1800 (Bogotá, 1983), pp. 531–533.

[16] The Jesuits established the Universidad Javeriana in Santafé in 1621 and in Quito, in 1622, the so-called Universidad de San Gregorio. These universities conferred degrees on civilians. In Caracas the Company were not responsible for studies nor was it present in the University.

[17] The Universidad de San Gregorio of Quito had 71 foreign professors that taught in the university and wrote up their subject in a manuscript volume. The local professors were 21, out of which 5 were from Loja, 4 from Quito, 3 from Guayas, 3 from Cuenca, 3 from Riobamba, 2 from Ibarra and 1 from Ambato. See J. M. Vargas, *Polémica universitaria en Quito colonial* (Quito, 1983), p. 11.

[18] La Condamine was not only linked to the Jesuits in Quito, but years later gave lectures in the Universidad de Lima. Besides Magnin's book, several Jesuit publications of this period are known, mainly geographical descriptions, See A. Lafuente and E. Estrella, "Una ciencia para el Estado: la expedición geodésica hispano-francesa al virreinato del Perú (1734–1743)", *Revista de Indias,* 172:549–629 (1983).

[19] Magnín had sent this work to Europe in 1744 and in 1747. (On the first occasion the book was lost when the ship sank.) Magnín was promoted to the rank of correspondent of the Royal

Academy of Sciences of Paris. See C. P. Escudero, "Estudio Introductorio", in *Pensamiento Ilustrado Ecuatoriano* (Quito, 1981), pp. 36–39.

[20] The *Physica specialis et curiosa* was writen in the year 1755; see O. N. Fajardo and D. S. Arango, "El debate sobre el sistema copernicano en la Nueva Granada durante el siglo XVIII", *Revista Lull* 7:53–75 (1984), For a Spanish translation of this course, see P. N. Ramírez (ed.) *Nueva filosofía natural. Physica specialis et curiosa.* Manuscrito colonial anónimo, 1755 (Bogotá, 1988). The author was probably a Spanish Jesuit or someone with Spanish philosophical education; *Ibid.*, p. 52. He displays a wide geographical knowledge of the Neogranadino Viceroyship, that leads us to believe he was probably a Spanish missionary who visited these territories.

[21] The expeditions to Peru and Lapland yielded observational verification of Newton's prediction of the shape of the earth, hastening the definitive acceptance of Newtonian physics and the Copernican world system. In the text of the *Physica especialis et curiosa* the debate about the shape of the Earth is reduced to the results of the two expeditions and is thoroughly "modern". See Ramírez, *Nueva Física Natural,* pp. 75–77.

[22] This course has been published in Francisco Aguilar S. J., "Curso de filosofía", in *Pensamiento Ilustrado Ecuatoriano* (Quito, 1981).

[23] Cf. Negrín and Soto, "El debate sobre el sistema copernicano en la Nueva Granada durante el siglo XVIII", pp. 50–51. It should be pointed out that Aguilar reproduced the generalized support of Tycho that took place after the prohibition of the Copernican system by the church in 1616.

[24] Cf. Paladines, "Estudio Introductorio", p. 34. When Aguirre explained sunspots, he was proving the rotation of the sun on its axis. This subject was also explained in the philosophy course of 1755 in the Universidad Javeriana. See J. M. Pacheco, *Ciencia, filosofía y educación en Colombia, siglo XVIII* (Bogotá, 1984), p. 11.

[25] Aguirre apparently had at least one disciple who went beyond his master, as can be concluded from the following quotation taken from the dissertation defended by José Mará Linati, under the direction of Father Juan Bautista de Aguirre, of the Company of Jesus, Public Professor of Philosophy in the Universidad de San Gregorio, Quito, 1759. "We defend and prove with purely philosophical reasons, leaving to the theologians the theological reasoning". The original is to be found in the Library and Archeological Museum "Aurelio Espinosa Pólit", Quito, Cotocallao. See Aguilar, "Curso de filosofía", pp. 129–130.

[26] Cf. C. F. Granizo, "El siglo XVIII en la Real Audiencia de Quito", in *Eugenio Espejo, conciencia crítica de su época.* (Quito, 1978), pp. 11–33. Paladines, "Estudio Introductorio", p. 34, says that the Jesuit, Juan Hospital, born in Bonolas, was possibly one of the first of the new generation of American Creoles that publicly served as disseminator of modern Astronomy, during the physics course he taught during 1761–1762 in Quito; see also Angel Nicanor Bedoya Mararu, *El doctor Francisco Xavier Eugenio de Santa Cruz y Espejo* (Quito, 1982).

[27] Cf. M. Carvajal, *Lo que se debe probar: el sistema de Copérnico como el más acorde con las observaciones astronómicas y las leyes de la física* (Quito, 1761). Theses defended under the guidance of Father Juan de Hospital, in Universidad de San Gregorio, Quito, 14th December 1761, with the approval of his superiors.

[28] Cf. E. Keeding, "Las ciencias naturales en la antigua Audiencia de Quito. El sistema copernicano y las leyes newtonianas", *Boletín de la Academia Nacional de Historia* 57:43–67 (1973).

[29] Mutis papers on the Copernican system are published, by example, in G. H. de Alba, *Pensamiento científico y filosófico de José Celestino Mutis* (Bogotá, 1982).

[30] Cf. L. C. Arboleda, "Sobre una traducción inédita de los *Principia* al castellano hecha por Mutis en la Nueva Granada *circa* 1770", *Quipu, Revista Latinoamericana de Historia de las Ciencias y la Tecnología* 4:291–313 (1987). The Arboleda study is an analysis of the manuscript that contains the 1770 Mutis translation of the *Principia Mathematica* . See also L. C. Arboleda, "Los *Principia* de Newton en la Nueva Granada", in C. A. Lértora,

E. Nicolaïdis, and J. Vandersmissen (eds.), *The Spread of the Scientific Revolution in the European Periphery, Latin America and East Asia* (Turnhout, 2000).

[31] Cf. G. H. de Alba, *Pensamiento científico y filosófico de José Celestino Mutis* (Bogotá, 1967). This is the text of the Mutis allocution on the Copernican system (Colegio del Rosario, Santafé, December 1773).

[32] There are not many bibliographical references on Valverde's activities. For instance, it is not known if he was Creole or Peninsular, and the type of appointment he had in Caracas. Caracciolo is the Venezuelan historian who provides most information on this philosopher. See Valverde, "Carta de Valverde al Conde de San Javier sobre filosofía", 7 de agosto de 1770, Archivo del General Miranda, sección Diversos, also Caracciolo Parra León, *Filosofía universitaria venezolana, 1788–1821* (Caracas, 1989).

[33] In this sense Valverde was situated at the complete opposite place of those English and European Newtonians who supported the Unitarian and Arian theses of the author of the *Principia.*

[34] Cf. A. Á. de Morales, *Inquisición e Ilustración, 1700–1834* (Madrid, 1982), pp. 130–131.

[35] The Jesuits and the Dominicans had carried on a long debate during the 17th century over academic privileges and the right to confer degrees. A critical stage was the *Real Cédula* of 2nd March 1655, which established that neither the Jesuits nor the Dominicans could govern universities or confer degrees. This conflict was resolved on 27th May 1702, when the Pope accepted the Royal Council's decision of 13th August 1700, to confer equal rights on both communities.

[36] Cf. "Expedientes sobre la Universidad Pública. Santafé, 30th June 1777", A.G.I., Audiencia de Santafé, Legajo 759, doc. 13, f. 15.

[37] "The king approved this decision in the *Cédula* of 4th April 1786, in which it was decreed to organize the university in accordance to the customs of Lima and Mexico, according to what was ordered in title 22 of the *Recopilación de las Indias*, alternating ecclesiastics and laymen in the chancellorship". Quoted by Á. R. Cruz, *Historia de las Universidades Hispanoamericanas,* Vol. 1, "Periodo hispánico" (Bogotá, 1973), p. 561.

[38] Francisco Antonio Moreno y Escandón (1736–1792), born in Mariquita and died in Santiago de Chile. He was a scholar and a teacher at the Colegio de San Bartolomé of the Jesuits in Santafé, but was a graduate from Universidad de Santo Tomás. As fiscal of the Real Audiencia and protector of Indians, and as a member of the Junta de Temporalidades, in charge of administering the property of the expelled Jesuits, Moreno proposed in the session of 9th May 1768, a project of reforms to the plan of studies and the creation of the Public University for the city of Santafé.

[39] Cf. "The first fiscal report protecting natives on the establishing of the Public University in Santafé, 9th May 1768". AGI, Fifth Section, Audiencia de Santafé, Legajo 75, doc. 8, ff. 3 and 10 (our numbering).

[40] The polemic between Mutis and the Dominicans on the Copernican system is analyzed in Negrín and Soto, "El debate sobre el sistema copernicano en la Nueva Granada en el siglo XVIII" (note 22 above).

[41] "Report of the Holy Office of Santafé on the debate on the Copernican system", 24th July 1774, Archivo del Jardín Botánico, Madrid (AJB), Mutis Section, legajo 25, f. 2 (our numbering).

[42] Cf. *Mutis y la Expedición Botánica-Documentos* (Bogotá, 1983); p. 57.

[43] According to the census carried out by Moreno y Escandón, the city of Santafé had 24,000 inhabitants. On 20th March 1630, the Consejo Real approved the University of Santo Tomás, which functioned until 3rd October 1826, when Santander suppressed it and created the Official University. Nevertheless, the Dominicans continued with the school, that was suppressed once more by Tomás Cipriano de Mosquera on 18th July, 1861. Finally, in 1942 the School and the University of Santo Tomás were reopened. See A. Ariza, *El Colegio Universidad de Santo Tomás de Aquino en Santa Fe de Bogotá* (Bogotá, 1980), pp. 67.

[44] Toward the second half of XVIII century, the city of Quito had 20,000 inhabitants, and in 1780, 28,451. The presidency of Quito was created in 1739, dependent on the Viceroyship of Nueva Granada. See Ernesto Cisneros Alfaro, *Eugenio el médico.* Quito, p. 69.
[45] The Colegio Real de San Fernando was founded in 1693 under the control of the Dominicans. This was the first school for laymen. The Universidad de Santo Tomás was created in 1686. The San Andrés school in 1555. Later, this was called Colegio Imperial de San Buenaventura; it was under the control of the Franciscans.
[46] The University of San Fulgencio of the Reformed Augustinians in Quito, had Papal Consent on 20th August 1586 to create the university, and was given the right to confer degrees in any of its faculties, on religious and lay people. The General of the order only allowed degrees to be conferred on the friars of the convent in 1602. In 1603 the university began functioning, catering for religious and lay people. In 1755 the reformer visitor Joaquin Izerta, suppressed degrees for laymen, and in 1786, Carlos III prohibited degrees in this university. Cf. R. Cruz, *Historia de las universidades,* pp. 417–418.
[47] The Convent of *La Merced* was created in 1670 and in 1837 it was suppressed and its property passed to the faculty of Medicine of the University. Cf. L. C. Lara, *Los mercedarios y la vida política y social de Caracas en los siglos XVII y XVIII*, Vol. I (Caracas, 1980), 332–333. The San Francisco convent, controlled by Franciscans, was established toward 1580. The convent of San Francisco was founded in 1597, under Dominican direction. See Parra (1954), pp. 111–122.
[48] Cf. C. P. León, *Obras* (Madrid, 1954), pp. 229. From 1673 the convents of San Francisco and San Jacinto opened their classrooms to laymen.
[49] The seminary of Santa Rosa de Lima was created in 1637, and in 1721 led to the foundation of the university. In 1826 it is no longer called Pontificia y Regia, but it was called Universidad Central De Venezuela.
[50] Francisco Javier Eugenio de Santacruz y Espejo was born in Quito in 1747 and died in 1795. He studied in San Fernando school, and was Hospital's student during the course 1761–1762, when he was 13 and 14 years old. He entered the Faculty of Medicine of Universidad de Santo Tomás and graduated in 1767, but was only able to practice as a doctor until 1772. In 1783 he escaped to Quito as he was about to be arrested, and in 1789 we find him in Santafé de Bogotá.
[51] Miguel Antonio Rodriguez (1769–1817); his father was a student of Father Hospital's in the philosophy course from 1759–1762. It should be said that in his physics course at the university, he introduced the teaching of anatomy and gave great importance to the teaching of mathematics. In 1801 he become a priest and in 1813 he published the rights of man. Because of this, he was expatriated to Panama and later to the Philippines. See Paladines, *op. cit.,* pp. 50–51. See the manuscript: Keeding, Ekkehard, "The revolutionary teacher of Universidad Colonial of Quito, Doctor Miguel Antonio Rodriguez". A transcript of this article was given to us by Doctor Carlos Paladines, in the city of Quito as it was not possible to find it in libraries. The article introduces some aspects of the life Miguel Antonio Rodríguez: his birthplace, his studies, the reforms to studies he introduced as teacher of philosophy and his participation in the independence movement.
[52] Cf. Paladines, "Estudio Introductorio", p. 50.
[53] There is a report of 1788 on the visits to the teachings that says: "The hateful singularity enjoyed by the religious Dominicans of the convent of San Jacinto of Caracas, of reading in the Real and Pontificia University the teachings of philosophy and Holy Scriptures, without the disputations and examinations that in observance of their Constitutions and Statutes they applied to other teachers". See Archivos de la Universidad Central de Venezuela (A.U.C.V.), *Provisiones y opositores a varias cátedras,* I, 7, book 128, reg. 12, doc.10, pp. 338–348. The report of Universidad de Caracas dated 22nd December 1786 says that: "the teachings of the Dominicans should conform to the new reforms (...) and in order to maintain uniformity in the provisions of all of them without exception, including the two that have been and are in charge of the convent of Santo Domingo". A.U.C.V., book I, Vol. I, t. I, No. 24, f. 271.

[54] Cf. Paladines, "Estudio Introductorio", pp. 50–52, and Rodríguez Cruz, *op. cit.,* p. 565.
[55] Cf. "The Philosophy Students of Colegio Santafé Require the Appointment, at Their Expense, of a Philosophy Teacher to Instruct Them in Physics, Mathematics, Botany and Natural History", Santafé, 179, Archivo Histórico Nacional de Colombia, (A.H.N.C.), Sección Colonia, Fondos Milicias y Marina, t. 128, pp. 200–201. See also L. C. Arboleda, "La ciencia y el ideal de ascenso social de los criollos en el virreinato de Nueva Granada", in A. Lafuente and J. S. Catalá (eds.), *Ciencia colonial en América* (Madrid, 1992).
[56] Cf. "Professor Vallecilla requires the King to fix a term for the Chancellor to present the report that had been requested from him". Cf. Santafé, 27th November 1790, A.H.N.C., Sección Colonia, Fondo Miscelánea, t. 31, p. 38. (Doc. No. 6, our numbering).
[57] Cf. "The Chancellor Martinez Caso Suspended the Literary Act in Which Vásquez Gallo was to Defend the Copernican System", Santafé 1795, A.H.N.C., Sección Colonia, Fondo Colegio, t. III, pp. 630–631.
[58] Cf. Guillermo Hernández de Alba, *Crónica del muy ilustre Colegio Mayor de Nuestra Señora del Rosario en Santa Fe de Bogotá,* Vol. II. (Bogotá, 1938), p. 301.
[59] The new curriculum that replaced scholasticism by modern "philosophy" in San Nicolás de Bari, was instituted by circular dated 18th October 1773. This circular was transcribed by Fray Bautista González, reformer of the Order of Nuestra Señora de la Gracia. It should be mentioned here that during the 1780s and 1790s, students of the Augustinian schools in Spain discussed modern ideas . See R. Herr, *España y la revolución del siglo XVIII* (Madrid, 1964), p. 13. The University of San Nicolás de Bari was authorized by a *Bula*, dated 24th April 1694, to confer academic degrees in philosophy and theology on the Augustinians of Province La Gracia. The academic course began in 1697, but the *pase regio* was only given on 22nd April 1703. Initially, the university operated within the Augustinian convent. From 1739 to 1775 it functioned in a separate building together with the Colegio de San Miguel. In 1775, the Reformer Visitor Juan Bautista González closed the school and donated the building to the Conciliar seminary; the university continued functioning in spite of this up to 1861, and its last chancellor was Felipe Bernal. See C. del Pozo, "Método y profesores de la Universidad de San Nicolás", pp. 200–201.
[60] The Peruvian priest Francisco Javier Vásquez, was appointed superior general of the Augustinian order in 1753. During his administration, the reform of ecclesiastical studies was carried out. His constitutions were not published, but it is known that he put into practice matters referring to studies by means of decrees. The friendship of Father Vásquez with the enlightened ministers of Carlos III, including Monino y Roda, was publicly known. It is believed that Vásquez collaborated closely with the representatives of Carlos III in order to carry out the abolition of the Jesuits. Cf. Campo del Pozo, *loc. cit.,* p. 60. See also "Método y profesores de la Universidad de San Nicolás", *Augustinian Archives*, 68(186):196; J. Sarrailh, *La España Ilustrada en la segunda mitad del siglo XVIII* (México, 1974), pp. 204, 700, and 703; and Herr, *op. cit.*, pp. 19 and 143.
[61] Communication of Fray Bautista González, reformer of the Augustinian order in the province of La Gracia, on Father Vásquez's reform to be applied in the University of San Nicolás de Bari, Santafé, 18th October 1773. Archivo Histórico Nacional de Colombia (A. H. N. C.), Sección Colonia, Fondo Conventos, t. 47, p.92v.
[62] The Creole José Francisco Padilla, a Reformed Augustinian, was in charge of the teaching of philosophy in San Nicolás de Bari from 1776 to 1782,. He imposed a new system that accepted the modern philosophy according to Father Vásquez's plan. In the first year he taught logic; in the second general and particular physics, including metaphysics, and in the third an integrated course of ethics in three parts, according to the new methodology. In 1786, he visited France, Italy, and Spain, where he got hold of several books on encyclopedic philosophy for the University. In 1788 he was appointed Regent of Studies, and together with the Chancellor, Father Bernardo Londoño, organized the philosophy studies based on the ideas of Descartes, Bacon, Newton, Locke, Montesquieu, Pascal, and other authors. Cf. Campo del Pozo, *op. cit.,* pp. 66, 70, and 71.

[63] In 1788, the philosophy class at Universidad de Bari, was attended by laymen and ecclesiastics, and toward 1800 the teaching of mathematics was started, where Copernicus and Newton were taught. Campo del Pozo, *op. cit.,* p. 71.
[64] Marrero was born in Caracas, on 6th January 1752 and died in 1809. He studied in Seminario de Santa Rosa and in 1779 he was ordered priest. In 1775 he became a teacher at the University of Caracas, lecturing in arts until 1776. Later he taught rhetoric until 1788, when he begun teaching modern philosophy. See Caracciolo Parra León, *op. cit.,* pp. 54–57 and Leal, p. 146.
[65] Cf. Parra León (1989), p. 57. See also "Expediente de la Real y Pontificia Universidad de la ciudad de Santiago de León de Caracas, capital de provincias de Venezuela, practicada en virtud de la Orden Real del 14 de mayo de 1815". A.G.I., Audiencia de Caracas, Sección V, legajo 446 (Doc. 10, our numbering).
[66] In Caracas several wills can be found in important libraries, with the works of a great number of illustrated Europeans. See Leal (1985), pp. 453–481.
[67] "On 27th July 1791, in Madrid, the final decision on the lawsuit between Dr Baltazar Marrero and Dr Cayetano Montenegro, the father of the erring student, was handed down. The decision was as follows: the lessons of algebra, geometry and arithmetic which were not customarily taught in the philosophy course and which were not included within the university statutes could only be received by students who expressed this voluntarily". The student José Cayetano was reinstated and Doctor Marrero was ordered to pay 793 pesos, "the costs of the legal dispute". The fine was extremely high; a university professor in 1803 earned a salary of 150 pesos per annum. See Leal, op. cit., p. 152, and Ildefonso Leal, *Historia de la Universidad de Caracas, 1721–1827* (Caracas, 1985), pp. 515–516.
[68] In 1785 de la Torre was the Academic Vice-Chancellor of the university and in 1789 he became Chancellor.
[69] "Archives of Proceedings of the creation of the Academy of Public Spanish Law, when Miguel José Sanz was elected President", Caracas, 1790, A.U.C.V., Reclamos Cátedras, vitrina I, tramo 5, libro 142, pp. 1–6.
[70] Miguel José Sanz is known in Venezuelan history as a famous legal consultant, founder of the College of Caracas Lawyers and of the Academy of Public Law, consultant to the royal consulate, accurate critic of the unequal system of colonial education, journalist of the *Seminario de Caracas* and precursor of an important political reform. See Leal, *op. cit.,* p. 268.
[71] See Arboleda, "La ciencia y el ideal de ascenso social de los criollos en la Nueva Granada", in connection with Zea and the enlightened elite of Santafé.
[72] Cf. J. A. De la Torre, "Discurso Económico. Amor a las letras en relación a la agricultura y el comercio" (Caracas, 1790); see also Leal, *op.cit.*, pp. 227, 229, and 240–241.
[73] Cf. Parra, *op. cit.*, p. 70, 113, and ff.